D1001325

VADOSE ZONE MONITORING FOR HAZARDOUS WASTE SITES

VADOSE ZONE MONITORING FOR HAZARDOUS WASTE SITES

WITHDRAWN

by

L.G. Everett

L.G. Wilson

E.W. Hoylman

Kaman Tempo
Santa Barbara, California

NOYES DATA CORPORATION

Park Ridge, New Jersey, U.S.A.

1984

Tennessee Tech. Library
Cookeville. Tenn.

387304

Library of Congress Catalog Card Number: 84-16509
ISBN: 0-8155-1000-4
ISSN: 0090-516X
Printed in the United States

Published in the United States of America by
Noyes Data Corporation
Mill Road, Park Ridge, New Jersey 07656

10 9 8 7 6 5 4 3 2

Library of Congress Cataloging in Publication Data

Everett, Lorne G.
Vadose zone monitoring for hazardous waste sites.

(Pollution technology review, ISSN 0090-516X ; no. 112)
Bibliography: p.
Includes index.
1. Hazardous waste sites--United States--Zone of aeration--Monitoring--Handbooks, manuals, etc. 2. Water, Underground--Pollution--United States--Handbooks, manuals, etc. 3. Sewage disposal--United States--Handbooks, manuals, etc. 4. Water quality--United States--Measurement--Handbooks, manuals, etc. I. Wilson, L.G. (Lorne Graham), 1929- . II. Hoylman, Edward W. III. Title. IV. Series.
TD811.5.E94 1984 628.4'456 84-16509
ISBN 0-8155-1000-4

Foreword

This book is an extensive state-of-the-art review and evaluation of vadose (unsaturated) zone monitoring methods and their applicability to hazardous waste disposal sites. More than 70 different sampling and nonsampling vadose zone monitoring techniques are described in terms of their advantages and disadvantages.

The vadose, or unsaturated, zone is that ground layer, beneath the topsoil and overlying the water table, in which water in pore spaces coexists with air, or in which the geological matter is unsaturated. A major concern at all hazardous waste disposal sites—abandoned, active, or planned—is potential pollution of the underlying groundwater system. Current federal regulations require both groundwater monitoring and vadose zone sampling at land treatment installations.

This book will serve as a compendium of monitoring techniques from which the user may select a method to develop a vadose zone monitoring program. Each of the methods presented is quantitatively described according to physical, chemical, geologic, topographic, hydrologic, and climatic constraints. The monitoring techniques are categorized for premonitoring, active, and postclosure site assessments. Waste disposal methods are categorized for piles, landfills, impoundments, and land treatment. Conceptual vadose zone monitoring approaches are developed for specific waste disposal method categories.

The information in the book is from *Vadose Zone Monitoring for Hazardous Waste Sites* prepared by L.G. Everett, L.G. Wilson, and E.W. Hoylman of Kaman Tempo for the U.S. Environmental Protection Agency under Contract No. 68-03-3090, October 1983.

The table of contents is organized in such a way as to serve as a subject index and provides easy access to the information contained in the book.

Advanced composition and production methods developed by Noyes Data Corporation are employed to bring this durably bound book to you in a minimum of time. Special techniques are used to close the gap between "manuscript" and "completed book." In order to keep the price of the book to a reasonable level, it has been partially reproduced by photo-offset directly from the original report and the cost saving passed on to the reader. Due to this method of publishing, certain portions of the book may be less legible than desired.

NOTICE

The information in this document has been funded wholly or in part by the United States Environmental Protection Agency under Contract Number 68-03-3090 to Kaman Tempo, Santa Barbara. It has been subjected to the Agency's peer and administrative review, and it has been approved for publication. The contents reflect the views and policies of the Agency.

Contents and Subject Index

1. Introduction

The purpose of this book is to describe the applicability of selected vadose zone monitoring methods to hazardous waste disposal sites. Based upon existing data, each of the monitoring methods presented is quantitatively described according to physical, chemical, geologic, topographic, hydrogeologic, and climatic constraints. In addition, each monitoring technique is evaluated in terms of preactive, active, and postclosure assessment applications. Various hazardous wastes and waste disposal methods are described. Vadose zone monitoring approaches that would be applicable at various sites using particular waste disposal methods are presented.

This document is a comprehensive state-of-the-art technical review and evaluation of vadose zone monitoring. All available literature sources were reviewed to develop this compendium of monitoring devices from which a user can make a selection to develop a vadose zone monitoring program.

The Resource Conservation and Recovery Act of 1976 (RCRA) was enacted to promote the protection of public health and the environment through various regulatory, technical assistance, and training programs. Subtitle C of RCRA specifically addresses regulation of hazardous waste sites.

A major concern at all hazardous waste disposal sites, including abandoned, active, and planned sites, is the possibility of polluting an underlying groundwater system. The Interim Status Hazardous Waste and Consolidated Permit Regulations, issued by the U.S. Environmental Protection Agency (EPA) on May 19, 1980, require a minimum of four groundwater sampling wells at sites containing impoundments, landfills, and waste piles, and at land treatment areas. The regulations also stipulate the location of such wells relative to site boundaries and specify the water sample parameters to be determined.

The final permitting standards issued on July 26, 1982 require both groundwater monitoring and vadose zone sampling at land treatment units. Insufficient information was available for EPA to consider vadose zone monitoring at waste piles, landfills, and impoundments. This document will provide technical background information but is not intended to interpret the regulations.

VADOSE ZONE DESCRIPTION

The topsoil is the region that manifests the effects of weathering of geological materials, together with the processes of eluviation and illuviation of colloidal materials, to form more or less well-developed profiles

(Simonson, 1957). Water movement in the topsoil usually occurs in the unsaturated state, where soil water exists under less-than-atmospheric pressures. A great deal of literature is available on the subject in periodicals and textbooks. Within the topsoil, saturated zones may develop over horizons of low permeability. A number of references are available on the theory of flow in perched water tables (e.g., Luthin, 1957; van Schilfgaarde, 1970). Soil chemists and soil microbiologists have also attempted to quantify chemical-microbiological transformations during soil-water movement (Rhoades and Bernstein, 1971; Dunlap and McNabb, 1973).

Weathered topsoil materials gradually merge with underlying materials such as unconsolidated alluvium, lacustrine deposits, eolian deposits, glacial or consolidated sedimentary deposits, and metamorphic or igneous rocks. The zone beneath the topsoil and overlying the water table, in which water in pore spaces coexists with air or in which the geological materials are unsaturated, is known as the vadose zone. Perched water tables may develop above interfaces between layers having greatly different textures. Saturated conditions may also develop beneath recharge sites as a result of prolonged infiltration. In contrast to the large number of studies on water movement in the topsoil, parallel studies in the vadose zone have been few. Meinzer (1942) coined the term "no-man's land of hydrology" to describe the limited knowledge of this zone.

A number of techniques have been developed for monitoring water content, water movement, and water quality in the topsoil. Many detailed descriptions, specifications, and methods of using these techniques are compiled in Methods of Soil Analysis (Black, 1965). Monitoring in the vadose zone requires an extension of topsoil monitoring technology.

The hypothetical flow paths of a pollutant beneath a waste disposal impoundment are shown in Figure 1-1. When groundwater underlying such a disposal site begins to manifest pollution by quality changes in pumped water, the aquifer may have already become badly contaminated. Monitoring the quality of wastes entering and within a disposal site (i.e., source monitoring), coupled with monitoring in the vadose zone, offers an early warning of groundwater pollution. Remedial measures implemented before the onset of severe pollution may eliminate or reduce the associated renovation costs. In addition to providing an early warning of pollutant movement in groundwater, if a vadose zone monitoring program fails to detect the movement of pollutants, the requirements for groundwater monitoring may be reduced or largely precluded. The cost savings for constructing monitoring wells could be significant.

Vadose zone monitoring is not a panacea for all geohydrologic conditions and for all waste disposal operations. The need for and extent of such monitoring should be tailored to site-specific conditions. For example, if the water table at a given site is relatively shallow, say within 10 feet* of the

* See Appendix A for conversion to metric units. English units are generally used in this book because of their current usage and familiarity in industry and the hydrology-related sciences. Certain units, (e.g., concentrations) more commonly expressed in the metric system are used where appropriate.

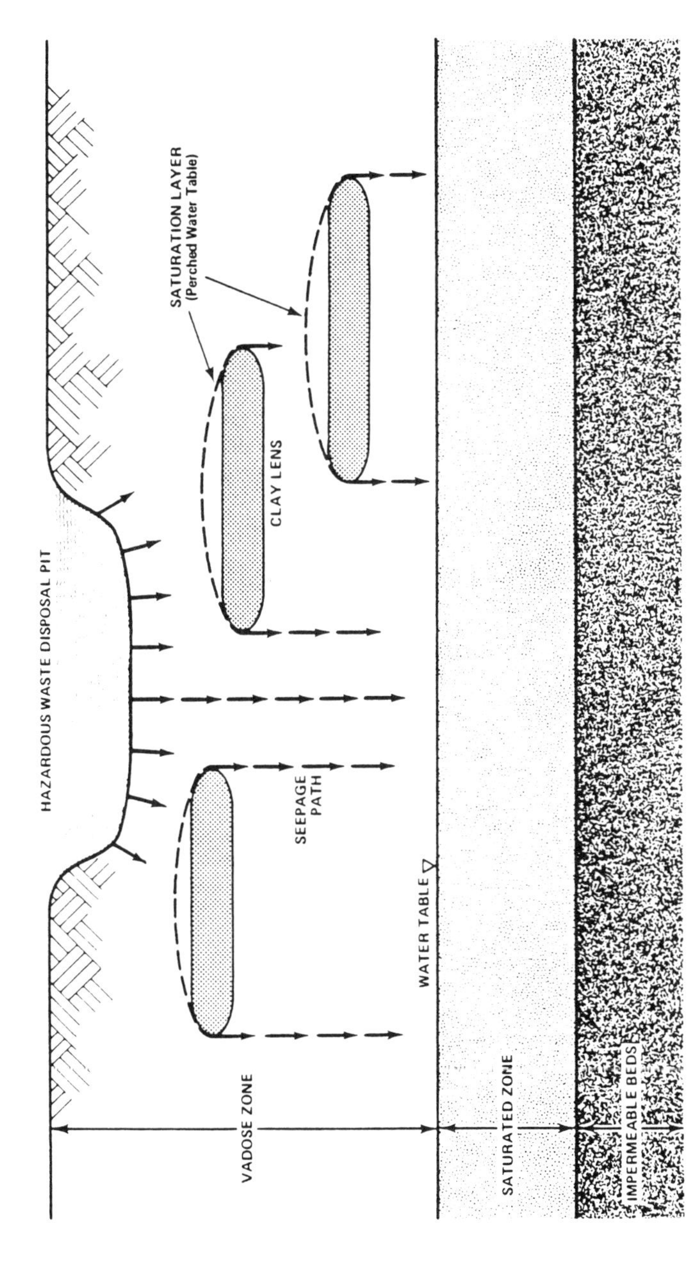

Figure 1-1. Vadose zone including saturated and unsaturated flow.

land surface, vadose zone monitoring may be minimal. Similarly, if the vadose zone consists of fractured media, flow occurs primarily in channels and the ameliorating interactions of the vadose zone and water-borne pollutants may be minimal. Monitoring activities for these two cases could be restrictec to source monitoring and groundwater monitoring. Installation of in-source monitoring facilities, including vadose zone units, may not be recommended for certain sources, such as landfills. This rationale is presented in the proposed guidelines for landfill disposal of solid waste (U.S. EPA, 1979e): "In no case should groundwater or leachate monitoring wells be installed through the bottom of the landfill proper since such installation could result in creation of a conduit for the direct passage of landfill leachate into underlying groundwater." Thus, it would be necessary to either install monitoring facilities around the periphery of the landfills or use angle-drilling methods to install such facilities below the base of the lowermost landfill cells.

CATEGORIZATION OF VADOSE ZONE MONITORING METHODS

A vadose zone monitoring program for a waste disposal site includes premonitoring activities followed by active and postclosure monitoring programs. Basic premonitoring activities consist of assessing hydraulic properties of the vadose zone, specifically storage and transmissive properties, and the geochemical properties relating to pollutant mobility (Table 1-1). The premonitoring program results will yield information on the potential mobility rates of liquid-borne pollutants through the vadose zone, the storage potential of the region for liquid wastes, and the likelihood that specific pollutants will become attenuated. Fuller (1979) discusses premonitoring activities in detail. A premonitoring program will also provide valuable information for the design of a vadose zone monitoring system. For example, if it appears that potential pollutants of concern will become attenuated in the near-surface horizon, shallow monitoring facilities (vacuum-operated lysimeters) could be installed.

The active and postclosure monitoring programs will comprise a package of sampling (Table 1-2) and nonsampling methods (Table 1-3) selected from an array of possible methods. Sampling methods provide actual liquid or solid samples from the vadose zone, whereas nonsampling methods provide inferential evidence about the movement of liquid-borne pollutants. Thus, nonsampling methods can be viewed as a backup or supplemental procedure for the preferred sampling techniques.

SUMMARY OF CRITERIA FOR SELECTING ALTERNATIVE VADOSE ZONE MONITORING METHODS

A guiding principle for selecting methods for a monitoring program (premonitoring/active/postclosure monitoring) is aptly stated by Vanhof, Weyer, and Whitaker (1979) as follows: "... for an efficient, long-term operation of an operational monitoring network, the devices to be used must be simple enough to be used by trained but not educationally skilled personnel." If a device meets these criteria and is also inexpensive (Everett et al., 1976), one need not look further. In practice, selection of a method from a group of alternatives is governed by additional site-specific and function-specific

TABLE 1-1. PREMONITORING OF THE VADOSE ZONE AT HAZARDOUS WASTE DISPOSAL SITES

Property	Purpose of Premonitoring	Approach	Alternative Methods
1. Storage	1. To determine overall storage capacity of the vadose zone	1. Relate storage capacity to depth of water table or depth to confining layer	1. Examine groundwater level maps 2. Measure water levels in wells 3. Examine drillers' logs for depth to water table or confining layer 4. Drill test wells
		2. Estimate available porosity	1. Estimate from grain-size data 2. Drill test wells and obtain drill cuttings 3. Neutron moisture logging (available porosity = total porosity-water content, by volume)
	2. To determine regions of potential liquid accumulation (perched groundwater)	1. Characterize subsurface stratigraphy	1. Examine drillers' logs 2. Drill test wells and obtain samples for grain-size analyses 3. Natural gamma logging
		2. Locate existing perched groundwater zones	1. Examine drillers' logs 2. Drill test wells 3. Neutron moisture logging
2. Transmission of liquid wastes a. Flux	1. To determine infiltration potential	1. Measure infiltration rate in the field	1. Infiltrometers 2. Test plots
	2. To estimate percolation rates in vadose zone	1. Measure unsaturated hydraulic conductivity for use in Darcy's equation	1. Instantaneous rate method 2. Laboratory column studies
		2. Measure or estimate saturated hydraulic conductivity for use in Darcy's equation a. Use core samples or grain-size data	1. Permeameters 2. Estimate from grain-size data using a catalog of hydraulic properties of soils

(continued)

TABLE 1-1. (continued)

Property	Purpose of Premonitoring	Approach	Alternative Methods
2. Transmission of liquid wastes (continued)			
a. Flux (continued)		b. Measure saturated hydraulic conductivity in shallow regions	1. Pump-in method 2. Air entry permeameter 3. Infiltration gradient 4. Double tube method
		c. Measure saturated hydraulic conductivity in deep regions	1. USBR open end casing test 2. USBR open hole method 3. Stephens-Neuman method
b. Velocity	1. To estimate the flow rate of liquid pollutants in the vadose zone	1. Estimate from field data on flux	1. Use flux values obtained as above; divide flux values by water content values at field capacity
		2. Tracer studies	1. Field plots, coupled with a depthwise sequence of suction samplers, using conservative tracer
3. Pollutant mobility.	1. To estimate the mobility of potential pollutants in the vadose zone.	1. Characterize solids samples for properties affecting pollutant mobility: cation exchange capacity, clay content, content of hydrous oxides of iron, pH and content of free lime, and surface area	1. Obtain solids samples (e.g., by drilling test holes) and conduct standard laboratory analyses
		2. Estimate from laboratory or field testing using liquid wastes	1. Batch testing 2. Column studies 3. Field plots

TABLE 1-2. SAMPLING METHODS FOR ACTIVE AND POSTCLOSURE VADOSE ZONE MONITORING PROGRAMS AT HAZARDOUS WASTE DISPOSAL SITES

Property	Purpose of Monitoring	Approach	Alternative Methods
1. Chemical and microbial properties of liquid wastes in unsaturated regions of the vadose zone	1. To obtain liquid samples for determination of pollutants in unsaturated regions	1. Obtain soil samples and drill cuttings, followed by field or laboratory extraction of pore fluid for analyses of inorganics; other extraction methods for organics and microorganisms	1. Hand auger 2. Power auger 3. Hydraulic squeezer for fluid extraction 4. Field saturated extract method 5. Laboratory saturated extract methods
		2. Obtain in situ samples of pore fluid	1. Suction cup lysimeters, vacuum-operated type 2. Suction cup lysimeters, vacuum-pressure type 3. Suction cup lysimeters, high-pressure-vacuum type 4. Filter candles 5. Hollow fiber filters 6. Membrane filters
2. Chemical and microbial properties of liquid wastes in saturated regions of the vadose zone (perched groundwater)	1. To obtain liquid samples for determination of pollutants in saturated regions	1. Obtain an integrated sample	1. Existing cascading wells or installed monitor wells 2. Collection pans and manifolds 3. Tile-drain outflow samplers
		2. Depth-wise samples	1. Profile sampler 2. Multilevel sampler 3. Piezometers

TABLE 1-3. NONSAMPLING METHODS FOR ACTIVE AND POSTCLOSURE VADOSE ZONE MONITORING PROGRAMS AT HAZARDOUS WASTE DISPOSAL SITES

Property	Purpose of Monitoring	Approach	Alternative Methods
1. Salinity	1. To determine overall salinity of pore fluids	1. Relate salinity to electrical conductance properties	1. Four-probe method 2. EC probe 3. Salinity sensors
2. Infiltration rate	1. To infer the movement of liquid-borne pollutants based on intake rates at the land surface	1. Measure infiltration in ponds	1. Water budget method 2. Instantaneous rate method 3. Seepage meters
		2. Measure infiltration on land treatment areas	1. Water budget method 2. Infiltrometers 3. Test plots
3. Flux	1. To infer the movement of liquid-borne pollutants based on changes in storage	1. Determine water content changes	1. Gravimetric 2. Neutron moisture logging 3. Gamma ray attenuation 4. Tensiometers 5. Hygrometer/psychrometer 6. Heat dissipation sensor 7. Resistor/capacitor type sensors 8. Remote sensing
	2. To infer the movement of liquid-borne pollutants based on changes in matric potential	1. Measure hydraulic gradient, use Darcy's law	1. Tensiometers 2. Hygrometer/psychrometer 3. Heat dissipation sensors 4. Resistor/capacitor type sensors
		2. Assume unit hydraulic gradient, measure water pressure at a point, and use Darcy's law	1. Tensiometers 2. Hygrometer/psychrometer 3. Heat dissipation sensors 4. Resistor/capacitor type sensors
	3. To directly measure flux	1. Install flowmeters	1. Direct flow measurement 2. Heat pulse method
4. Velocity	1. To estimate the velocity of liquid-borne pollutants in the vadose zone	1. Introduce a tracer into liquid waste at land surface and collect depthwise samples	1. Suction cup lysimeters 2. Vadose zone wells
		2. Estimate from information on infiltration rates	1. Assume steady-state and unit hydraulic gradients; divide infiltration rate value by estimated or measured water content values (volumetric) at field capacity
		3. Estimate from field determined flux values	1. Assume steady-state and unit hydraulic gradients; divide flux values by estimated or measured water content values (volumetric) at field capacity

requirements. Table 1-4 lists 14 possible criteria that should be considered. Each of these criteria is briefly reviewed.

TABLE 1-4. CRITERIA FOR SELECTING ALTERNATIVE VADOSE ZONE MONITORING METHODS

Item	Criteria
1	Applicability to new, active, or abandoned sites
2	Applicability to laboratory or field usage
3	Power requirements
4	Depth limitations
5	Multiple use capabilities
6	Data collection system
7	Continuous or discrete sampling
8	Point and bulk volume samplers
9	Reliability and life expectancy
10	Degree of complexity
11	Direct versus indirect sampling measurements
12	Applicability in alternate media
13	Effect of sampling/measurement on flow region
14	Effect of hazardous waste on sampling/measurement

Applicability to New, Active, and Abandoned Sites

A vadose zone monitoring program for sites under construction will differ from those for active and abandoned sites. Premonitoring activities at a new site lead to the design of a full-scale monitoring scheme, with a spectrum of possible methods. In contrast, premonitoring and monitoring activities merge in a program for an existing site because the primary concern is determining the extent that pollutants have already moved into the vadose zone. Direct sampling procedures, such as obtaining solids samples for saturated extracts or sampling perched groundwater, are used at the outset for an existing site. If pollution is not detected, plans can be developed for a long-term monitoring program and facilities installed during the initial effort can be used. For example, casing can be installed in boreholes left during the presampling effort for use in water sampling and moisture logging.

Obviously, some methods can be employed at new sites but not at existing facilities. For example, suction cup lysimeters or filter candle samplers could be installed below the base of impoundments before laying down a liner. Installing these samplers beneath liners in existing ponds is impractical.

Applicability to Field Usage

Some of the standard monitoring techniques that have been modified for laboratory studies may not be suitable for field operation. For example, experimental transducerized tensiometers offer a distinct advantage over presently used tensiometers because they can be linked to an automatic data collection system. Unfortunately, existing units are plagued with problems such as difficulties in purging air from the system (Brakensiek, Osborn, and Rawls, 1979). A related problem is that some devices may be easily broken during field installation. Implantable sensors such as hygrometer/psychrometers fall into this category.

Power Requirements

From the viewpoint of power requirements, monitoring units can be classified into two groups: those that require power and those that can be operated manually or mechanically. A source of AC and/or DC electricity is advantageous when using automatic samplers and recorders and transmitting data. A disadvantage is that if a power failure occurs or a battery fails, the operation of associated units ceases and data acquisition is stopped. The designer of a monitoring program will need to choose between the convenience and flexibility of power-driven units for a particular need and the greater reliability of mechanical or manually operated units.

Depth Limitations

As stated by Wilson (1980), many of the techniques for vadose zone monitoring are adaptations of methods developed by soil scientists. Consequently, certain units are more suitable for monitoring in shallow rather than deeper depths. If the vadose zone is relatively shallow, this limitation may not be a problem. In regions with deeper vadose zones, the problems become magnified. For example, successful measurement of water content changes in a porous medium by installing in-place sensors of matric potential depends greatly on three requirements: (1) good contact between the sensor and the surrounding media, (2) good calibration relationships between soil matric potential and the response signal from the sensor, and (3) good calibration relationships between water content and matric potential. Success in achieving these requirements diminishes as the installation depth increases. In contrast to the questionable reliability and dependability of point sensors in deep vadose zones, reasonably good estimates of water content changes can be obtained using the neutron moisture method, an alternative technique used by soil scientists.

Multiple-Use Capabilities

Because of the great expense in purchasing and installing monitoring units, they should be used whenever possible for multiple purposes. For instance, test holes near waste disposal facilities could be cased and used for neutron moisture logging, obtaining groundwater samples, and observing water-level fluctuations.

Data Collection System

Data output monitoring techniques can be divided into three groups: (1) those requiring manual recording of data, (2) those that automatically record onsite data, and (3) those that transmit data to remote stations. A simple example is the tensiometer. The basic unit, with mercury manometers or bourdon-type gages, requires manual recording of the readings. A more sophisticated version, the pressure transducer type, can be used for either onsite automatic recording of data or transmission to remote stations (Watson, 1967). Another example is the neutron moisture logger. Most commercial units developed by soil scientists require stepwise manual recording of data. One commercial unit is available, however, that automatically records water content values as the probe is lowered continuously down the casing. Manual recording of neutron moisture data has been simplified with the recent development of portable data acquisition systems. These systems offer an additional advantage in that the memory package can be read into a computer program for tabulating or graphing the output (Warrick, University of Arizona, personal communication, 1981).

Continuous or Discrete Sampling

When monitoring liquids, samples are required either on a continuous or discrete time basis. For example, 24-hour composite samples from suction cup lysimeters depict daily changes in constituents of interest. In contrast, discrete samples from suction cup lysimeters are useful in characterizing the breakthrough of a particular constituent at the sampling depth. Similarly, a sequence of samples from a monitoring well in a perched groundwater region may show the arrival of a pollutant front.

The simplest approach for obtaining continuous or discrete samples is to manually collect the samples using field technicians. This approach, however, is tedious, time-consuming, and expensive. Alternatively, most sampling devices can be adapted for either continuous or discrete sampling by appropriate mechanical devices. Both manual and automatic sampling have advantages and disadvantages that a project manager must weigh.

Point and Bulk-Volume Samplers

Sampling units may be categorized into two groups depending on sample size: point samplers or bulk-volume samplers. For example, suction cup lysimeters provide point samples, but pumping of perched groundwater provides an integrated bulk sample. Measurement techniques also may be categorized from the viewpoint of region. Tensiometers manifest water content at a single point, whereas neutron loggers are used to scan water content along a vertical line in the vadose zone.

Reliability and Life Expectancy

Two desirable characteristics to be considered when selecting sampling or measurement units for hazardous waste sites are reliability and life expectancy. That is, the units should provide accurate data for a prolonged period of time. Thus, when choosing between alternate units for measuring, for example, soil water content, units that will not drift in calibration should be given preference. Similarly, some types of sampling or measurement units may become ineffective if clogged. Suction cup lysimeters may fall into this category.

Degree of Complexity

Whenever possible, a guiding principle in the selection of sampling or measurement techniques should be "the simpler, the better." More complex units requiring an array of equally complex auxiliary components will have a greater potential for failure than simpler units. An example is the simple tensiometer versus the transducerized version. Naturally, remote recording of data offers numerous advantages over onsite measurements, but the possibility of equipment failure may be greater.

A related consideration is that equipment that is easy to operate and maintain will not require highly trained technicians.

Direct Versus Indirect Sampling Measurements

Point and bulk sampling units are grouped into two classes: direct sample or measurement units and indirect measurement units. Suction cup samplers measure soil salinity directly, whereas measurements using the four-probe method must be converted to salinity via a calibration relationship. Tensiometers measure matric potential directly, whereas the output from heat dissipation sensors must be converted to matric potential using a calibration curve. Obviously, the error is reduced when using direct rather than indirect techniques.

Applicability in Alternate Media

Techniques developed by soil scientists for sampling and measuring storage changes are primarily suited for granular, porous media. The applicability of some of these techniques is questionable for media in which fractures and solution cavities are the principal avenue for fluid transport. For instance, installing a point sampler such as a suction cup lysimeter in a fractured rock requires skill and luck in positioning the unit within a crack. For such media, a more appropriate choice would be to install clustered units within a common borehole (see, for example, Hounslow et al., 1978).

Problems may arise even in porous media. Tensiometers and suction cup sampling units require that the porous cups function as an extension of the surrounding porous matrix. Such continuity may be difficult to obtain when installing cups in coarse sand or gravels. Data from neutron moisture loggers may be affected by side leakage when access tubes are installed in very coarse media.

Effect of Sampling/Measurement on Flow Region

When obtaining liquid samples or measuring storage changes, it is important that the sample/measurement reflects actual conditions in the exterior medium. Thus, if the process of obtaining a sample affects exterior flow paths, the resultant analyses may not represent actual quality. A suction cup lysimeter may represent actual quality and may influence solute flow upon application of a vacuum at the point. In contrast, obtaining solids samples and extracting the pore fluid will not immediately affect water movement. (The presence of the hole will, of course, cause a subsequent change in unsaturated flow paths in the vicinity of the cavity, or serve as a conduit for leakage from saturated regions.) Indirect techniques such as buried electrical conductivity (EC) probes will not affect flow if the porosity and bulk density of the material placed in the cavity are similar to the surrounding medium.

Effect of Hazardous Waste on Sampling/Measurement

Liquid hazardous wastes, alone or mixed with water, may have different storage and flow characteristics than water alone. The surface tension of liquid wastes may alter storage patterns in a porous matrix because of its effect on capillarity. Similarly, the viscosity of the liquid waste may cause a difference in flow velocity compared to that of water alone. Thus, the velocity of gasoline is about 1.5 times that of water (Schwille, 1967). These effects should be recognized when selecting monitoring equipment. For example, neutron moisture loggers operate on the principle that the high-energy neutrons are thermalized by hydrogen in water molecules in the soil. The presence of hydrogen in hydrocarbon-type waste may affect the results by overestimating storage changes. In contrast, the presence of excessive quantities of neutron absorbers such as boron or chlorine in the waste or surrounding media will result in an underestimation of storage changes.

Vadose zone monitoring techniques for premonitoring at proposed hazardous waste disposal sites are summarized in Table 1-5. Vadose zone sampling and nonsampling monitoring methods for active and postclosure monitoring programs are summarized in Tables 1-6 and 1-7, respectively.

TABLE 1-5. SUMMARY, VADOSE ZONE MONITORING TECHNIQUES--PREMONITORING PHASE

Property	Type of Site		Applicability		Power Requirements				Depth Limit			Mult Use Capability		Data Collection System			Continuous Sampling	
	New	Old	Lab	Field	ac	dc	Eng	No	Surf	<30 ft	>30 ft	Samp	Meas	Man	Auto	Remote	Yes	No
Storage																		
I. Overall storage capacity																		
A. Depth to water table																		
1. Water level maps	X	X		X				X	X	X	X	X	X	X			N/A	N/A
2. Sounding in wells	X	X		X		X		X	X	X	X		X		X	X	X	
3. Examine drillers' logs	X	X	Office					X	X	X	X	N/A	N/A	X			N/A	N/A
4. Drill test wells	X	X		X			X			X	X	X	X	X				X
B. Available porosity																		
1. Grain-size data	X	X	Office				X		X	X	X	N/A	N/A	X			N/A	N/A
2. Test wells, cuttings	X	X		X			X		X	X	X	X	X	X				X
3. Neutron logging	X	X		X	X	X	X		X	X	X	X	X	X	X		X	
II. Perched groundwater																		
A. Stratigraphy																		
1. Drillers' logs	X	X	Office					X	X	X	X	N/A	N/A	X			N/A	N/A
2. Test wells	X	X		X			X		X	X	X	X	X	X				X
3. Nat gamma	X	X		X	X	X				X	X	X	X	X	X			X
B. Existing perched groundwater																		
1. Drillers' logs	X	X	Other					X	X	X	X	N/A	N/A	X			N/A	N/A
2. Test wells	X	X		X				X	X	X	X	X	X	X				X
3. Neutron logs	X	X		X	X	X		X	X	X	X	X	X	X	X			X
Transmission																		
I. Flux																		
A. Infiltration																		
1. Infiltrometers	X			X				X	X				X	X	X		N/A	N/A
2. Test plots	X	X		X				X	X				X	X			N/A	N/A
B. Percolation rates																		
1. Meas unsat K																		
a. Last rate method	X			X				X	X				X	X	X		N/A	N/A
b. Lab vol	X	X	X					X	X	X	X		X	X			N/A	N/A
2. Meas sat K																		
a. Permeametric	X	X	X					X	X	X	X		X	X			N/A	N/A
b. Est from grain size	X	X	Other					X	X	X			X	X			N/A	N/A
c. Shallow field methods	X	X		X				X	X				X	X			N/A	N/A
d. Deeper field methods	X	X		X			X			X	X		X	X				X
C. Velocity																		
1. From field value of flux	X	X	Office					X	N/A	N/A	N/A	N/A	N/A	N/A	N/A	N/A	N/A	N/A
2. Tracer studies	X	X		X				X	X	X	X	X	X	X				
Pollutant Mobility																		
(1) Chemical properties	X	X	X					X	X	X	X	X		X			N/A	N/A
(2) Batch tests	X	X	X		X				X	X	X	X		X			N/A	N/A
(3) Columns	X	X	X					X	X	X	X	X	X	X			N/A	N/A
(4) Field plots	X	X		X				X	X	X		X	X	X			N/A	N/A

(continued)

TABLE 1-5. (continued)

Property	Sample Meas Vol		Reliability and Life Expectancy		Degree of Complexity		Direct/ Indirect		Media		Effect on Flow Region		Effect of Haz Waste Type on Res	
	Point/Line	Bulk	Good	Quest.	Simple	Complex	Dir	Ind	Porous	Fract.	Yes	No	Yes	No
Storage														
I. Overall storage capacity														
A. Depth to water table														
1. Water level maps	N/A	N/A	X		X		X		X	X		X		X
2. Sounding in wells	X		X		X		X		X	X		X		X
3. Examine drillers' logs	X		X		X		X		X	X		X		X
4. Drill test wells	X		X		X		X		X	X		X		X
B. Available porosity														
1. Grain-size data	N/A	N/A	X		X			X	X	X		X		X
2. Test wells, cuttings	X		X		X			X	X			X		X
3. Neutron logging	X		X			X		X	X	X		X	X	
II. Perched groundwater														
A. Stratigraphy														
1. Drillers' logs	X		X		X		X		X	X		X		X
2. Test wells	X		X		X	X	X		X	X		X		X
3. Nat gamma	X		X					X	X	X		X		X
B. Existing perched groundwater														
1. Drillers' logs	X		X		X	X		X	X	X		X		X
2. Test wells	X		X		X	X	X		X	X		X		X
3. Neutron logs	X		X		X	X		X	X	X		X		X
Transmission														
I. Flux														
A. Infiltration														
1. Infiltrometers	X		X		X		X		X	X	X		X	
2. Test plots	X		X		X		X		X	X	X		X	
B. Percolation rates														
1. Meas unsat K														
a. Last rate method	X		X			X		X	X	X	X		X	
b. Lab vol	X		X		X			X	X			X	X	
2. Meas sat. K														
a. Permeametric	X		X		X		X		X			X	X	
b. Est from grain size	X		X		X			X	X			X		X
c. Shallow field methods	X		X			X		X	X			X	X	
d. Deeper field methods	X		X			X		X	X	X		X	X	
C. Velocity														
1. From field value of flux	N/A	N/A	N/A	N/A	X		N/A	N/A	X	X	N/A	N/A	X	
2. Tracer studies	X		X		X		X		X	X	N/A	N/A	X	
Pollutant Motility														
(1) Chemical properties	X		X			X	X		X	X		X		X
(2) Batch tests	X		X			X	X		X	X		X	X	
(3) Columns	X		X		X		X		X			X	X	
(4) Field plots		X	X		X		X		X	X		X	X	

TABLE 1-6. SUMMARY, VADOSE ZONE MONITORING TECHNIQUES--SAMPLING METHODS

	Type of Site		Applicability		Power Requirements				Depth Limit			Mult Use Capability		Data Collection System			Continuous Sampling	
Property	New	Old	Lab	Field	ac	dc	Eng	No	Surf.	<30 ft	>30 ft	Samp	Meas	Man	Auto	Remote	Yes	No
Pollutant Mobility																		
I. Chemical/microbial properties w/liquid wastes in unsat region																		
A. Solids samples/extraction of pore fluid																		
1. Obtaining solids samples																		
a. Hand auger	X	X		X					X	X		X	X	N/A	N/A	N/A	N/A	N/A
b. Power auger	X	X		X			X		X	X	X	X	X	N/A	N/A	N/A	N/A	N/A
2. Obtaining solution samples																		
a. Hydraulic squeezer	X	X		X				X	X	X	X	N/A	N/A	N/A	N/A	N/A	N/A	N/A
b. Field sat. extract.	X	X		X				X	X	X	X	N/A	N/A	N/A	N/A	N/A	N/A	N/A
c. Lab sat. extract.	X	X		X				X	X	X	X	N/A	N/A	N/A	N/A	N/A	N/A	N/A
B. In situ pore water samples																		
1. Lysimeters, vac oper	X			X				X	X	X		X	X	N/A	N/A	N/A	X	
2. Lysimeters, vac pres	X			X	X				X	X	X	X	X	N/A	N/A	N/A	X	
3. Lysimeters, high-pres vac	X			X					X	X	X	X	X	N/A	N/A	N/A	X	
4. Filter candle	X			X	X				X			X	X	N/A	N/A	N/A	X	
5. Hollow fiber filters	X		X	X				X	X			X		N/A	N/A	N/A	X	
6. Membrane filters	X		X	X				X	X			X		N/A	N/A	N/A	X	
II. Chemical/microbial properties w/liquid wastes in sat. regions																		
A. Integrated samples																		
1. Wells	X	X		X	X		X	X	X	X	X	X	X	N/A	N/A	N/A	X	
2. Collection pans	X	X		X	X		X		X	X		X		N/A	N/A	N/A	X	
3. Tile outflow	X	X		X				X		X	X	X		N/A	N/A	N/A	X	
B. Depthwise samples																		
1. Profile samples	X	X		X	X		X			X	X	X		N/A	N/A	N/A	X	
2. Multilevel sampler	X	X		X	X		X			X		X	X	N/A	N/A	N/A	X	X
3. Piezometers	X	X		X			X			X	X	X	X	N/A	N/A	N/A	X	

(continued)

TABLE 1-6. (continued)

Property	Sample Meas Vol		Reliability and Life Expectancy		Degree of Complexity		Direct/ Indirect		Media		Effect on Flow Region		Effect of Haz Waste Type on Res	
	Point/Line	Bulk	Good	Quest.	Simple	Complex	Dir	Ind	Porous	Fract.	Yes	No	Yes	No
Pollutant Mobility														
I. Chemical/microbial properties w/liquid wastes in unsat region														
A. Solids samples/extraction of pore fluid														
1. Obtaining solids samples														
a. Hand auger	X		X		X		X		X	X	X			X
b. Power auger	X		X		X		X		X	X	X			X
2. Obtaining solution samples														
a. Hydraulic squeezer	X		X	X	X		X		X			X	X	X
b. Field sat. extract.	X		X	X	X		X		X			X		X
c. Lab sat. extract.	X		X	X	X		X		X			X		X
B. In situ pore water samples														
1. Lysimeters, vac oper	X		X		X		X		X	X	X			X
2. Lysimeters, vac pres	X		X		X		X		X	X	X			X
3. Lysimeters, high-pres vac	X		X		X		X		X	X	X			X
4. Filter candle	X		X		X		X		X	X	X			X
5. Hollow fiber filters	X			X	X		X		X	X	X			X
6. Membrane filters	X			X	X		X		X	X	X			X
II. Chemical/microbial properties w/liquid wastes in sat. regions														
A. Integrated samples														
1. Wells	X		X		X		X		X	X	X			X
2. Collection pans	X		X		X		X		X	X		X		X
3. Tile outflow		X	X		X		X		X	X		X		X
B. Depthwise samples														
1. Profile samples	X		X		X		X		X	X	X			X
2. Multilevel sampler	X		X		X		X		X	X	X			X
3. Piezometers	X		X		X		X		X	X	X			X

TABLE 1-7. SUMMARY, VADOSE ZONE MONITORING TECHNIQUES--NONSAMPLING METHODS

Property	Type of Site		Applicability		Power Requirements				Depth Limit			Mult Use Capability		Data Collection System			Continuous Sampling	
	New	Old	Lab	Field	ac	dc	Eng	No	Surf	<30 ft	>30 ft	Samp	Meas	Man	Auto	Remote	Yes	No
Pollutant Mobility																		
I. Salinity																		
A. Four-probe method	X	X		X		X			X				X	X			N/A	N/A
B. EC probe	X			X		X				X			X	X	X	X		
C. Salinity sensor	X			X		X				X			X	X	X	X	N/A	N/A
II. Infiltration rates																		
A. Ponds																		
1. Water budget	X	X		X				X	X			X	X	X	X	X	N/A	N/A
2. Int rate	X	X		X				X	X			X	X	X	X	X	N/A	N/A
3. Seepage meter	X	X		X				X	X				X	X			N/A	N/A
B. Land treatment areas																		
1. Water budget	X	X		X				X	X			X	X	X	X	X	N/A	N/A
2. Infiltrometer	X	X		X				X	X				X	X			N/A	N/A
3. Test plots	X	X		X				X	X			X	X	X			N/A	N/A
III. Flux																		
A. Gravimetric	X	X		X			X	X	X	X	X	X	X	X			N/A	N/A
B. Neutron moist. log	X	X		X	X	X			X	X	X	X	X	X	X		N/A	N/A
C. Gamma ray att	X	X		X	X	X				X			X	X	X		N/A	N/A
D. Tensiometers																		
1. Bourdon tube type	X	X		X			X			X			X	X			N/A	N/A
2. Transducer type	X	X		X	X					X	X		X		X	X	N/A	N/A
E. Hydrometer/psych	X	X		X		X				X		X	X	X	X	X	N/A	N/A
F. Heat diss sensor	X	X		X		X				X		X	X	X	X	X	N/A	N/A
G. Resistor/capac	X	X		X		X				X		X	X	X	X	X	N/A	N/A
H. Remote sensing	X	X		X		X	X		X			X	X	X	X	X	N/A	N/A
I. Flowmeters	X			X		X				X		X	X	X	X	X	N/A	N/A

(continued)

TABLE 1-7. (continued)

Property	Sample Meas Vol		Reliability and Life Expectancy		Degree of Complexity		Direct/ Indirect		Media		Effect on Flow Region		Effect of Haz Waste Type on Res	
	Point/Line	Bulk	Good	Quest.	Simple	Complex	Dir	Ind	Porous	Fract.	Yes	No	Yes	No
POLLUTANT MOBILITY														
I. Salinity														
A. Four-probe method		X	X		X			X	X	X		X	X	
B. EC probe	X		X		X			X	X	X		X	X	
C. Salinity sensor	X			X	X			X	X	X		X	X	
II. Infiltration rates														
A. Ponds														
1. Water budget		X	X		X			X	X	X		X	X	
2. Int rate		X	X		X			X	X	X		X	X	
3. Seepage rate	X		X		X			X	X	X		X	X	
B. Land treatment areas														
1. Water budget		X	X		X			X	X	X		X	X	
2. Infiltrometer	X		X		X			X	X	X		X	X	
3. Test plots		X	X		X			X	X	X		X	X	
III. Flux														
A. Gravimetric	X		X		X			X	X	X		X		X
B. Neutron moist. log	X		X		X			X	X	X		X	X	
C. Gamma ray att		X	X		X			X	X	X		X		X
D. Tensiometers														
1. Bourdon tube type	X		X		X			X	X			X	X	
2. Transducer type	X			X		X		X	X			X	X	
E. Hydrometer/psych	X		X	X		X	X	X				X	X	
F. Heat diss sensor	X		X	X		X	X	X	X			X	X	
G. Resistor/capac	X		X	X		X	X	X	X			X	X	
H. Remote sensing		X	X	X		X	X	X	X	X		X	X	
I. Flowmeters	X		X	X		X	X	X	X	X		X	X	

2. Geohydrologic Framework of the Vadose Zone

The geological profile extending from ground surface to the upper surface of the principal water-bearing formation is called the vadose zone. As pointed out by Bouwer (1978), the term "vadose zone" is preferable to the often-used term "unsaturated zone" because saturated regions are frequently present in some vadose zones. The term "zone of aeration" is also often used synonymously.

Davis and De Wiest (1966) subdivided the vadose zone into three regions designated as: the soil zone, the intermediate vadose zone, and the capillary fringe. The surface soil zone is generally recognized as that region that manifests the effects of weathering of native geological material. The movement of water in the soil zone occurs mainly as unsaturated flow caused by infiltration, percolation, redistribution, and evaporation (Klute, 1969). In some soils, primarily those containing horizons of low permeability, saturated regions may develop during surface flooding, creating shallow perched water tables.

The physics of unsaturated soil-water movement has been intensively studied by soil physicists, agricultural engineers, and microclimatologists. In fact, copious literature is available on the subject in periodicals (Journal of the Soil Science Society of America, Soil Science) and books (Childs, 1969; Kirkham and Powers, 1972; Hillel, 1971). Similarly, a number of published references on the theory of flow in perched water tables are available (Luthin, 1957; van Schilfgaarde, 1970). Soil chemists and soil microbiologists have also attempted to quantify chemical-microbiological transformations during soil-water movement (Bohn, McNeal, and O'Connor, 1979; Rhoades and Bernstein, 1971; Dunlap and McNabb, 1973).

Weathered materials of the soil zone may gradually merge with underlying deposits, which are generally unweathered, comprising the intermediate vadose zone. In some regions, this zone may be practically nonexistent, the soil zone merging directly with bedrock. In alluvial deposits of western valleys, however, this zone may be hundreds of feet thick. Figure 2-1 shows a geologic cross section through a vadose zone in an alluvial basin in California. By the nature of the processes by which such alluvium is laid down, this zone is unlikely to be uniform throughout, but may contain micro- or macrolenses of silts and clays interbedding with gravels. Water in the intermediate vadose zone may exist primarily in the unsaturated state, and in regions receiving little inflow from above, flow velocities may be negligible. Perched groundwater, however, may develop in the interfacial deposits of regions containing

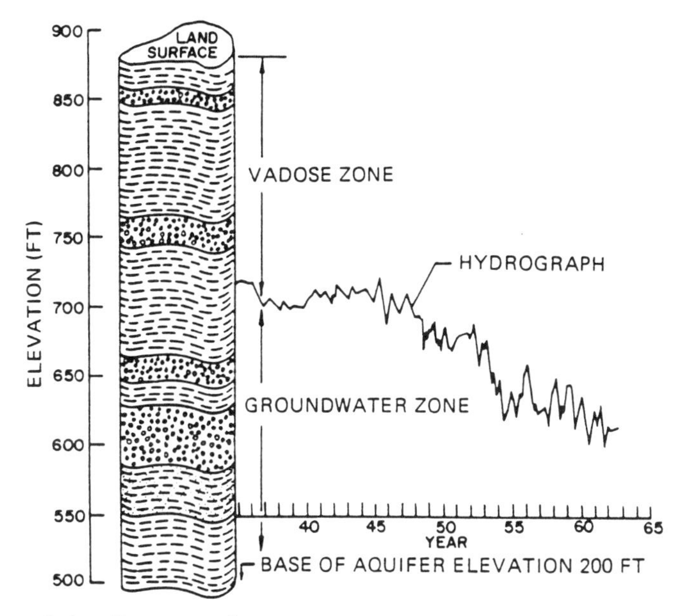

Figure 2-1. Cross section through the vadose zone and groundwater zone (after Ayers and Branson, 1973).

varying textures. Such perching layers may be hydraulically connected to ephemeral or perennial stream channels so that, respectively, temporary or permanent perched water tables may develop. Alternatively, saturated conditions may develop as a result of deep percolation of water from the soil zone during prolonged surface application. Studies by McWhorter and Brookman (1972) and Wilson (1971) have shown that perching layers intercepting downward-moving water may transmit the water laterally at substantial rates. Thus, these layers serve as underground spreading regions transmitting water laterally away from the overlying source area. Eventually, water leaks downward from these layers and may intercept a substantial area of the water table. Because of dilution and mixing below the water table, the effects of recharge may not be noticeable until a large volume of the aquifer has been affected.

The number of studies on water movement in the soil zone greatly exceeds the studies in the intermediate zone. Reasoning from Darcy's equation, Hall (1955) developed a number of equations to characterize mound (perched groundwater) development in the intermediate zone. Hall also discusses the hydraulic energy relationships during lateral flow in perched groundwater. More recently, Freeze (1969) attempted to describe the continuum of flow between the soil surface and underlying saturated water bodies.

The base of the vadose zone, the capillary fringe, merges with underlying saturated deposits of the principal water-bearing formation. This zone is not characterized as much by the nature of geological materials as by the presence of water under conditions of saturation or near saturation. Studies by Luthin and Day (1955) and Kraijenhoff van deLeur (1962) have shown that both the hydraulic conductivity and flux may remain high for some vertical distance in the capillary fringe, depending on the nature of the materials. In general, the thickness of the capillary fringe is greater in fine materials than in coarse deposits. Apparently, few studies have been conducted on flow and chemical transformations in this zone. Taylor and Luthin (1969) report on a computer model to characterize transient flow in this zone and compare results with data from a sand tank model. Freeze and Cherry (1979) indicate that oil reaching the water table following leakage from a surface source flows in a lateral direction within the capillary fringe in close proximity to the water table. Because oil and water are immiscible, oil does not penetrate below the water table, although some dissolution may occur.

The overall thickness of the vadose zone is not necessarily constant. For example, as a result of recharge at a water table during a waste disposal operation, a mound may develop throughout the capillary fringe extending into the intermediate zone. Such mounds have been observed during recharge studies (e.g., Wilson, 1971) and efforts have been made to quantify their growth and dissipation (Hantush, 1967; Bouwer, 1978). Sampling techniques in the groundwater zone also modify the overall thickness of the vadose zone, as is apparent in Figure 2-2.

As already indicated, the state of knowledge of water movement and chemical-microbiological transformations is greater in the soil zone than elsewhere in the vadose zone. Renovation of applied wastewater occurs primarily in the soil zone. This observation is borne out by the well-known studies of McMichael and McKee (1966), Parizek et al. (1967), and Sopper and Kardos (1973). These studies indicate that the soil is essentially a "living filter" that effectively reduces certain microbiological, physical, and chemical constituents to safe levels after passage through a relatively short distance (e.g., Miller, 1973; Thomas, 1973). As a result of such favorable observations, a certain complacency may have developed with respect to the need to monitor only in the soil zone. However, in one reported instance at a disposal operation for cannery wastes, an apparent breakdown of the living filter occurred because of excessive soil loading with sucrose and fructose (W.R. Meyer, personal communication, 1979). There is growing concern, moreover, that certain toxic substances, such as chlorinated hydrocarbons, other refractory organics, and microorganisms may escape the soil zone, eventually passing into underlying groundwater (Walker, 1973; Ongerth et al., 1973; Shuval and Gruener, 1973; Robertson, Toussaint, and Jorque, 1974). For example, when soils are thin and underlain in the intermediate zone by fractured rocks, short circuiting through these materials may occur directly into groundwater. Thomas and Phillips (1979) have shown that for well-structured soils, rapid water movement occurs in macropores between soil peds. Consequently, entrained pollutants may not interact with the bulk of the soil matrix.

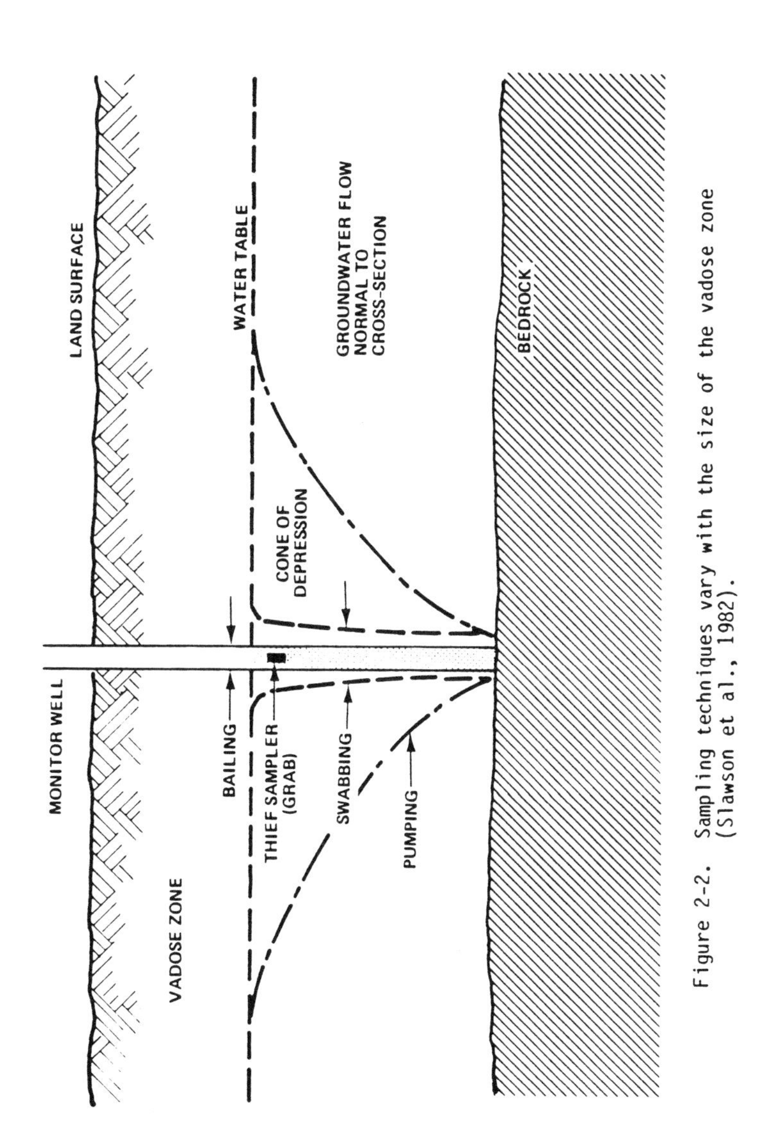

Figure 2-2. Sampling techniques vary with the size of the vadose zone (Slawson et al., 1982).

Dunlap and McNabb (1973) also point out that microbial activity may be significant in the regions underlying the soil. They recommend that investigations be conducted to quantify the extent that such activity modifies the nature of pollutants travelling through the intermediate zone.

For the soil zone, numerous techniques were compiled by Black (1965) into a two-volume series entitled "Methods of Soil Analyses." Monitoring in the intermediate zone and capillary fringe will require the extension of technology developed in both the soil zone and in the groundwater zone. Examples are already available where this approach has been used. For example, Apgar and Langmuir (1971) successfully used suction cups developed for in situ sampling of the soil solution at depths up to 50 feet below a sanitary landfill. W.R. Meyer (personal communication, 1979) reported that suction cups were used to sample at depths greater than 100 feet below land surface at cannery and rock phosphate disposal sites in California.

REQUIREMENTS OF MONITORING IN THE VADOSE ZONE

The basic requirement of a monitoring program in the vadose zone is to characterize three properties: (1) storage properties, (2) flow rates, and (3) spatial and temporal changes in pollutants. Storage properties are important in that pollutants are placed in temporary storage within the void space of the vadose zone before movement in groundwater. In some cases, such as in western valleys, the storage space may be sufficient to preclude movement into productive aquifers. In fact, Winograd (1974) suggests using the vadose zone for the storage of low-grade radioactive wastes. Mann (1976) recommends using the storage space of the vadose zone in arid regions to augment the above-ground storage of tailings ponds. Wilson, Osborne, and Percious (1968) showed that by locating a disposal pond for blowdown effluent near an ephemeral stream, periodic recharge events diluted the effluent stored in the vadose zone.

Information on the flux of water (the quantity of water flowing through a unit area per unit time) may indicate the macroscopic velocity of water in the vadose zone. Thus, if a piston flow model is assumed, the arrival time of pollutants at the water table may be determined from estimates of flux and the thickness of the vadose zone. (As will be shown later, a more realistic estimate of particle velocity is the average linear velocity, which accounts for variations in the cross sectional areas of the flow path and degree of saturation.)

Pollutants may not necessarily travel at the same rate as water because of attenuation in the vadose zone. Consequently, the travel time of water represents an upper limit on the mobility of most pollutants. Some pollutants may move at a higher rate than water because of physical and chemical properties. For example, based on differences in viscosity, the velocity of gasoline should be about 1.5 times that of water (Schwille, 1967). In considering rate, it is important to quantify both the infiltration rate at the ground surface and the percolation rate through the vadose zone. Recharge refers to the movement of water and pollutants across the water table.

Information on the third element, change in pollutants during transit within the vadose zone, is needed to quantify the mobility and spatial distribution of specific pollutants. The vadose zone may attenuate certain pollutants sufficiently to exclude them as pollution sources. For example, Runnells (1976) demonstrated that soil from Sulfur Spring, New Mexico, has an enormous capacity to remove copper from mill water. The additional observation that copper removal was irreversible indicates that thousands of years would be required before groundwater (located at about 100 feet) at the site would be affected.

PRELIMINARY MONITORING OF THE VADOSE ZONE

Before initiating a vadose zone monitoring program, existing monitoring efforts should be assessed and relevant background information collected. Such a "nonsampling" approach will aid in the selection of methods to offset monitoring gaps and generally increase the chances of designing an effective system. Assessing existing monitoring includes determining sample collection methods for the source, defining the network of existing monitoring wells including the locations and types of wells, and determining methods for sampling groundwater. Background information should be obtained on the following items: (1) nature of pollutants, (2) pollutant source loading, (3) existing groundwater usage and quality, and (4) a description of the geohydrologic framework, including infiltration potential and hydraulic/geochemical properties of the vadose and groundwater zones.

The existing source monitoring program could be evaluated by comparing the ongoing sampling techniques with recommended approaches described in recent published guidelines, such as the report of Huibregtse and Moser (1976). Similarly, the groundwater monitoring program could be compared with approaches recommended by Mooij and Rovers (1976) and Everett (1979, 1980). Approaches for collecting background quality information on a source and existing groundwater usage and quality have been described in detail by Todd et al. (1976) and Fenn et al. (1977). The quality data should be examined for completeness; for example, gaps may exist in analyses. Specific pollutants should be delineated and ranked.

As a first step in characterizing the geohydrologic framework with particular emphasis on the vadose zone, all relevant data should be collected. For example, comprehensive soil survey data may be available through State agricultural experiment stations, the Soil Conservation Service, or other agencies. Such surveys generally include information on depths of soil horizons, presence of hardpan, clay pan, color, porosity, structure, texture, organic matter content, soil pH, and infiltration rates.

For deeper horizons in the lower vadose zone, data may be available from drillers' well logs, Water Supply Papers of the U.S. Geological Survey (USGS), university theses, dissertations, and special reports. In general, well logs from drilling companies mainly describe conditions below the water table. However, they should be consulted for possible clues on stratification. In some areas, the USGS collects drill cuttings from drilling sites. These samples are evaluated in the laboratory for particle size distribution, color, and source of origin. Some universities also collect samples as part of their hydrology or agricultural engineering programs. Again, the purpose of these sampling programs is to characterize the saturated zone, but valuable clues on the vadose zone may also be included (Matlock, Morin, and Posedly, 1976). Historic trends in groundwater levels should be determined by examining existing water level contour maps.

In addition to collecting available data, wells in the area of a monitoring site should be used to the fullest extent. For example, static water levels in a network of wells can be plotted to give the water table configuration and thus an approximate idea of the thickness of the vadose zone. Also, the presence of cascading water in wells will indicate the possible presence of perching layers. Finally, abandoned wells, or wells in which the pumps have been removed for servicing, may be used with borehole geophysical methods (described later) to provide further clues on the vadose zone.

After exhausting all existing sources for clues on the nature of the vadose zone at a site, a test drilling program may be considered necessary to obtain more detailed information. The extent and thoroughness of the program will depend on the availability of funds. At any rate, the program should be carefully planned to ensure the maximum value of the data. For example, a careful grid should be laid out for the project area. Also, test well locations should recognize the possibility of their eventual use as observation wells, piezometers, or access wells during an actual monitoring program.

In general, samples from the soil zone are obtained to: (1) incrementally characterize the average soil texture, water content, and chemistry (say, 6-inch increments); (2) observe the precise depth-distribution of soil texture; or (3) determine the bulk density or water-release curves of soil increments. For the first purpose, samples obtained using post-hole augers, screw or sleeve-type augers, or power-driven augers are useful. For the second purpose, cores are obtained by driving small-diameter tubes into the soil to the desired depth. For the third purpose, larger-diameter core samplers are used. These may be hand-driven or forced into the soil by power-driven hydraulic soil coring equipment. With these methods, cores of a specific volume are obtained. More information on coring is presented by Blake (1965).

Sampling throughout the vadose zone may require drilling deep wells using standard techniques. Such techniques include jetting, rotary, cable-tool, augering, and air drilling. Of these methods, perhaps augering, using continuous flights, and air drilling provide the most usable samples. Problems develop in characterizing the distribution of indigenous salt and water content with cable-tool and rotary methods because of water additions during the drilling process.

One air drilling technique involves driving a double-wall tube with a pile hammer while forcing air under pressure down the annulus of the pipe. Air and entrained material cut by the bit return to the surface through the inside pipe. The continuously available sample is diverted into a cyclone sampler where it is bagged for laboratory analyses. According to the manufacturer, formation changes can be determined within a few inches. Furthermore, water seams can be determined immediately. This feature is advantageous in locating the depth and thickness of perching layers. Whichever technique is used, samples should be taken in increments throughout the vadose zone.

Drill cuttings could be examined in the laboratory for particle-size distribution and other geologic parameters. Knowing the vertical distribution of grain size in the hole, the location of potential perching zones may be delineated. The collected samples can also be used to estimate storage and transmissive properties of the vadose zone.

Following construction of a test well or wells, geophysical logging should be initiated. First, the test hole(s) should be logged. The results of logging in the test well should be correlated with drill cutting analyses to delineate stratigraphy. Subsequent logs from other wells can thus be interpreted and the lateral and vertical extent of various layers (e.g., those favoring perching) defined.

Among the common borehole logging techniques, nuclear logging is the most useful within the vadose zone. The spontaneous potential resistance and acoustic logs require an uncased fluid-filled well. These conditions are generally not available in the vadose zone unless the wells penetrate confining beds with a high piezometric surface or water is added to the borehole.

Common nuclear logging techniques include natural gamma logging, gamma-gamma logging, and neutron logging. The principles of and procedures in applying these techniques in groundwater investigations are discussed thoroughly by Keys and MacCary (1971). Briefly, natural gamma logging comprises the detection of natural gamma activity of rocks. Keys and MacCary (1971) list the following sequence of sedimentary rocks in ascending order of natural gamma activity: anhydrite, coal, salt, dolomite, sandstone, sandy shale, shale, organic marine shale, and potash beds. Natural gamma logs are particularly advantageous in characterizing the vertical extent of sediments and tracing them laterally from well to well (Keys and MacCary, 1971; Norris, 1972).

Gamma-gamma logs record the intensity of radiation from a source in a down-hole probe after it has backscattered and attenuated in the well and surrounding media (Keys and MacCary, 1971). The down-hole probe contains a source of gamma photons, such as cobalt-60 or cesium-137, and a sodium iodide detector. The principal uses of gamma-gamma logs are to identify the lithology and to estimate bulk density and porosity of rocks (Keys and MacCary, 1971).

Neutron logging comprises lowering a down-hole probe containing a source of high-energy neutrons and a detector for thermalized neutrons into a well. Neutron logs detect the hydrogen content, and consequently water content, of sediments within the vadose zone and the porosity of sediments below the water table. Neutron logs are useful in detecting the presence of perched groundwater. For example, Keys (1967) used a combination of gamma logs and neutron logs to delineate a clay perching bed at the National Reactor Testing Station in Idaho. Neutron logging between wells may indicate the lateral extent of perched groundwater bodies.

3. Premonitoring of Storage at Disposal Sites

STORAGE

Physical properties of the vadose zone associated with storage of water include: (1) total thickness, (2) porosity, (3) bulk density, (4) water content, (5) soil-water characteristics, (6) field capacity (specific retention), (7) specific yield, and (8) fillable porosity. Technical information on each of these properties may be found in reference works by Childs (1969); Hillel (1971); Davis and DeWiest (1966); Cooley, Harsh, and Lewis (1972); Freeze and Cherry (1979); Bouwer (1978); and Brakensiek, Osborn, and Rawls (1979).

Thickness

The storage capacity of a vadose zone is related to the overall thickness, i.e., depth from ground surface to the water table. As shown in Figure 2-1, the thickness is not constant but changes in response to pumping or recharge. The theoretical depth of water that could be stored in a vadose zone assuming 100-percent saturation is expressed by the following relationship:

$$d_w = \left(\frac{\Delta P_w}{100}\right)\left(\frac{D_b}{D_w}\right) d_{vz} \tag{3-1}$$

where d_w = depth of water applied during surface flooding

D_b = bulk density of soil

D_w = density of water

d_{vz} = depth of vadose zone to be wetted

ΔP_w = difference in the percentage of water on a mass basis at saturation and after oven drying (105°C for 24 hours).

As an example of the use of this equation to estimate the theoretical maximum storage capacity of a vadose zone, consider the following hypothetical conditions:

1. 100-foot thick vadose zone, uniform and unlayered, e.g., d_{vz} = 100 feet

2. Bulk density of media = 81.1 lb/ft^3 = 1.3 gm/cm^3

3. Density of water = 62.4 lb/ft^3 = 1.0 gm/cm^3

4. Water content of drained vadose zone material = 10 percent (dry weight basis)

5. Water content of vadose zone material assuming 100-percent saturation = 30 percent (dry weight basis).

Using equation 3-1, the depth of water, d_w, that could be stored in the 100-foot thick hypothetical profile would be:

$$d_w = \frac{30 - 10}{100} \times 1.3 \times 100$$

$$= 26 \text{ feet.}$$

In other words, 26 acre-feet of water per acre of surface area would theoretically be required to saturate the 100-foot vadose zone. In reality, the vadose zone would never attain 100-percent saturation, even for an ideal medium, because of entrapped air and because of drainage of the medium to field capacity.

In multilayered media, the individual bulk densities of the various layers must be considered in using equation 3-1 to calculate total storage. Note also that only vertical storage is considered by this relationship. For point and line sources, lateral storage should also be accounted for.

The thickness of the vadose zone underlying the pollution source may be estimated directly by determining water levels in the area using either existing wells or specifically constructed wells. Table 3-1 lists constraints on determining the depth to water. If the area is underlain by perched groundwater that is a source of potable water, the depth to the perched water table should be used as the depth of the vadose zone (Silka and Swearingen, 1978). For confined leaky aquifers, the depth to the overlying unconfined water table should be used. For nonleaky confined aquifers, the distance to the base of the uppermost confining layer should be used. When estimating storage for pollutants, historic minimum depth to groundwater should be used to avoid overestimating vadose zone storage.

Indirect methods have been employed to detect the presence of shallow perched groundwater and estimate the depth of the vadose zone. For example, Estes et al. (1978) report on using three imagery techniques to detect perched groundwater: visible and reflected infrared imagery, thermal infrared imagery, and microwave techniques. They evaluated the potential of these techniques for detecting perched water tables within 5 to 10 feet of the surface in the lower San Joaquin Valley, California. Thermal infrared imagery appeared to be of greatest value in water table detection because of unique soil and water thermal characteristics. In particular, the flux of heat through soils underlain by shallow water tables differs from that through dry soils. Consequently, a diurnal surface anomaly occurs that appears warmer at night

TABLE 3-1. CONSTRAINTS ON DETERMINING DEPTH TO WATER TABLE

Category	Constraint
Geohydrologic	1. If water levels are moving either upward or downward in the aquifer, water level measurements will not be exact. 2. Cascading water may cause higher water levels inside well than outside. 3. In composite aquifer systems, the water level is affected by varying head differentials, permeabilities, and recharge rates (Saines, 1981). Thus, water levels may not be the same as for the top of the first aquifer.
Chemical	Movement of saltwater into a casing may disperse salts throughout the well, lowering the water level below the true value (Kohout, 1960). Thus, thickness of vadose zone may be overestimated.
Geologic	1. If the aquifer is confined by tight geologic formations, the water levels in wells will represent the pressure surface of water in the aquifer. Generally, the pressure surface is higher in elevation than the confining beds. If the geology is not known or understood and if the water level is assumed to be that of a water table, then the thickness of the vadose zone will be underestimated. 2. Faults may act as flow barriers or as drains, resulting in differing water levels on opposite sides of the fault (Freeze and Cherry, 1979). If water levels are obtained only in a well on one side of the fault, the thickness of the vadose zone will be either underestimated or overestimated on the other side.
Topographic	Slope, geology, and pore water pressures are related (Freeze and Cherry, 1979). Main effect on water levels is the presence of faults in slope areas. (See Geologic Constraints.)
Physical	1. Type of well associated with data must be known, e.g., deep perforated observation wells, shallow observation wells, piezometers. Water levels will be different in the three types of wells in recharge and discharge areas, but not in areas of lateral flow (Saines, 1981). (See Geohydrologic Constraints.) 2. Water level measurements may not have been exact at time of measurement because of air pressure changes in vadose zone or barometric pressure changes in a piezometer (Bouwer, 1978). 3. If wells are being pumped, or if there are nearby pumping wells at time of measurement, water levels will not represent static conditions and the thickness of the vadose zone will be overestimated.
Climatic	If the barometric pressure is unusually affected by climate at time of measurements, unrepresentative water levels may be obtained, resulting in false estimation of vadose zone thickness.

and cooler during the day than surrounding areas that do not have perched water tables.

Bulk Density and Porosity

As shown by equation 3-1, bulk density values are required in order to estimate total water storage in the vadose zone. Bulk density represents the density of a soil or rock, including solids plus voids, after drying (Bouwer, 1978). To determine the bulk density of a material, field cores of a precise volume are oven-dried at $105^{o}C$ until constant weight is obtained. The bulk density is the oven-dry mass divided by the sample volume. Blake (1965) and Paetzold (1979) also reviewed the alternative "clod method" for determining bulk density of disturbed samples. Constraints on using gravimetric methods are given in Table 3-2.

Bulk density values range from 68.6 lb/ft^3 (1.1 gm/cm^3) for clay soils to 100 lb/ft^3 (1.6 gm/cm^3) in sandy soils (Hillel, 1971).

As defined by Bouwer (1978), the porosity of a soil or rock material is the percentage of the total volume of the medium occupied by pores or other openings. Thus, the total porosity of a soil is a measure of the amount of water which could be stored under saturation. In actuality, saturation may not be reached until after a long period of time because of entrapped air.

According to Vomicil (1965), total porosity may be calculated by the following expression:

$$S_t = 100 \left(1 - (D_b/\rho_b)\right) \qquad (3\text{-}2)$$

where S_t = total porosity, the percentage of the bulk volume not occupied by solids

D_b = bulk density

ρ_b = particle density, approximately 165.4 lb/ft^3 (2.65 gm/cm^3) for mineral soil.

Typical values of porosity for representative earth materials are presented in Table 3-3 and Figure 3-1.

Laboratory estimation of the total porosity of a soil core requires determining the bulk density and particle density. Methods for determining the bulk density have been reviewed (equation 3-1). Particle density is measured precisely using the pycnometer method (Vomicil, 1965).

Although a knowledge of the total porosity is of value when considering storage in the vadose zone, the pore-size distribution is also important. For example, sandy soils have primarily large pores of nearly equal size permitting easy drainage (U.S. EPA et al., 1977). Medium-textured loamy soils, however, have a greater porosity than sands and also a wider pore-size distribution. "Thus, more water is held at saturation in soils than in sands and it

TABLE 3-2. CONSTRAINTS ON GRAVIMETRIC METHODS

Category	Constraint
Geohydrologic	If fluid is moving rapidly through the vadose zone, water content values will be changing within the wetting profile. Thus, the water distribution profile will change during sampling and a plotted profile will not be a true "snapshot."
Chemical	There may be a health hazard in collecting samples for measuring storage changes at some chemical waste disposal sites. Problems may also develop from the release of noxious or toxic gases during oven drying of samples.
Geologic	1. Natural heterogeneities in geologic deposits mean that water holding (storage) properties vary throughout the vadose zone. For example, water content at interfaces between layers of differing texture is greater than within the layers. Similarly, the presence of organic matter increases water holding capacity. Thus, a large number of samples are required to obtain an estimate of the average storage content. 2. Method is primarily intended for granular media. Not suitable for other types such as fractured media, carbonate rock, etc. 3. May be difficult to obtain samples in media with indurated layers (e.g., caliche), particularly with hand augers.
Topographic	Samples will be difficult to obtain using power equipment on very steep terrain.
Physical	1. Several physical factors affect the reliability of gravimetric methods for characterizing water content changes in the vadose zone. These factors include (Reynolds, 1970): a. Water content--an increase in water content of a sample is related to an increase of the variance of sample means b. Sampling method affects results, e.g., screw auger versus barrel auger c. Successive samples cannot be taken from the same hole d. Presence of stones in samples causes an underestimation of water content. 2. Samples cannot be obtained immediately beneath existing disposal facilities 3. Samples are difficult to obtain when very dry or very wet, e.g., samples will not remain in a bucket auger.
Climatic	May be difficult to obtain samples in frozen soils.

TABLE 3-3. RANGE OF POROSITY VALUES IN UNCONSOLIDATED DEPOSITS AND ROCKS (after Freeze and Cherry, 1979)

	Porosity (percent)
Unconsolidated Deposits	
Gravel	25 to 40
Sand	25 to 50
Silt	35 to 50
Clay	40 to 70
Rocks	
Fractured basalt	5 to 50
Karst limestone	5 to 50
Sandstone	5 to 30
Limestone, dolomite	0 to 20
Shale	0 to 10
Fractured crystalline rock	0 to 10
Dense crystalline rock	0 to 5

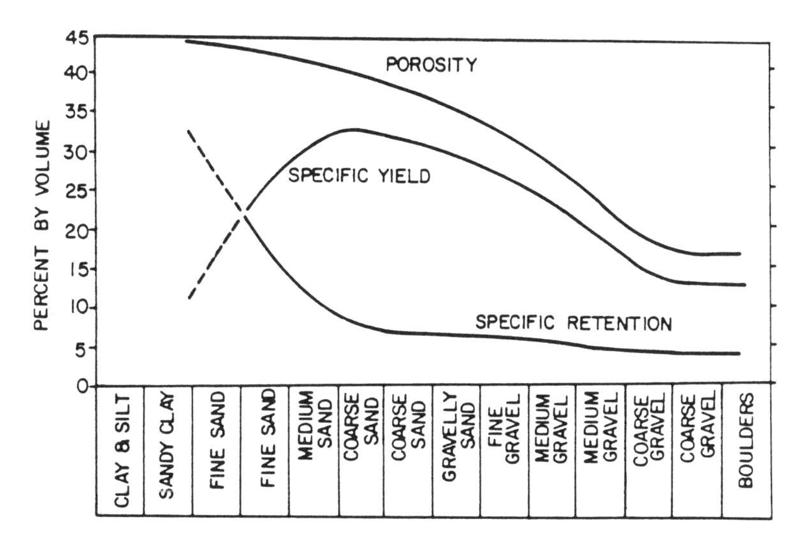

Figure 3-1. Variation of porosity, specific yield, and specific retention with grain size (after Scott and Scalmanini, 1978).

is removed much more gradually as matric potential becomes greater" (U.S. EPA et al., 1977).

The effect of macropores on retention and drainage of water in soils is receiving renewed attention (Thomas and Phillips, 1979). The effect of macropores is particularly important in well-structured soils. Water movement occurs rapidly in the interstices between soil peds and little storage may occur within the pores of the soil blocks. Pore-size distribution is commonly found by means of water characteristic curves (see Soil-Water Characteristics, this section).

The bulk density of soil may be determined in the field as well as in the laboratory. For in situ measurements, a technique developed by soil engineers is the excavation method, also called the sand cone method, described by Blake (1965). In this method, a hole is excavated to a desired depth. The hole is filled with sand from a container using a double-cone arrangement. Ottawa sand is commonly used because the bulk density is accurately known. Thus, knowing the mass of sand to fill the cavity, the associated volume is calculated. The mass of soil removed from the hole is determined by oven drying. The bulk density of the soil is then determined by dividing the dry soil mass by the calculated volume.

Alternatively, a balloon is placed in the cavity and subsequently filled with water until the cavity is completely filled. The volume of water to fill the hole is measured. The soil excavated from the hole is oven-dried and weighed. The dry mass of the soil divided by the water volume is equal to the bulk density.

An indirect method for determining bulk density in-place is the gamma radiation method. The gamma radiation method is discussed next under the heading of "Water Content."

Once the bulk density has been determined, the total porosity may be calculated from the dry weight bulk density using the relationship given in equation 3-2.

Water Content

The total porosity of vadose zone materials represents the upper limit for water storage in the vadose zone. Unless the vadose zone has not been wetted since deposition, a certain residual amount of water remains in the pores. Thus, the available storage is diminished by this residual water content. Water content is expressed on either a mass or a volume basis using equations presented by Brakensiek, Osborn, and Rawls (1979):

$$P_w = \frac{W_w - W_d}{W_d} \times 100 \qquad (3\text{-}3)$$

and

$$P_v = \frac{V_w}{V_s} \times 100 \qquad (3\text{-}4)$$

where P_w = moisture percent on a dry weight basis

W_w = wet weight of soil

W_d = dry weight of soil

P_v = volumetric moisture percentage

V_w = volume of water in sample

V_s = bulk volume of soil.

As pointed out by Bouwer (1978), the volumetric water content is more useful for hydrological purposes than is the mass water content. When expressed as a fraction, the volumetric water content is denoted by the symbol θ. The basic laboratory technique for determining water content of a soil consists of (1) weighing the sample, (2) drying the sample for a set period of time (e.g., at 105°C for 24 hours), and (3) reweighing the dried sample. These relationships are then used to calculate P_w and P_v. An alternative method entails placing a soil sample in a sealed container with calcium carbide reagent. The calcium carbide reacts with the water in the sample to produce a gas. The gas pressure, registered on a gauge, is converted into water content on a wet weight basis. Combined drying and weighing units are also commercially available for laboratory determination of water content. Gardner (1965) reviewed other methods, including the alcohol drying technique.

Principle of the Neutron Moderation Method--

The method of water content evaluation by neutron moderation depends on two properties related to the interaction of neutrons with matter: scattering and capture (Gardner, 1965). High-energy neutrons emitted from a radioactive source may be slowed down or thermalized by collisions with atomic nuclei. The statistical probability for such thermalization depends on the "scattering cross section" of various nuclei. The scattering cross section of hydrogen causes a greater thermalizing effect on fast neutrons than many elements commonly found in soils. This forms the basis for detecting the concentration of water in a soil (van Bavel, 1963). The second property of interest in the neutron moderation method is capture of slow neutrons by elements present in the soil with the release of other nuclear particles or energy. Cadmium and boron have extremely large capture cross sections compared with hydrogen. The property of energy release during capture serves as a means of detecting the concentration of slow neutrons.

In operation, when a source of fast neutrons is lowered into a soil through a suitable well or casing, a cloud of thermalized neutrons is established (Gardner, 1965). This cloud reflects the moderating effect of scattering cross sections of nuclei in the soil mass on fast neutrons. If a suitable calibration is made to isolate the moderating effects of soil nuclei other than hydrogen, changes in the volume of the thermalized cloud will reflect changes in water content. In general, the wetter the soil, the smaller the volume of thermalized neutrons. Finally, a detector that relies on capture of thermalized neutrons, in conjunction with suitable electronic circuitry, may be used to measure the water volume.

Instrumentation used to measure the water content by neutron thermalization requires three principal components: (1) a source of fast neutrons, (2) a detector of slow neutrons, and (3) an instrument to determine the count rate from the detection equipment (Figure 3-2).

Early moisture loggers relied on radium-beryllium as a fast neutron source. Because of radiation health problems with radium-beryllium, bulky shielding was necessary. A common source in today's units is americium-beryllium. Smaller units, such as those used by agriculturists, contain sources in the millicurie range (e.g., 50, 100, 200 millicuries) for small-diameter casing (e.g., 2-inch). Larger units with multicurie sources, such as those used in the petroleum industry for porosity determination, are used to log wells with larger casings.

Detectors of slow neutrons comprise materials with high capture cross sections (e.g., boron). A charged particle is emitted during capture, which is then detected by a solid or gaseous counting device (van Bavel, 1963). The predominant detector is lithium-enriched BF_3.

Both the source and detector are located in a cylindrical tool that is lowered via a cable into access tubing. The depth of measurement is determined either by graduations on the cable or by a counter. The relative positions of source and detector in the design of the tool are important. This question is reviewed in detail by van Bavel (1963).

The pulse emitted during the nuclear reaction resulting from capture is amplified by a preamplifier within the down-hole tool and transmitted through the cable to an aboveground meter for counting. Two portable meters are commonly used: rate meters or scalers. Gardner (1965) discusses the relative advantages of each. Count rate is converted to water content via suitable calibration curves.

The type of meter available for soils work usually involves placing the down-hole tool at a discrete depth, taking a count or number of counts, and then moving the tool by hand to another discrete depth. This process is not troublesome for shallow soils studies. However, for deeper access wells in the intermediate zone, an inordinate amount of time would be involved. A motorized unit is available that permits lowering the tool in the well at a constant rate. Concurrently, internal electronic components translate and record pulse rate as water content.

Agriculturists have employed the principle of neutron moderation or thermalization to measure the volumetric water content of soils in situ for many years. Recently, the technique has been used by geohydrologists to monitor water storage in the intermediate zone, particularly to delineate perching layers and mounds (Keys, 1967), and also to estimate flow rates. Figure 3-3 is a sequence of moisture logs from a well near the Santa Cruz River in Tucson showing the growth and dissipation of mounds during an ephemeral recharge event.

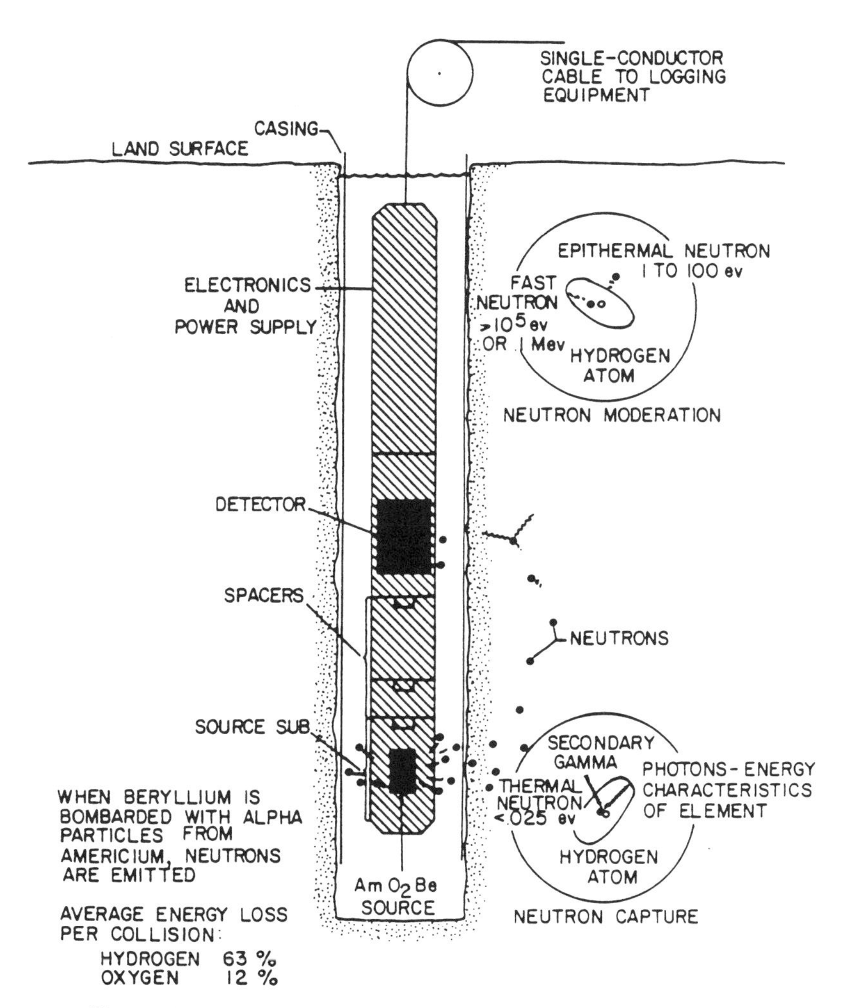

Figure 3-2. Equipment and principles of neutron moisture logging (after Keys and MacCary, 1971).

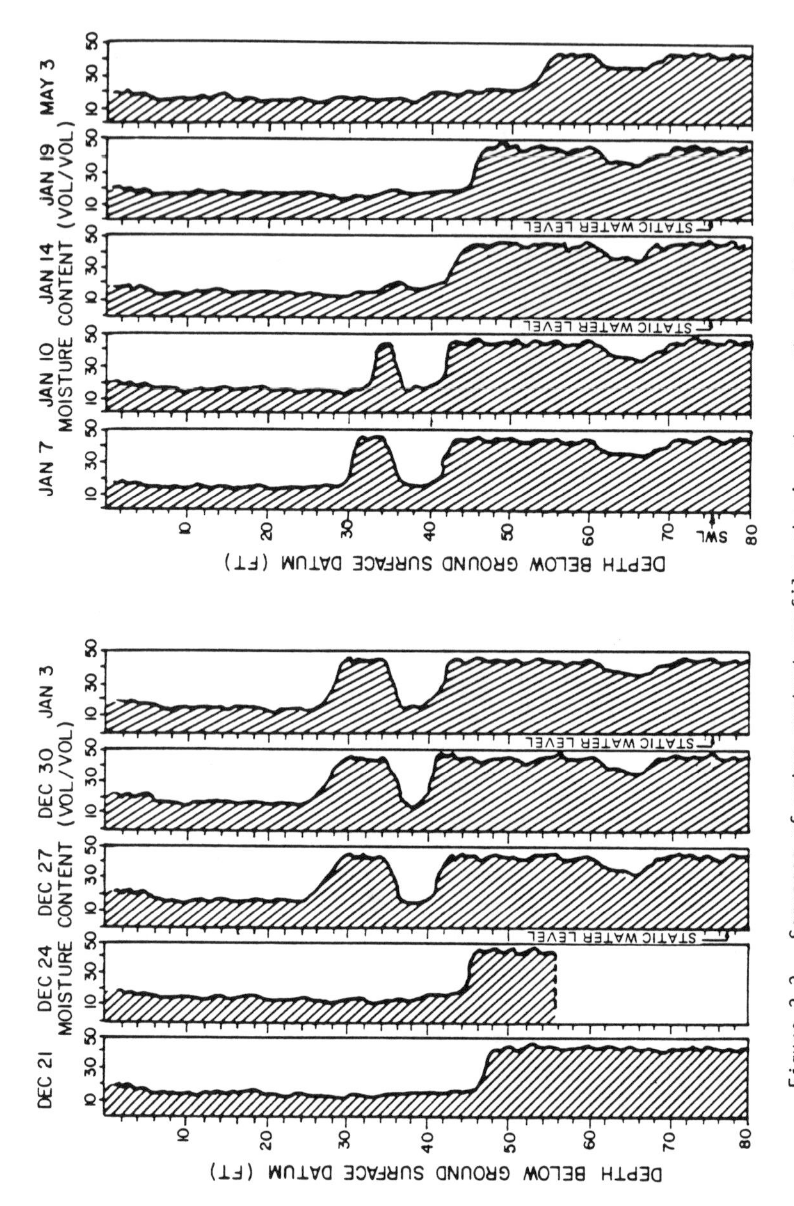

Figure 3-3. Sequence of water content profiles showing the growth and dissipation of perched groundwater during recharge from an ephemeral stream (after Wilson and DeCook, 1968).

Field implementation--Access tubes for neutron moisture logging are usually constructed of seamless steel or aluminum. Plastic wells may cause difficulty in that hydrogen or chlorine atoms in the tubing may moderate the thermal neutrons, interfering with soil moisture evaluation. Aluminum tubes may deteriorate in saline water.

The inside diameter of wells should be as close as possible to the outside diameter of the probe. Work by Ralston (1967) showed that for a 100-mc Am-Be source in a 1.5-inch tool, water content could not be accurately evaluated in wells greater than 4 inches. Brakensiek, Osborn, and Rawls (1979) recommend that the radial air gap does not exceed 0.02 inch.

Wells installed for shallow water content monitoring can be easily installed by successively augering within and driving a tube. Myhre, Sanford, and Jones (1969) report on a simple power-driven auger for installing wells to about 5-foot depths.

For deeper wells, standard drilling techniques are necessary, e.g., percussion drilling, rotary drilling, etc. During installation by drilling techniques, a tight fit between the borehole and the well casing is essential to minimize vertical leakage of water. Drilling mud is not recommended for the drilling operation because it inhibits caving around the well casing and also interferes with water content observations. Some workers have attempted to backfill the gap between the borehole wall and the casing with fine material (Brakensiek, Osborn, and Rawls, 1979) or with grout. Field studies by Halpenny (personal communication, 1979) show that bound water within grout markedly attenuates the flux of fast neutrons from a source, with attendent sensitivity reduction of a logger in detecting water content changes.

Achieving a tight fit of the logging probe access tube in the borehole may be difficult in rocky material or in deep consolidated vadose zones. Placement of the access tube to intersect permeable zones that transmit a large percentage of the groundwater flow is of importance. In well indurated, highly fractured media, the primary flow paths occur in cracks. If the access well does not intercept these cracks, no movement of water or waste fluids will be indicated and a potential problem could go undetected. In general, nonuniform conditions (e.g., stones and cracks) reduce the reliability of the method (Gairon and Hadas, 1973).

Installation of the access tube for the neutron moderation method requires a tight fit between the tube and borehole. This fit is especially important for installations at existing landfills or impoundments because of the possibility of providing a direct path for pollutants along the access well, thereby short-circuiting the waste through the vadose zone. Side leakage it will also cause erroneous readings (Gairon and Hadas, 1973).

Neutron loggers should be calibrated against samples of known water content. Field calibration for the neutron logger is described by Hsieh and Enfield (1974). This method utilized 55-gallon oil drums fitted with aluminum tubing down the center similar to that used for the field installations. The drums were filled with local soil. If drill cuttings are available for the depth range to be surveyed by the probe, these should be used in developing the calibration curve. Water content of the test soils was adjusted by adding a known volume of water to some samples or drying others through spreading on the ground. A calibration curve for the neutron probe was developed from multiple readings within the drum containing test soils.

Range of applications and limitations--The principal advantage of neutron moisture logging is that water content profiles are obtained in situ without disturbing the soil. Thus, a history of profiles can be established at the same site during a monitoring program. Moisture logs clearly show the presence of perched water tables, together with their growth and dissipation (see Figure 3-3).

Although the neutron thermalization technique is used mainly to determine changes in volumetric water content of materials in the vadose zone, the method may also be used to infer water movement. In particular, if a soil-water characteristic curve is available for incremental depths throughout the vadose zone, it may be possible to relate water content values to water pressure. Hydraulic head gradients, and therefore flow direction, could then be inferred. The accuracy of the method may not be great enough, however, to detect slight water content changes, particularly in the dry range.

Wilson and DeCook (1968) and Wilson (1971) used moisture logs from a network of 100-foot-deep access wells to infer the rates of lateral movement of recharge waves in the vadose zone during river recharge and during artificial recharge. The arrival of such waves was inferred by the change in water content in a perched mound at about 33 feet.

The neutron moderation method determines water content and changes in water content when flow conditions prevail. Some regions, however, may transmit water without detectable changes in water content, thereby leading to an erroneous conclusion that water and waste material are not flowing. It should be kept in mind that the method indicates change in storage capacity but tells nothing about the energetics of water movement.

Because interpretation and inference of water content from neutron logging is based on moderation by hydrogen, the presence in excessive concentrations of other fast neutron moderators (e.g., boron, chlorine, and hydrocarbons) may cause erroneous water content determinations. Further errors can occur when water content values are converted to standard units (soil-water suction values) via a soil-water characteristic curve. Chemicals may also cause deterioration of some access tubes (e.g., aluminum).

Water Content Measured by Gamma Radiation Methods--

Gamma radiation methods are also used for indirectly determining bulk density and estimating water content. Two variations are used: (1) the two-hole transmission technique (bulk density and water content), and (2) the scattering technique. Gamma photon sources include cobalt-60 and cesium-137. Sodium iodide detectors are commonly used. Details of the two-hole transmission method are described by Blake (1965). Basically, the technique entails lowering probes containing a source of gamma radiation in one of two parallel tubes (see Figure 3-4). A probe with a radiation detector is simultaneously lowered in a second tube located a fixed distance from the first tube. In this method, the density is measured in a collimated beam a few inches in vertical extent.

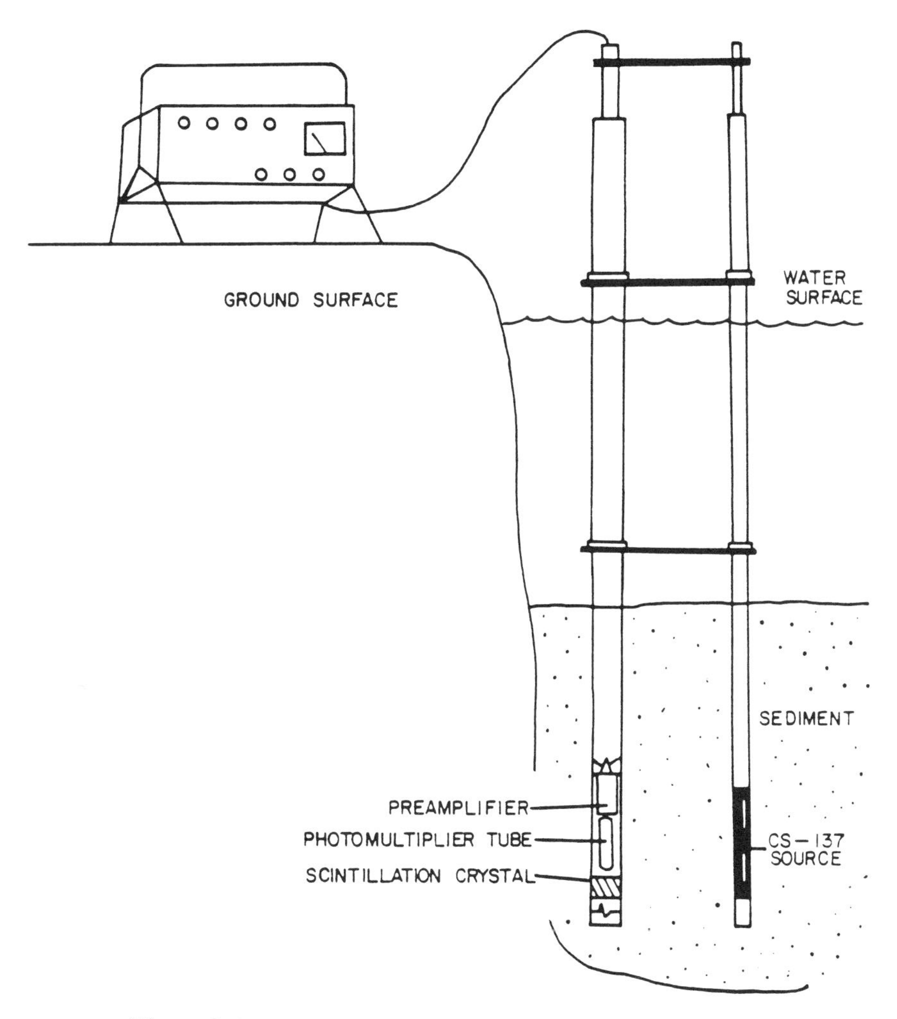

Figure 3-4. Dual probe used to measure water content by gamma ray attenuation (after Brakensiek, Osborn, and Rawls, 1979).

Keys and MacCary (1971) discuss the principles of the scattering technique. In practice, a probe containing both a source and a detector separated by shielding is lowered into a well. In contrast to the transmission method, the scattering method examines density in a spherical volume. According to Paetzold (1979), the scattering method requires a source of higher strength but is less accurate than the transmission method. The double-tube transmission method is primarily suitable for determining bulk density in shallow soils rather than deeper regions because of the difficulty of installing precisely aligned tubes. Consequently, the scattering method is used to measure bulk density in the vadose and groundwater zones.

An alternative technique involves placing a source of gamma photons on the soil surface and measuring the density in an underlying hemispherical volume.

Inasmuch as bulk density measured in the field also measures the density of water present in the medium, a correction is necessary to convert the measured wet bulk density to a dry weight basis. In particular, a calibration curve is prepared by measuring count rates in a number of wet soils and subsequently determining the corresponding wet bulk density using laboratory techniques (e.g., the calcium carbide method).

Water content values of the soil in the field are determined at the same time that a count rate is obtained. Using the calibration curve and knowing the water content, the following relationship is used to determine the dry weight bulk density (Blake, 1965):

$$D_b = D_{bw}/(1 + P_w/100) \qquad (3\text{-}5)$$

where D_b = dry weight bulk density

D_{bw} = wet weight bulk density

P_w = moisture percent on a dry weight basis.

The water content of the soil may be determined by collecting field samples and oven drying or using the neutron logging method. (When the latter method is employed, water content is determined on a volumetric basis and the above equation must be modified.) Corey, Peterson, and Wakat (1971) described a method for simultaneously measuring soil density and water content using a dual source containing americium-241 and cesium-137. Commercial units with dual sources are now available.

Gamma ray transmission can also be used to indirectly determine water content. Principles of the method for water content evaluation are reviewed by Gardner (1965), Bouwer and Jackson (1974), and Reginato and van Bavel (1964). As stated by Gardner (1965): "The degree to which a beam of monoenergetic gamma rays is attenuated ... in passing through a soil column depends upon the overall density of the column. If the density of the soil less its water content is constant, then changes in attenuation represent changes in water content."

The basic components for the method include a source of monoenergetic gamma rays such as cesium-137 and a detector such as a NaI (TI) scintillation crystal. Accessories include a high-voltage supply, amplifier, scaler, timer, spectrum analyzer and pulse height analyzer, and a photomultiplier tube (Bouwer and Jackson, 1974).

For field usage, two parallel access wells are required, one for the source and one for the detector (see Figure 3-4). Both source and detector must be properly positioned with respect to each other during sampling.

Once the bulk density has been determined, the total porosity may be calculated from the dry weight bulk density using the relationship given in equation 3-2, i.e., $S_t = 100\ (1 - (D_b/\rho_b))$.

Field implementation--Gamma ray transmission techniques are being used to measure water content and bulk density of soils in situ. These measurements are made between a detector and source of gamma radiation as discussed above. Two parallel holes are bored to the desired depth and standard 2-inch-diameter aluminum access tubes are installed. Care must be taken to keep the access tubes as near parallel as possible. The count rate is then measured through a standard material, e.g., magnesium or aluminum, with the detector and source in the access tubes extending above the ground surface. The source and detector are then lowered into the soil and readings are taken at the desired depth. After the calibration reading has been completed, the detector must not be exposed to unattenuated radiation. Industrial literature (RCA Electronic Components staff, personal communication, 1970) indicates that exposing the detector to the full intensity of the gamma ray source may cause fatigue and/or hysteresis in the multiplier and affect the operating stability of the photomultipliers, resulting in a gain change with time. If the unattenuated count rate is excessive, the phototube stability may be lost permanently. Three ways to prevent this exposure are suggested by Reginato (1974):

1. Put the standard material on the soil surface

2. Take a standard count somewhere in the soil bed where the bulk density is constant

3. Turn off the high voltage when moving the detector and source from the standard to the soil.

The author chose the second alternative to establish bulk densities at a depth where the density remains constant.

When using the scattering technique, a hole is driven into the test material to the maximum desired depth. A single probe containing both gamma source and detector separated by shielding is lowered to the desired depth. Field personnel must remain shielded from the high-energy gamma photons. The count-rate is measured and divided by the standard count-rate. Standard count-rate can be obtained through individual manufacturer's instructions, by using the count-rate obtained when the probe is within its carrying shield, or by preparing a standard source as described by Blake (1965). Dry weight bulk density is then calculated using equation 3-5.

Range of applications and limitations--The gamma radiation technique has a wide range of applications for in situ measurements and has been used extensively in defining density and water content. Radiation hazard exists with both methods however. Field personnel using this equipment should be knowledgeable in means of checking radiation levels under actual field conditions. Regular use of film badges and other radiation detectors as an added safety precaution is recommended. An additional health hazard can occur when installing access wells in existing sites containing toxic wastes.

According to Brakensiek, Osborn, and Rawls (1979), advantages of the method include the following:

> Measurements ... by gamma ray attenuation provide a nondestructive method of determining soil moisture and soil density. The site sampling can be repeated as often as desired. The measurements can be made vertically as close as one inch. Measurements at one inch from an interface (surface) are valid. The precision is high and accuracy of the results is excellent. Errors in measurement can usually be detected and corrected at time of measurement. No Particular health hazard is involved in this type of measurement provided simple safety precautions are observed and the radioactive source is shielded when out of the access tube.

Disadvantages of the method include: changes in bulk density in shrinking and swelling materials may affect calibration, equipment is expensive, and instabilities in count rate may occur (Reginato and Jackson, 1971; Reginato, 1974). Also, the need to maintain access wells a constant distance apart is impractical for deep sampling, particularly in rocky materials. Access holes for the scattering method need to be straight with a very small annular space between the access tube and the formation. Access tubes are difficult to install on steep terrain.

Accuracy of the method requires independent measurements of dry bulk density. In deep, highly layered material, it may be difficult to determine bulk density distribution, thus affecting accuracy. Brakensiek, Osborn, and Rawls (1979) found that valid measurements cannot be obtained in soils that shrink and swell during drying-wetting cycles.

Interpretation of the gamma ray attenuation method requires a thorough understanding of the hydrologic setting. Leakage of water from perched layers along the wall of the casing may cause erroneous logs. The method is basically the capacitance type, e.g., measuring water content and changes in water content. If water is moving through the sampling area at a rate equal to that being delivered, water content may not change and be erroneously interpreted as static in the soil profile. Stratified soils may exhibit large variations in water content (and bulk density), thus limiting the spatial resolution of the logs (Schmugge, Jackson, and McKim, 1980).

The specific gravity of many industrial organic wastes differs from the specific gravity of water. If calibration curves are constructed only with water and the fluid moving through the vadose zone is a mixture of water and organic wastes, a true picture of the storage change may not be obtained from the gamma ray log.

The primary physical constraint is ensuring a tight fit between the well casing and the borehole. This problem increases with depth. The annular space should be sealed to prevent water migration along the casing. Special equipment may be required when installing access tubes near existing landfills and impounds, e.g., hollow stem augers or driven casing as described by Kaufmann et al. (1981). The temperature during installation can hinder implacement of the access tube (e.g., frozen ground) or affect electronic components during sampling (e.g., extremely high temperature). In addition, freezing and thawing changes the bulk density of soils, thereby affecting water content determinations.

Water Content Using Tensiometers--

Tensiometers are used to measure soil-water negative pressures during unsaturated flow. Basically, a tensiometer consists of a porous ceramic cup cemented to a rigid plastic tube, small-diameter (spaghetti) tubing leading to a bourdon gauge or a manometer terminating in a reservoir of mercury, and a filler plug in the rigid plastic tube (see Figure 3-5). Except for the portion of the small-diameter tubing filled with mercury, the internal volume of the system is completely filled with water. When properly installed, the

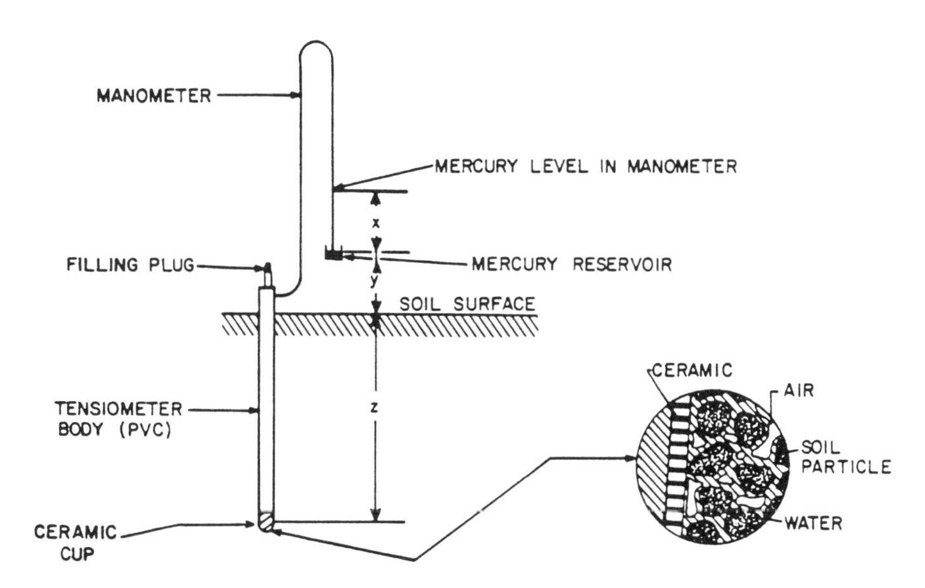

Figure 3-5. Schematic representation of tensiometer and section through the ceramic cup (after Nielsen, Biggar, and Erh, 1973; and Bouwer, 1978).

pores in the cup will form a continuum with the pores in the soil. Water will move either into or out of the tensiometer system until an equilibrium is attained across the ceramic cup. The mercury level in the manometer tubing will correspondingly adjust.

With care, tensiometers can be used to estimate soil-water content via suitable soil-water characteristic curves. In particular, knowing the soil-water pressure in the field via tensiometer data, a soil-water characteristic curve (e.g., Figure 3-6) for the field soil can be constructed to determine the corresponding water content value. From Figure 3-6, the soil-water pressure head, h, will be (Nielsen, Biggar, and Erh, 1973):

$$h = -(12.55\,x - y - z) \qquad (3\text{-}6)$$

where x = height of mercury column

y = distance from soil surface to mercury reference level

z = depth below land surface to ceramic cup.

Field implementation--Field installation of a tensiometer requires filling the instrument with water fluid and implanting it in the soil horizon or area of the vadose zone to be tested. Soilmoisture, Inc. recommends that its tensiometers be filled with a fluid prepared to inhibit algae growth and tinted (blue) to show air bubble accumulations. After filling the tensiometer body tube, it should remain in a vertical position until fluid completely saturates the sensing tip and drips from it for about 5 minutes (to ensure thorough wetting). The tensiometer body is again completely filled and, if it is a bourdon gauge type, evacuated using a vacuum hand pump (applying a vacuum to the tensiometer body is not required for the mercury manometer type instrument). This will cause air to bubble out from the stem of the dial gauge. The tensiometer is then refilled and the pumping/refilling operation cycle repeated four or five times or until no further air is seen to bubble from the stem of the dial gauge. The tensiometer is now ready for installation. If there is to be a delay in installing the tensiometer, the sensing tip should be covered with plastic to prevent evaporation of fluid.

For near-surface installation, an access hole is driven to the desired depth. Some companies provide an insertion tool that is tapered to provide a snug fit between the ceramic sensing tip and the soil. If this tool is not available, a length of pipe with an outside diameter equal to that of the tensiometer tube can be used to create an insertion hole. If impediments are encountered, the instrument should be moved to an adjacent location to avoid possible damage to the tensiometer during emplacement. For shallow depths, a hole can be dug to accept the tensiometer. If this method of installation is used, the soil must be packed firmly around the tensiometer after it is emplaced.

Several hours may be required after installation before the tensiometer reads the correct soil suction values because of installation disturbance to the soil. The correct reading will be reached more quickly in moist soil than in dry soil.

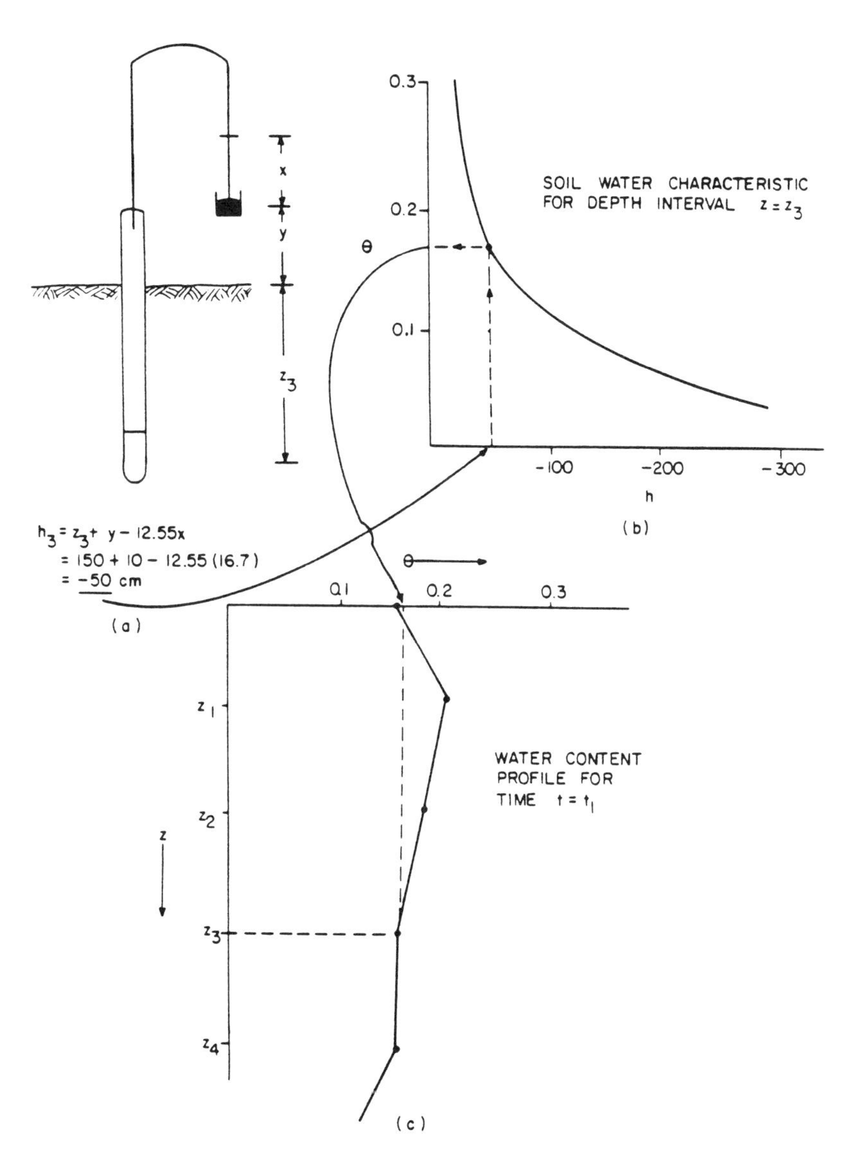

Figure 3-6. Schematic representation of procedure for converting pressure values from tensiometer data to water content values on a water content profile for a given time.

Tensiometers are well sealed, weatherproof, and require very little servicing other than the removal of accumulated air within the tensiometer tube. If the soil in which the instrument has been placed is moist, the soil suction readings will be low and very little air will accumulate in the body tube. However, if the tensiometer has been installed in relatively dry soil, suction values will be in the 40- to 60-centibar range and air will accumulate rather quickly for the first few days after installation because of coming out of solution and detaching from the internal walls of the tensiometer when exposed to high vacuums for the first time. After initial installation, the tensiometer must be checked every day or two and accumulated air removed. After the first few air removal servicing operations, the rate of air accumulation will drop off markedly and servicing will be required only infrequently.

Tensiometers should be removed from the soil before the onset of freezing conditions if they contain water (systems with ethylene glycol/water solutions can operate successfully in freezing weather). If allowed to remain in the ground during a deep penetrating freeze, expansion caused by ice formation can cause breakage of the porous ceramic sensing tip and/or distortion and rupture of the thin-walled bourdon tube within the dial gauge. If the bourdon tube is ruptured, the gauge will have to be replaced. However, if the tube is distorted but not ruptured, it may be possible to reset the dial gauge to correct for the change in calibration caused by freezing.

Holmes, Taylor, and Richards (1967) and Brakensiek, Osborn, and Rawls (1979) note the precautions to be taken during installation of tensiometers. The individual units should be filled with deaired water and the cups immersed in water during transport to the field. For installation at deeper depths, a hole slightly larger in diameter than the tensiometer is dug to the desired depth. A slurry is prepared from native soil and poured into the bottom of the hole. The tensiometer cup is then forced into the slurry and the tube carefully backfilled with soil to prevent surface water leaking down the tube.

Holmes, Taylor, and Richards (1967) point out that to ensure minimum effect of the tensiometer on the soil-water system, the sensitivity of the unit should be as great as possible. Furthermore, the response lag of the system should be kept to a minimum, particularly in rapidly changing soil-water systems.

Bianchi (1967) presents the design of a tensiometer coupled with strain gauges instead of the mercury system to permit conversion of soil-water pressure into an equivalent electrical resistance. His system allows deep emplacement of tensiometers and the use of automatic recording equipment. Watson (1967) presents a design for a tensiometer-pressure transducer system that also allows for the recording of soil-water pressure at depth. Figure 3-7 illustrates this system. He recommends such a device for recharge basin studies, and indicates that the unit can also be used for automatic recording of soil-water pressures at remote stations. In lieu of employing one transducer per unit, Brakensiek, Osborn, and Rawls (1979) describe a strain gauge/recorder combination using one transducer to service several tensiometers in turn.

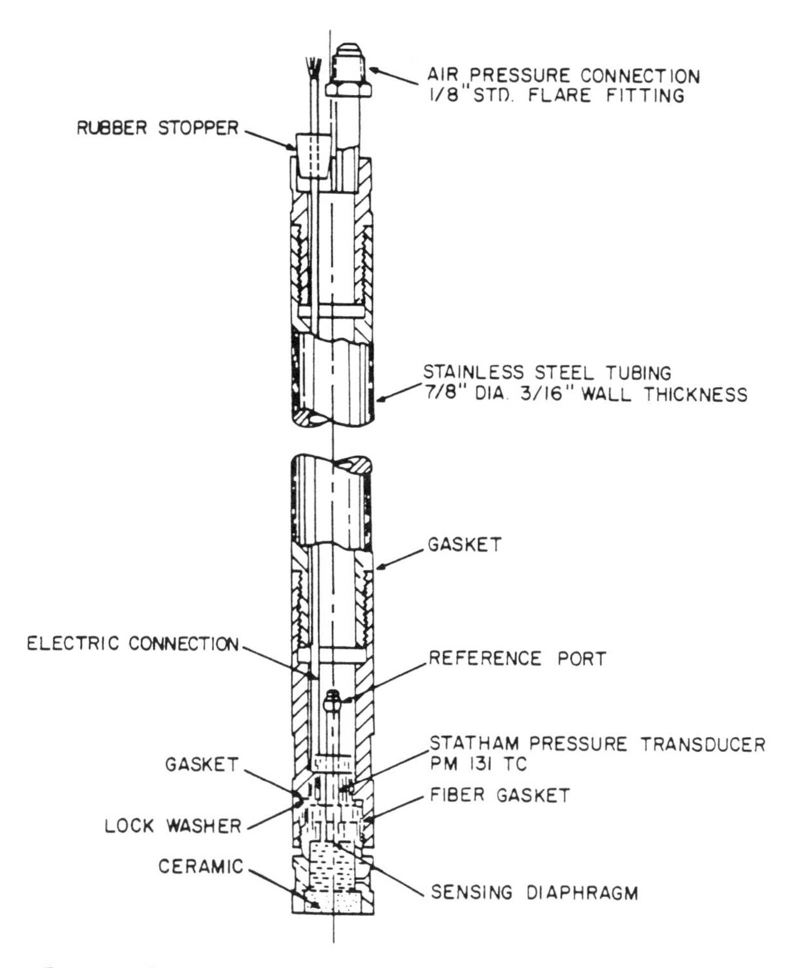

Figure 3-7. Cross section of tensiometer-pressure transducer assembly (after Watson, 1967).

Range of applications and limitations--Tensiometers can be used to characterize the direction of unsaturated water movement in the vadose zone. For this application, a battery of tensiometers must be installed with individual cups terminating at varying depths throughout the region of interest. Using the principles of hydraulics, the total hydraulic head, H, at a point in a flow system is the sum of the positional (gravitational) head, z, plus the pressure head, h. The positional head is the distance between the measuring point (i.e., tensiometer cup) and an arbitrary datum. That is, positional head equals zero at the datum. The pressure heads are readings taken from the tensiometer manometer system as shown in Figure 3-5. Positional heads are positive and pressure heads are negative during unsaturated flow. Differences in the total head between tensiometer cups may be used to estimate the direction of flow, which occurs in the direction of decreasing total head.

Figure 3-8 illustrates three hypothetical flow cases in a soil profile overlying a water table. This example is an adaptation of one presented by Reeve (1965). In the example, the datum is taken as the base of the aquifer. Three tensiometers are located in the soil zone above the water table. The cups are connected to water manometers for illustrative purposes only. In a real case, the tensiometers would be connected to aboveground mercury manometers as shown in Figure 3-5.

In the first case, water is not moving upward or downward in the profile, as evident by calculating the total hydraulic heads. The water levels in the open ends of the three manometers are equal and correspond to the water table. For unit #1, the total hydraulic head, H, is the sum of the positional head, z, and the pressure head, $-h_1$. That is, $H_1 = z_1 - h_1$. But this is simply the distance from the datum to the water table. For unit #2, $H_2 = z_2 - h_2$. Again, this equals the distance from the datum to the water table. It can be shown that the total hydraulic head at unit #3 also equals the distance from the datum to the water table. In order for flow to occur in the soil between units, a difference in total head must exist. But $H_2 - H_1 = 0$ and $H_3 - H_2 = 0$. In other words, the system is in equilibrium. Incidentally, for such a system, measurements in tensiometers afford a means of defining the water table position.

In the second case, the elevation of the water table is different and water is flowing vertically downward in the profile. In this example, the pressure heads are also zero because the level of the meniscus is at the same level as the cups. For all cups, the positional heads remain equal to the distance from the datum to the cups. In this case, the total hydraulic head at each cup is just the value for the positional heads, z_1, z_2, and z_3. Inasmuch as the positional head (and thus total head in this case) decreases downward, flow also occurs in a downward direction.

Finally, in the third case, evaporation is occurring at the soil surface. By applying the same reasoning as above, it can be shown that the total hydraulic head decreases vertically upward in the profile so that upward movement occurs.

More than one array of tensiometers is necessary to detect horizontal flow. Thus, if individual arrays terminate at varying depths, differences in total hydraulic heads between corresponding units in successive lateral arrays may suggest that horizontal movement is occurring in the unsaturated state. Because of the heterogeneity in soil properties, however, definitive conclusions on lateral flow may be tenuous.

An array of tensiometers may be useful in detecting the presence of clogging layers and in manifesting the head loss by such layers (Bouwer, 1978).

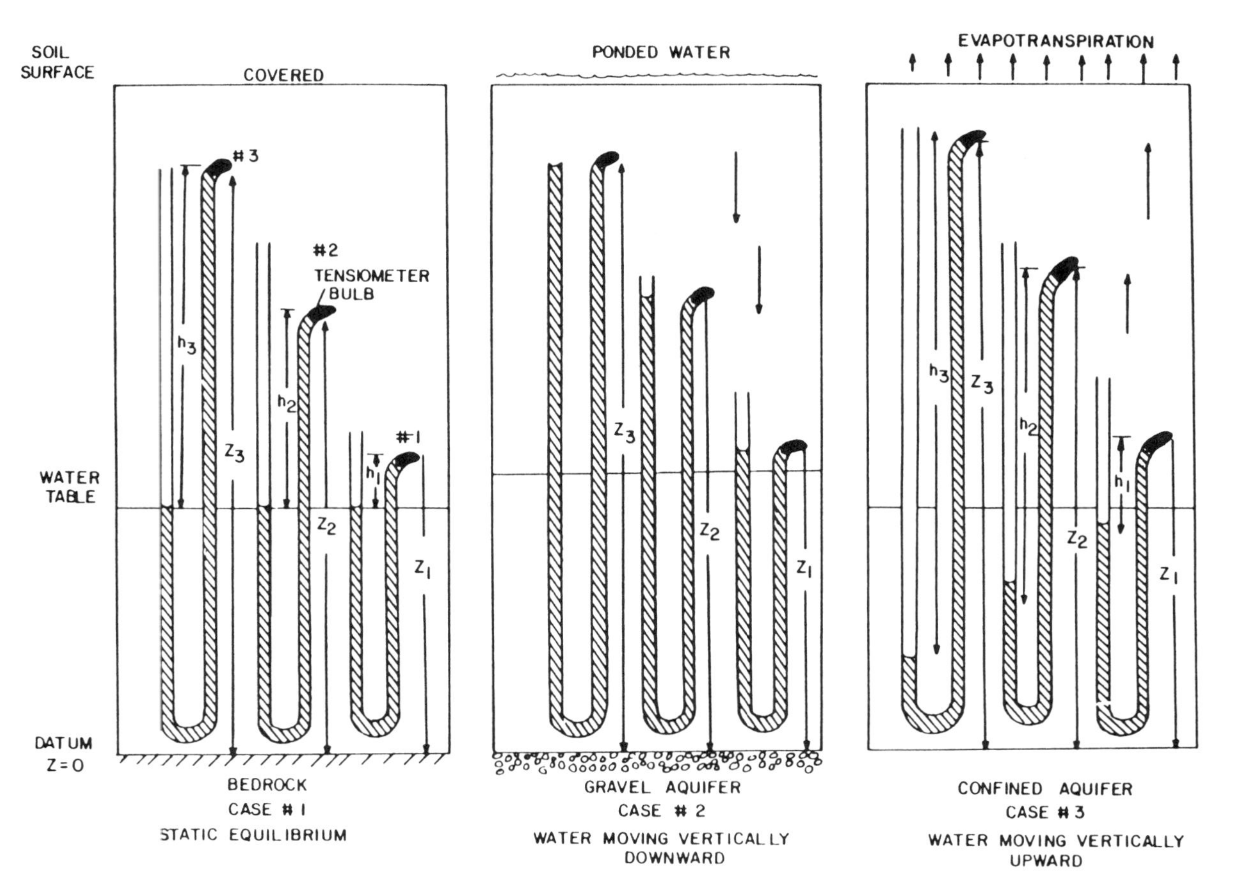

Figure 3-8. Distribution of hydraulic heads for three unsaturated flow cases (after Reeve, 1965).

For example, because of dispersion in water-spreading operations, fine solids may migrate in the profile and filter out at the interface of soil materials of differing textures. A tensiometer above this layer will gradually show an increase in soil-water pressure, while a unit below the layer will shift to the more negative side.

Time lag in response to a rapidly changing system is also a problem with tensiometers. The tensiometer design of Watson (1967) involving a pressure transducer system for measuring soil-water pressure has more rapid response characteristics than mercury-water manometer systems.

Varying geologic conditions can impede the use of tensiometers. In coarse materials, ensuring good contact between the tensiometer and formation to be tested is difficult. In fractured regions, locating the cups near fractures is difficult. Similar problems arise in calcareous strata where primary permeability is controlled by solution cavities. In well indurated areas, the units are difficult to install and the potential for damaging the ceramic tip is high.

The principal advantage of tensiometer units is in-place continuous readings of soil-water pressure, which may be translated into water content values (Gairon and Hadas, 1973). Tensiometers are useful only in measuring soil-water pressures in excess of about -0.8 bar. The most serious hydrologic problem with the method is that, because of hysteresis, knowing whether the soil is wetting or drying when the tensiometer is read is necessary. In particular, during the redistribution of soil water following infiltration, portions of the profile may be wetting while others are drying. Thus, a tensiometer reading could represent the value of a wetting curve, a drying curve, or one of an infinite number of "scanning curves," which represent the change in the water content-negative pressure head relationship if the soil is suddenly wetted during drainage, or vice-versa.

If tensiometers are used at waste disposal sites, soil-water pressure should be interpreted cautiously. Surface tension characteristics of some chemical wastes in the vadose zone may differ from those of water. This will affect the operation of the tensiometer.

For proper operation, tensiometers need to be calibrated to the water characteristics of the sediments being tested. In layered strata, this means multiple calibration curves. In addition, spatial variation in the physical properties of the vadose zone produce uncertainty when results are extrapolated from the point of measurement. Reliable estimation of regional water characteristics requires installation of a large number of units.

The ceramic tip of the tensiometer is fragile and easily broken during installation. This limitation is being overcome in part by units that use porous PVC tips, which substantially increase the durability of the equipment. The more durable units generally facilitate tighter packing of the tensiometer during field installation, thus reducing the occurrence of side leakage of surface waters.

In addition to freezing conditions causing tensiometer failure, freezing and thawing soils can produce vertical movement of the equipment in the soil profile, thus changing the reference elevation of the cup. Abnormally high temperatures can also affect tensiometer operation.

Water Content Using Electrical Resistance Blocks--

Electrical resistance blocks, used to measure either soil-water content or soil-water pressure, consist of electrodes embedded in suitable porous material. Plaster of paris, fiberglass, and nylon cloth have been used. The principle of operation of these blocks is that water content (or negative pressure) within the blocks responds to the water content (or suction) of the soil with which the blocks are in intimate contact (Holmes, Taylor, and Richards, 1967). The electrical properties of the block change correspondingly as reflected by wheatstone bridge resistance measurements.

Moisture blocks (Figure 3-9) are calibrated in soil from the site at which they are to be installed. Such calibration involves evaluating resistance readings against a range of soil-water contents or negative pressures.

Phene, Hoffman, and Rawlins (1971) present an alternative to the standard gypsum block comprising a heat source and sensor embedded in a ceramic body. This device measures both soil-water content and matric potential. The basis of the units is that air is a good thermal insulator with respect to water. With drying and entry of additional air into the media, the remaining water films become thinner. Consequently, the flow path for heat conduction increases and a larger temperature gradient is needed to dissipate heat. Phene, Rawlins, and Hoffman (1971) also discuss laboratory and field tests with their sensor. Apparently, the accuracy of the sensor is as good as or better than other techniques used to measure matric potential.

The precision of determining soil moisture should not be expected to be greater than ±2 percent water content and errors as great as 100 percent are possible (Gardner, 1965).

Electrical resistance blocks can also be calibrated to measure soil moisture tension, which allows considerably better accuracy (Gardner, 1965). According to Bouyoucas (1960), the calibration procedure for moisture tension or suction consists of initially saturating the blocks in a soil suspension and then plastering the blocks with mud. The plates are placed on a pressure membrane and are allowed to set for a period of time for equilibrium moisture condition to be attained between the blocks and the soil. The resistance is then read at various atmospheric pressures attained in the pressure chamber (Richards, 1949) and a calibration curve is plotted.

Field implementation--Before installing the electrical resistance blocks in the field, each block must be calibrated in soil typical of the site to be monitored. Gardner (1965) describes the calibration procedure when measuring percent available moisture:

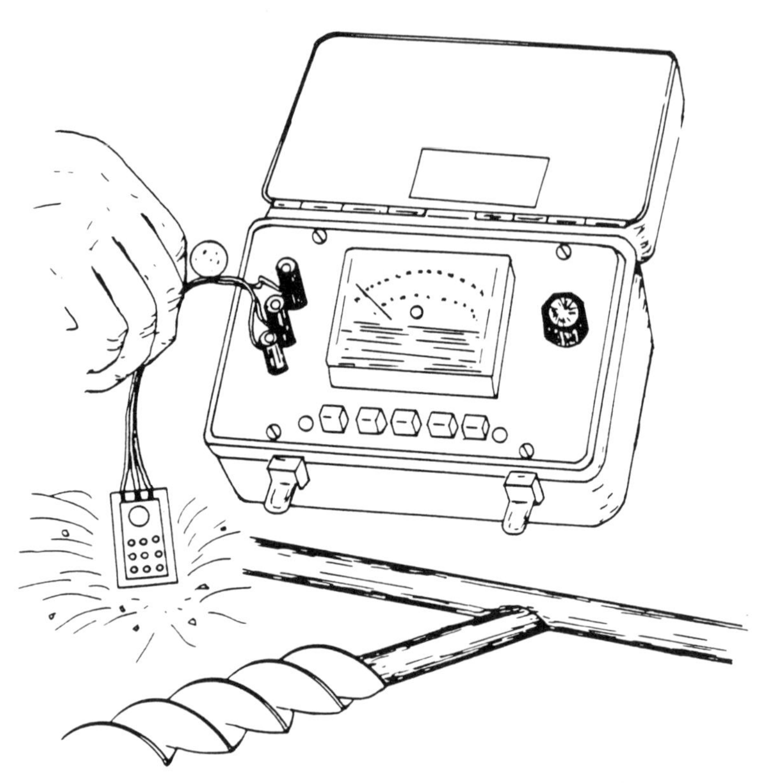

Figure 3-9. Thermistor soil cell, meter, and hand auger. Cell is being placed in a soil cavity (after Soiltest Inc., 1976).

To carry out the calibration, saturate the block with water, preferably with vacuum soaking, and weigh the wet block, its attached leads, and the screen box together to obtain a tare weight. Moisten the soil to be used to a state where it can be packed around the block to approximately its field density, mix the soil thoroughly, and take a sample for water content determination. Then pack the soil around the block in the screen box and weigh the entire apparatus. Using the water content determined independently by gravimetric methods (see section 7-2.2.2.1 or 7-2.3.2.1), compute the dry mass of the soil in the box. This will be:

$$\text{dry mass of soil} = \frac{(\text{mass of moist soil + tare}) - (\text{tare})}{\text{percent water content}/100 + 1},$$

where the water content figure is on a dry-mass basis. At all subsequent calibration points, the water content percentage will be

$$\text{water content, dry-mass basis} = \frac{100\big((\text{mass of tare + wet soil}) - (\text{mass of tare + dry soil})\big)}{\text{mass of dry soil}}.$$

Wet the soil in the screen box to near saturation, weigh the entire assembly, and then measure the block resistance to determine the first calibration point. Allow water to evaporate from the apparatus in the air until the desired weight for the next calibration point is reached. After the desired weight is reached, place the entire apparatus in a closed container (such as a desiccator without desiccant) in the dark at uniform temperature overnight, or longer, to permit the water to equilibrate in the block and soil. Then measure the resistance. Plot the resistance as a function of water content (3-cycle semilogarithmic paper is convenient).

When installing the moisture block in the field, a hole slightly larger than the block is dug to the zone to be monitored. The block is thoroughly wetted and then is placed in the desired monitoring location. The remaining loose soil is packed around the block (to assure good contact with the soil mass) and the leads are extended to the ground surface. Following installation of the moisture block, the apparatus should be left overnight to equilibrate the moisture in the block and surrounding soil before making a soil moisture reading.

Range of applications and limitations--Holmes, Taylor, and Richards (1967) discuss the advantages of resistance blocks and indicate that (1) they appear to be best suited for general use in study of soil-water relations, (2) they are inexpensive, and (3) they can be calibrated for either suction or water content. Generally, blocks are used for soil-water pressures less than -0.8 atmosphere.

Phene, Hoffman, and Rawlins (1971) reviewed the problems with gypsum blocks together with alternatives to circumvent these problems:

> In a gypsum block, the water content is measured by the electrical resistance between two imbedded electrodes. Several problems are usually encountered when using gypsum blocks: first, salinity affects electrical conductivity independently of water content; second, the gypsum used in attempting to mask variations in soil salinity eventually dissolves, resulting in an unstable matrix for the sensor; third, contact resistance between the porous body and the soil can restrict the exchange of water between the block and the soil; and fourth, the hysteresis in the water content-matric potential relation of the porous body can cause errors in the interpretation of the data, depending on whether the measurements are taken during a drying or wetting cycle. Elimination of the first

> problem requires a measurement of the water content of the porous body that is independent of salinity. This, in turn, would eliminate the need for the buffer material so that a stable matrix could be used. Contact resistance between the block and the soil can be minimized by packing the dry soil carefully around the sensor and then irrigating. Generally, the status of soil water is only of interest during drying conditions, since after a rain or an irrigation, the soil water content changes rapidly as the wetting front passes. Thus, hysteresis is of little concern and only a desorption calibration curve is needed.

Blocks are rather insensitive to moisture changes in the wet range (Gairon and Hadas, 1973). Hence, blocks are commonly used as an adjunct to tensiometers to monitor moisture content values at pressures less than -0.8 atmosphere.

Electrical resistance blocks do not perform well in sandy, coarse soils or in soils that experience considerable shrinking and swelling. This is primarily due to the problem of maintaining good contact with the native soils.

The gypsum type resistance blocks are not as sensitive as the fabric type in wet soils. In addition, the calibration can change with successive wetting and drying cycles (Brakensiek, 1979). Periodic recovery of the blocks to check the calibration may be necessary to alleviate this concern.

Soil moisture blocks can be calibrated to account for moisture conditions occurring during either wetting or drying cycles. Hysteresis makes it difficult and time consuming to calibrate the porous blocks for both conditions. Generally, for monitoring applications the status of soil water is of interest only during wetting conditions, since leachate detection is a primary concern. Calibrating the porous blocks for wetting cycles eliminates the hysteresis problem.

During installation of the moisture blocks, soil around their leads must be firmly packed to eliminate moisture movement along this path. If they are to be installed in saturated/unsaturated transition zones, units should be made of durable materials. In these conditions, gypsum blocks tend to deteriorate rapidly.

The electrical resistance measured by soil moisture blocks is dependent upon the ion concentration of the soil water as well as the amount of water present. If the moisture block is made of inert fabric material, a small change in ion concentration of the soil solution will influence the resistance readings. Moisture blocks made of gypsum read a saturated solution of calcium sulfate, which buffers the resistance readings from changes in the electrolyte concentration in the soil solution.

Iodic soils have been found to accelerate the deterioration of gypsum blocks. A porous plastic covering for the gypsum blocks has been developed to slow the degeneration due to chemical alteration.

A time lag occurs between the advancement of the moisture front through the soil and reaching equilibrium in the porous block. An overnight wait is usually sufficient to attain equilibrium between the moisture in the soil and the porous block and allow an accurate resistivity reading. Installation of moisture blocks is difficult in consolidated materials and zones composed of coarse debris. Care is needed to assure good contact between the blocks and surrounding media. In soils experiencing considerable expansion and contraction during freeze/thaw cycles, the blocks may be moved vertically in the profile, changing the zone being monitored.

Soil-Water Characteristics

The theoretical saturation value of a soil is the total porosity. As soil drains, a range of soil-water content values is possible, depending on various forces acting on the solids-water system. According to Day, Bolt, and Anderson (1967), these forces include matric forces, osmotic forces, and body forces (e.g., gravity). Matric forces include forces related to the surface tension of water, the cohesion of water molecules, the adhesion of water molecules to the surface of soil particles, and electrical forces at the molecular level (U.S. EPA et al., 1977). When expressed on a unit volume basis, the energy associated with these forces is termed soil-water pressure. The most common method for expressing the energy is on a unit weight basis, and the energy is termed soil-water pressure head, or head (U.S. EPA et al., 1977). The relationship between soil-water pressure and water content is called the soil-water characteristic. According to Klute (1969), the change in water content per unit pressure head change ($d\theta/dh$) is called the water capacity.

As pointed out by the U.S. EPA et al. (1977), "The force by which water is held in the soil is approximately inversely proportional to the pore diameter. Thus, the larger the pore, the less energy is required to remove water. As soil dries or drains, water is removed from the larger pores first. The water remains in the smaller pores because it is held more tightly. Thus, as soil-water content decreases, soil-water potential increases." Hillel (1971) points out that the amount of water retained at low negative head values depends on capillary effects and pore-size distribution and, consequently, is strongly affected by soil structure. In contrast, at higher negative heads, water retention is influenced less by structure and more by texture and specific surface.

Figure 3-10 shows the effect of soil texture on the shape of the soil-water characteristic curve for a clay soil and a sandy soil. The curve for the clay soil is more gradual and the water content at a given negative head is greater than for the sandy soil. In the sandy soil, however, most of the pores are large and, once drained, the amount of water retained is small (Hillel, 1971).

To obtain soil-water characteristic curves in the laboratory down to negative pressures of -800 cm of water, the modified Haines method may be used for individual core samples. This method employs equipment shown in Figure 3-11. Soil cores are carefully placed on the fritted glassbead plate and saturated. After ensuring that the samples are saturated, excess water is removed and the hanging water column is adjusted through a desired range of negative heads. The cores are weighed after the soil water has equilibrated with each successive negative head. At the end of the test, the soil cores are oven-dried and intermediate masses are converted to volumetric water content values.

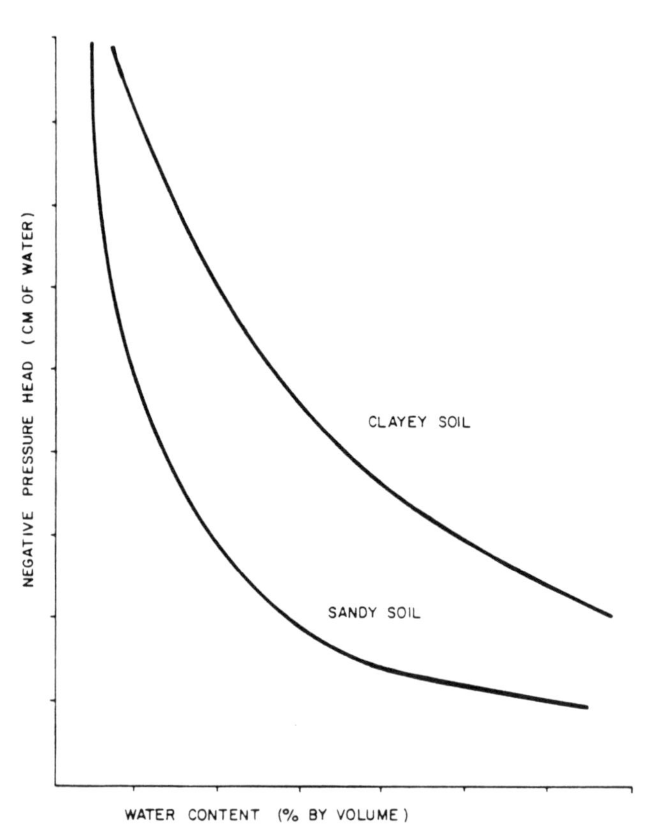

Figure 3-10. Soil-water characteristic curve for a clayey soil and sandy soil (after Hillel, 1971).

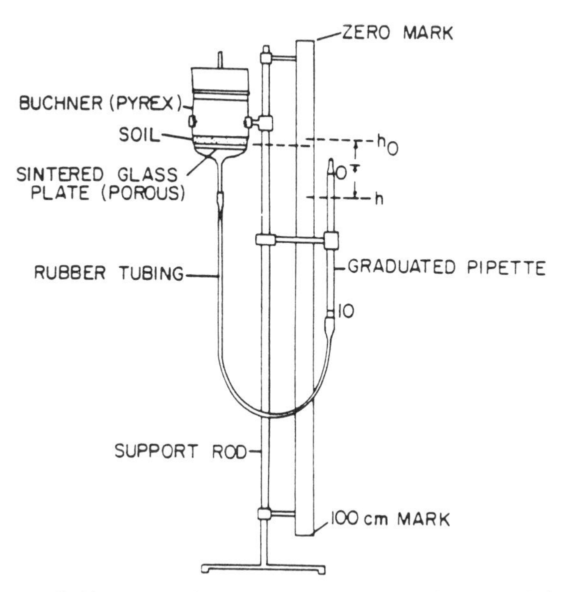

Figure 3-11. Modified Haines apparatus for obtaining soil-water characteristic curves (after Day, Bolt, and Anderson, 1967).

A plot of water content versus applied negative head for a desorption cycle (obtained via the Haines method) is different from the curve obtained during a wetting cycle, as shown in Figure 3-12. Water content values at a given negative head are generally less during the wetting cycle than those during the drying cycle. Such hysteresis is caused by entrapment of air in pore spaces during wetting. The figure also exhibits scanning curves.

Topp and Zebchuck (1979) present a variation of the hanging column method that permits the simultaneous measurement of soil-water desorption curves from a number of large-diameter cores. The soil cores contact a "tension medium" in a tank. The medium comprises glass beads with a narrow pore-size distribution, permitting high hydraulic conductivity and high air-entry values. Negative heads up to -100 cm of water are applied to the tension medium via a hanging water column. Heads from -100 cm to -500 cm are obtained using a regulated vacuum. After equilibrating at successive pressures, the cores are weighed. At the end of the extraction period, the cores are oven-dried and intermediate masses are converted to volumetric water content values.

Inasmuch as the air-entry value of the tension medium for the above methods is about 0.8 atmosphere of negative pressure, other methods are required to obtain water content versus head relationships for greater negative head values. Commonly, the pressure-plate method is employed. Details of the

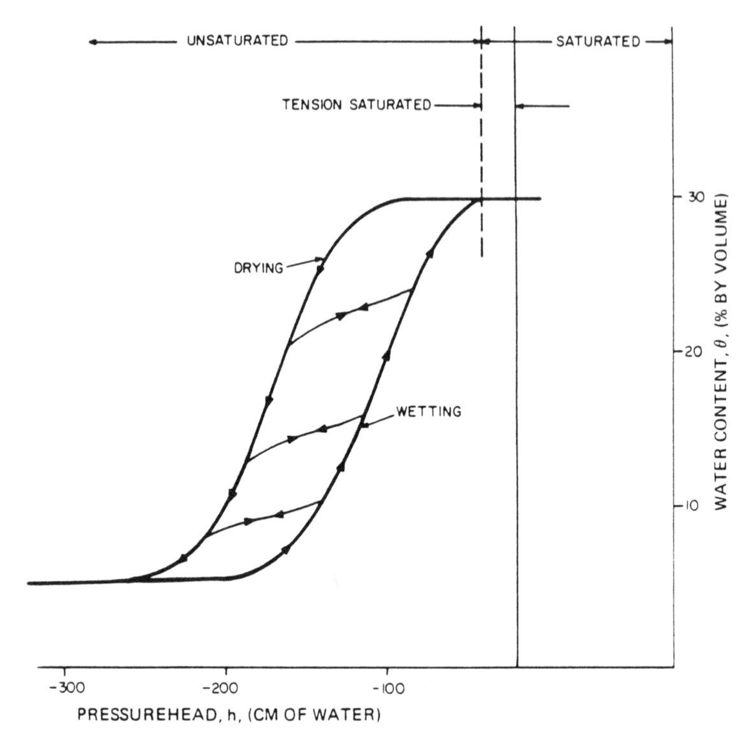

Figure 3-12. Soil-water characteristic curves for a sandy soil during wetting and drying showing the effects of hysteresis (after Freeze and Cherry, 1979).

method are presented by Richards (1965). As described by Paetzold (1979), the soil sample is wetted and placed in a pressure chamber on a ceramic plate or a cellulose-acetate membrane. A positive air pressure is applied to the soil within the chamber causing water to flow from the sample through the membrane. The air pressure value corresponds to the negative pressure head retaining water in the sample. The various air-entry values of membranes permit a range of possible pressures (for example, up to 15 atmospheres for ceramic plates). After the water content of the sample equilibrates at the applied air pressure, the sample is removed and weighed.

Soil-water characteristic curves may be used to construct the associated pore-size distribution--water content curves. The relationship between suction, h, and effective pore-size diameter is approximated by:

$$h = 2T \cos \alpha/\rho g r \tag{3-7}$$

where T = surface tension of the liquid

α = contact angle

ρ = density of liquid

g = acceleration of gravity

r = effective radius.

As discussed by Bouwer (1978), the negative pressure head in the vadose zone equals the vertical distance above a water table provided vertical flow is not occurring. Consequently, for uniform conditions, a plot of the change in volumetric water content with distance above a water table represents the water characteristic curve of the vadose zone material, again assuming no vertical flow. In practice, the volumetric water content could be determined using the neutron moisture logger.

Bouwer (1978) describes the nature of characteristic curves for layered conditions. The negative pressure head still equals the vertical distance above the water table. The water content-head relationship, however, depends on the soil at the measurement point. In other words, if an irregular water content distribution occurs above the water table, although head changes occur continuously with vertical distance, discontinuities may occur in water content distribution. Bouwer (1978) indicates that certain fine-textured soils may actually be saturated (e.g., contain perched groundwater lenses), whereas coarser-textured material above and below may be unsaturated.

Field Capacity (Specific Retention)

Field capacity may be defined in a general sense as the volume of water that a unit volume of soil will retain against the force of gravity during drainage. The concept of field capacity was developed many years ago by agriculturists concerned with quantifying the amount of water to apply to irrigated fields. The original premise was that field capacity is a fixed value representing the amount of water stored in a soil a certain time after drainage has "essentially ceased." By the same token, it is usually assumed that during recharge (wetting), water movement will not occur until the medium has been wetted to field capacity. Although these concepts of field capacity are useful in an applied sense, they have certain technical limitations. Hillel (1971) discusses such limitations in detail. Briefly, one limitation is that the simplistic concept of field capacity fails to account for the dynamic nature of soil-water movement. In particular, drainage does not really cease at field capacity but may continue at a slower rate for a prolonged period of time. That is, "The redistribution process is in fact continuous and exhibits no abrupt 'breaks' or static levels. Although its rate decreases constantly, in the absence of a water table the process continues and equilibrium is approached, if at all, only after very long periods" (Hillel, 1971). The modern conception of field capacity is that it is not a unique soil property; instead, a range of values is possible.

For sandy soils, field capacity may be reached in a few hours. For soils finer than sandy soils (e.g., sandy loams), 2 or 3 days may be required to reach field capacity, and for medium- to fine-textured soils, a week may be required. For poorly structured clays, the time will be much greater (U.S. EPA et al., 1977). Approximate values of field capacity (on a mass basis) vary from 4 percent in sands to 45 percent in heavy clays and up to 100 percent or more in organic soils (Hillel, 1971). In terms of matric potentials, field capacity values for sands range from 0.1 to 0.15 bar (1 bar = 0.9869 atmosphere). For medium- to fine-textured soils, the corresponding range is 0.3 to 0.5 bar (U.S. EPA et al., 1977). The value of 0.3 bar is chosen as an average value.

Knowing the water content values of a given soil at field capacity and the observed water content value at a given time, the depth of water applied at the land surface to bring the soil to field capacity may be calculated from equation 3-1. For layered soil, it is necessary to account for the sum of the water contents of individual layers (see Brakensiek, Osborn, and Rawls, 1979).

Among the factors affecting the apparent field capacity are (Hillel, 1971): (1) soil texture, (2) type of clay (e.g., clays predominantly comprised of the montmorillonite type exhibit a higher water-holding capacity at field capacity), (3) organic matter content (the higher the organic matter level, the higher the field capacity), (4) antecedent water content, (5) presence of impeding layers, and (6) evapotranspiration. Soil structure is also an important factor in evaluating field capacity inasmuch as large interpedal cracks permit more rapid drainage than the micropores within the soil blocks.

The water content of a soil sample at 0.3 bar is obtained in the laboratory using the pressure membrane method discussed in Soil Water Characteristics, this section. An alternative method to estimate field capacity is to assume that field capacity equals one-half of the percentage of water content at saturation; that is, $F_C = SP/2$. Saturation percentage is measured in the laboratory by determining the number of grams of water to saturate 100 grams of air dry soil (U.S. EPA et al., 1977).

The above discussion relates to the concept of field capacity as employed by agriculturists. The parallel term used by geohydrologists is "specific retention," defined as the "quantity of water per unit total volume which will not drain under the influence of gravity" (Cooley, Harsh, and Lewis, 1972). Specific retention may be visualized as the water remaining in the dewatered region of the vadose zone after recession of the water table (Figure 3-1).

Specific Yield

"Specific yield" is a term employed by geohydrologists to characterize storage in an unconfined aquifer. That is, specific yield is "... the volume of water that an unconfined aquifer releases from storage per unit surface area of aquifer per unit decline in the water table" (Freeze and Cherry, 1979). Figure 3-1 shows the conceptual relationship between specific yield and specific retention. As shown in Figures 3-1 and 3-13, the specific yield for a medium equals the porosity value minus the value of specific retention.

Representative specific yield values for valley sediments in California are listed in Table 3-4.

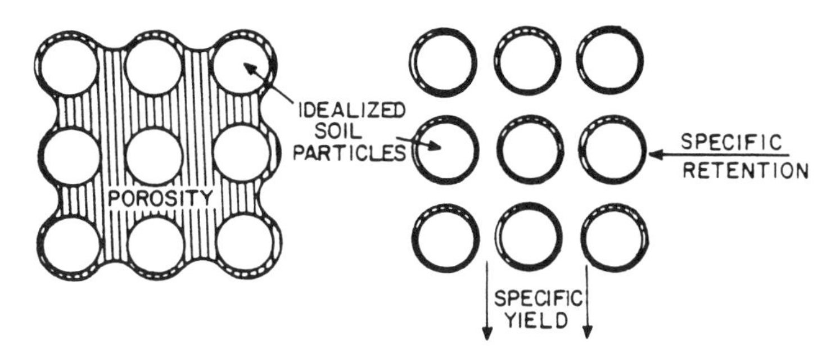

Figure 3-13. Schematic representation of porosity, specific yield, and specific retention (after Scott and Scalmanini, 1978).

TABLE 3-4. COMPILATION OF SPECIFIC YIELD VALUES FOR VARIOUS MATERIALS IN CALIFORNIA VALLEYS (after Cooley, Harsh, and Lewis, 1972)

Material	Average Specific Yield (percent)
Clay	2
Silt	8
Sandy clay	7
Fine sand	21
Medium sand	26
Coarse sand	27
Gravelly sand	25
Fine gravel	25
Medium gravel	23
Coarse gravel	22

A number of techniques are available for estimating the specific yield of the groundwater zone. If the storage properties of material above and below the water table are similar, the specific yield value determined for a groundwater zone can be assumed to approximate the specific yield in the vadose zone. In actuality, this assumption may be very tenuous because of the marked variations in lithology noted in the field.

The most common technique for estimating specific yield is to conduct pumping tests on wells. (Such tests also provide data on the transmissivity, T, of the aquifer.) Common testing techniques are described by Lohman (1972). Alternatively, if long-term groundwater withdrawals are known together with the corresponding head changes, specific yield values of the regional groundwater system can be estimated (e.g., Matlock and Davis, 1972).

As indicated by Stallman (1967) and Bouwer (1978), specific yield values determined from short-term pumping tests may underestimate the true value because of delayed drainage. Instead of relying on results from short-duration tests, Stallman (1967) suggests that a more realistic estimate of specific yield can be obtained by observing variations in water content profiles during the decline of a water table. In particular, the specific yield can be calculated by using a sequence of water content profiles to determine the volume of water drained near the water table during a decline in the water table (Bouwer, 1978). Meyer (1963) used this technique to estimate temporal variations of the apparent specific yield near pumping wells.

The use of neutron moisture logs appears to be the most suitable approach for estimating the specific yield of vadose zone sediments near the water table. Such logs can also be used to estimate the specific yield of deposits in other regions of the vadose zone where perched groundwater bodies are generated during cyclic recharge events.

Fillable Porosity

The volume of water that an unconfined aquifer stores during a unit rise in water table per unit surface area is called the fillable porosity (Bouwer, 1978). As shown in Figure 3-14, the amount of water placed into storage (fillable porosity) during the rise of a water table is less than the corresponding amount released during drainage (specific yield). The difference reflects hysteresis caused by air entrapped in pore sequences during the rise of the water table.

Neutron moisture logs also allow estimates of the fillable porosity of sediments near the water table, i.e., the volume of water placed into storage per unit rise in water level. Because of entrapment of air during the rise of a water table, the fillable porosity will initially be less than the specific yield.

Specific Retention

Specific retention, or the volume of water retained against the pull of gravity during drainage, can also be determined from neutron moisture logs obtained during the recession of a water table.

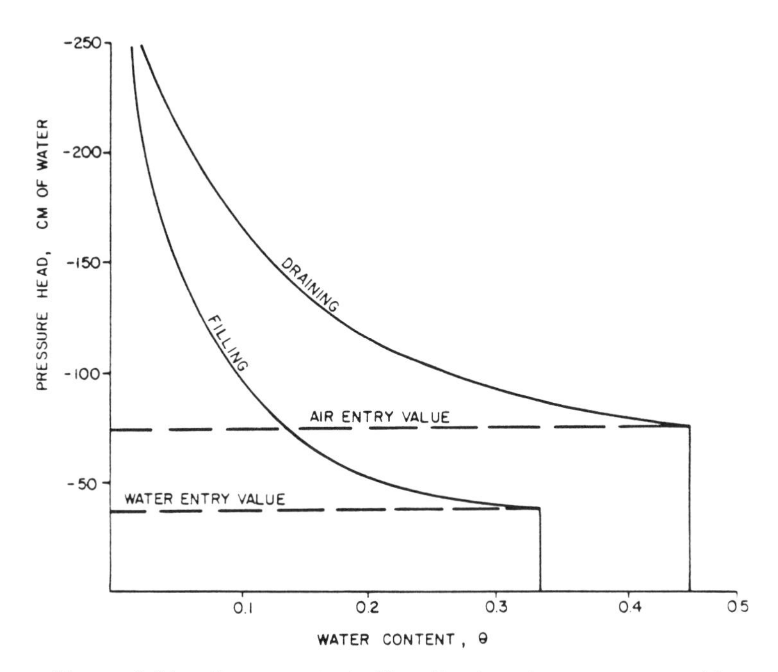

Figure 3-14. Water content distribution above a water table during a fall (draining) and rise (filling) in the water surface (Wilson, 1980).

UTILIZATION OF MONITORING METHODS FOR OBSERVING STORAGE CAPACITY BENEATH A POLLUTION SOURCE

The methods described in this section are examples of the available technology for estimating the capacity of a given volume of vadose zone underlying a pollution source to store water and water-borne pollutants. The example given in the subsection entitled Thickness showed the maximum (theoretical) capacity. In light of the discussion on field capacity, however, it is apparent that the actual storage capacity would be less than the maximum because of drainage. Another example using equation 3-1 illustrates this point.

Repeating equation 3-1:

$$d_w = \left(\frac{\Delta P_w}{100}\right)\left(\frac{D_b}{D_w}\right) d_{vz}$$

In practice, bulk density values (D_b) required to solve equation 3-1 can be evaluated using the gamma ray transmission method for shallow depths and the scattering method for deeper regions. Water content values under background conditions and at field capacity can be estimated from such alternative methods as neutron moisture logging, gamma ray transmission or scattering, tensiometers, and moisture blocks.

Again assuming (1) a 100-foot thick vadose zone, (2) a bulk density of 8.1 lb/ft^3, (3) water content of dry vadose zone material equal to 10 percent of weight, and assuming also that the water content of the material comprising the vadose zone at field capacity is 20 percent by weight, then

$$d_w = \frac{20 - 10}{100} \times 1.3 \times 100$$

$$= 13 \text{ feet.}$$

For this hypothetical case, 13 acre-feet of water could be theoretically stored in the vadose zone per acre of surface area when the media are at field capacity.

4. Premonitoring of Water Movement at Disposal Sites

TRANSMISSION OF LIQUID WASTES

Methods for estimating the capacity of the vadose zone to store water and water-borne pollutants are reviewed in Section 3. Although a knowledge of storage capacity is important, individuals charged with monitoring for water-borne pollutants are more likely to be concerned with the rate of movement through the vadose zone to the water table. This section examines the movement of water alone. Section 5 reviews relationships affecting the mobility of pollutants.

In practice, measuring the rate of water movement across a soil-water interface at the land surface is a fairly simple task. For example, seepage in a pond can be estimated by determining water level changes and using appropriate depth-volume-surface area relationships. In contrast, the process of water movement through the soil and lower vadose zone underlying the source is staggeringly complex. Elements contributing to the complexity of flow include variations in the state of water saturation and spatial variations in the physical and hydraulic properties of the vadose zone.

The difficulties in attempting to describe water movement in the vadose zone preclude presenting exact techniques for estimating transit time of water and water-borne pollutants through this region. A number of indirect methods are available, however, that are reviewed in this section.

For ease of discussion, the following categorizations are used: (1) infiltration at the land surface, (2) unsaturated flow in the vadose zone, and (3) flow in saturated regions of the vadose zone.

Infiltration

Infiltration is the process by which water enters a soil. The maximum rate at which water enters a soil is the infiltration capacity. The Soil Conservation Service has classified soils on the basis of infiltration rates and transmission rates as follows (U.S. EPA et al., 1977):

Class	Transmission Rate (in/hr)
Very slow	<0.06
Slow	0.06 to 0.2
Moderately slow	0.2 to 0.6

Class	Transmission Rate (in/hr)
Moderate	0.6 to 2.0
Moderately rapid	2.0 to 6.0
Rapid	6.0 to 20.0
Very rapid	>20.0

The infiltration rate of soil wetted continuously is initially high but decreases steadily with time, approaching an asymptotic value. The asymptotic value is approximately that of the saturated hydraulic conductivity of the soil. The physical reason for the typical decrease in infiltration with time is described by Hillel (1971):

> When infiltration takes place into an initially dry soil, the suction gradients are at first much greater than the gravitational gradient .... As the water penetrates deeper and the wetted part of the profile lengthens, the average suction gradient decreases, since the overall difference in pressure head (between the saturated soil surface and the unwetted soil inside the profile) divides itself along an ever-increasing distance. This trend continues until eventually the suction gradient in the upper part of the profile becomes negligible, leaving the constant gravitational gradient as the only remaining force moving water downward in this upper or transmission zone. Since the gravitational head gradient has the value of unity ... it follows that the flux tends to approach the hydraulic conductivity as a limiting value.

The distribution of water as a function of soil depth during infiltration is generally called a water content profile. As first described by Bodman and Coleman (1944), water content profiles may be divided into three distinct zones: (1) a transmission zone in which the upper few inches of soil may be saturated and below which the soil is very nearly saturated--this zone continually lengthens during infiltration; (2) a wetting zone exhibiting rapid changes in water content with depth and time; and (3) the wetting front, the visible limit of water penetration. Figure 4-1 shows water content profiles during infiltration into a soil that is continuously flooded and into the same soil when water is applied slowly by sprinkler. Because of this slow application rate, the water content values in the profile during sprinkling at all times remain below the values for the ponded case.

Following the cessation of surface flooding during a water-spreading operation, water becomes redistributed in the soil profile. The upper portion of the profile drains, while the lower portion continues to wet. The physics of redistribution is complex, particularly because of hysteresis.

For years, hydrologists have attempted to model infiltration using (1) algebraic and empirical equations, (2) approximations of rigorous flow equations, and (3) equations derived from simplifications of the physical system (Brakensiek, 1979). That the work still continues is apparent by examining the papers presented at a recent workshop on infiltration research, published by the Sciences and Education Administration (1979).

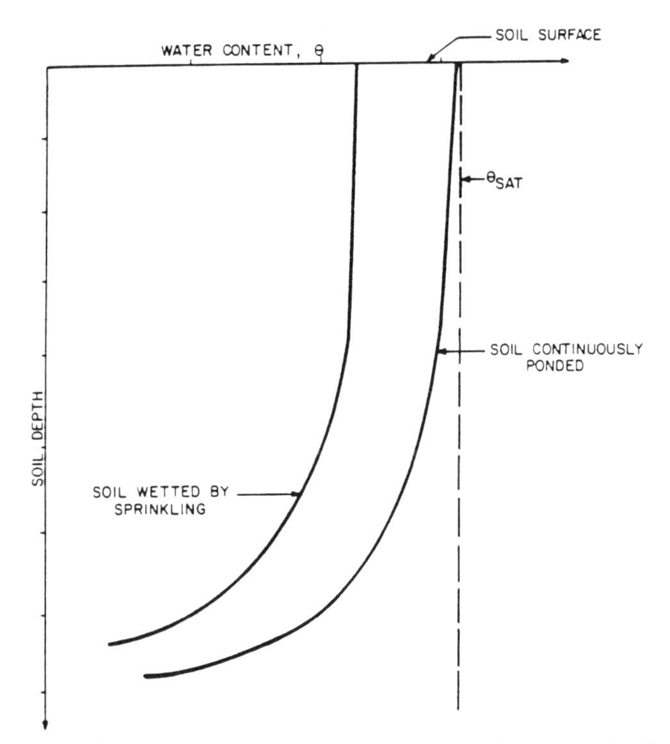

Figure 4-1. Water content profiles of a soil wetted slowly by sprinkling and the same soil continuously ponded (after Hillel, 1971).

Of the algebraic/empirical equations for describing infiltration, Brakensiek (1979) recommends the Green and Ampt (1911) equation and the Philip (1969) two-parameter equation. The Green and Ampt equation is discussed in detail by Childs (1969), Philip (1969), Hillel (1971), and Bouwer (1978).

Assumptions of the Green and Ampt approach are (Hillel, 1971): "... there exists a distinct and precisely definable wetting front, and that the matrix suction at this wetting front remains effectively constant, regardless of time and position." (Matrix suction is a measure of the sorptive capacity of a soil for water.) The Green and Ampt equation has been found to be satisfactory for describing infiltration into initially dry coarse-textured soils.

As presented by Bouwer (1978), the Green and Ampt equation is:

$$V_i = K \frac{H_w + L_f - h_{cr}}{L_f} \tag{4-1}$$

where V_i = infiltration rate

K = hydraulic conductivity of wetted zone

H_w = depth of water above soil

L_f = depth of wetting front

h_{cr} = critical pressure head of soil for wetting.

Brakensiek (1979) itemized parameter estimation procedures for the Green and Ampt equation.

Philip (1969) derived a two-parameter algebraic equation from physical principles to describe vertical one-dimensional infiltration by solving the basic partial differential equation for unsaturated flow. Philip's solution described the time dependence of infiltration in terms of a power series that, when t is not too large, reduces to:

$$I_t = St^{1/2} + At; \quad v_i = 1/2\ St^{-1/2} + A \tag{4-2}$$

where I_t = cumulative infiltration

v_i = infiltration rate at time t

S = sorptivity

A = constant.

According to Bouwer (1978), the sorptivity depends on the pore configuration of the soil, the initial water content, and the depth of applied water. Approaches for estimating the parameters in the Philip two-parameter equations are summarized by Brakensiek (1979). For short infiltration events, A is taken to equal K/2, where K is the saturated hydraulic conductivity. For long infiltration events, A is equal to K (Bouwer, 1978).

The initial infiltration rate of a soil is theoretically infinite (Bouwer, 1978). Both equations 4-1 and 4-2 comply with this requirement. In the case of equation 4-1, the value of the initial rate approaches infinity because the depth to the wetting front is zero (L_f = 0). For equation 4-2, the infiltration rate, V_i, at time t = 0 is also infinite because the first term on the right-hand side overwhelms the second (A) term.

An empirical infiltration curve commonly used by hydrologists is that of Horton (1935). This equation may be written as:

$$V_i = V_\infty + (V_0 - V_\infty)e^{-Kt} \tag{4-3}$$

where V_i = infiltration rate at time t

V_0 = initial infiltration rate

V_∞ = final infiltration rate

e = base of natural logarithms

K = a constant depending on soils and vegetation.

Values for K are found from field infiltration data by expressing equation 4-3 in logarithmic form. A basic problem with Horton's equation is that it does not satisfy the theoretical requirements that the initial infiltration must be of infinite value (Bouwer, 1978, p 256). According to Bouwer, the equation is best suited to describing infiltration when water is applied by rain or sprinkling for short periods.

Factors affecting the infiltration capacity (and thus the potential amount of pollutants entering the lower vadose zone) of a soil include soil texture, soil structure, initial water content, presence of shallow impeding layers, entrapped and confined air, biological activity, entrained sediment, and salinity of applied water. The effect of soil texture on infiltration is complex. For example, a clay soil may possess a rather high sorptivity value but water movement is retarded by the energy losses in the fine pores. Inasmuch as sorptivity decreases with time, the long-term infiltration rate depends on the hydraulic conductivity of a soil. Thus, a sandy soil should have a greater long-term infiltration rate than a clay soil. Soil structure has a profound effect on infiltration rate, particularly at the beginning of infiltration. Water moves preferentially very rapidly through the larger interpedal cracks at the beginning of infiltration. Later, as the soil swells and closes the cracks, water movement occurs through the soil blocks. Hillel (1971) points out that the long-term infiltration rate of a highly structured soil approaches the rate of a uniform soil because of the control imposed by the hydraulic conductivity of the lower soil zone.

The initial water content of a soil affects the infiltration rate because of the concurrent effect on sorptivity and available porosity. Thus, the dryer the soil, the greater will be the initial infiltration rate compared to a wetter soil (see Figure 4-2). The final, or long-term, infiltration rate of a soil, however, is independent of the initial water content. Shallow impeding layers within the soil zone retard the infiltration rate compared to a nonstratified soil. As pointed out by Hillel (1971), the effect is the same whether a fine soil is underlain by a lens of coarser material, or vice-versa. Air entrapped in the soil pores during infiltration effectively behaves as solid grains in restricting water movement. Eventually, however, the air may dissolve in the water, causing an increase in intake rates. Confined air ahead of the wetting front may also restrict infiltration because of the buildup in air pressure due to a restricting lens or shallow water table (Wilson and Luthin, 1963). In fact, the air pressure may increase to the point that the air-entry value of the soil is exceeded and air escapes at the surface.

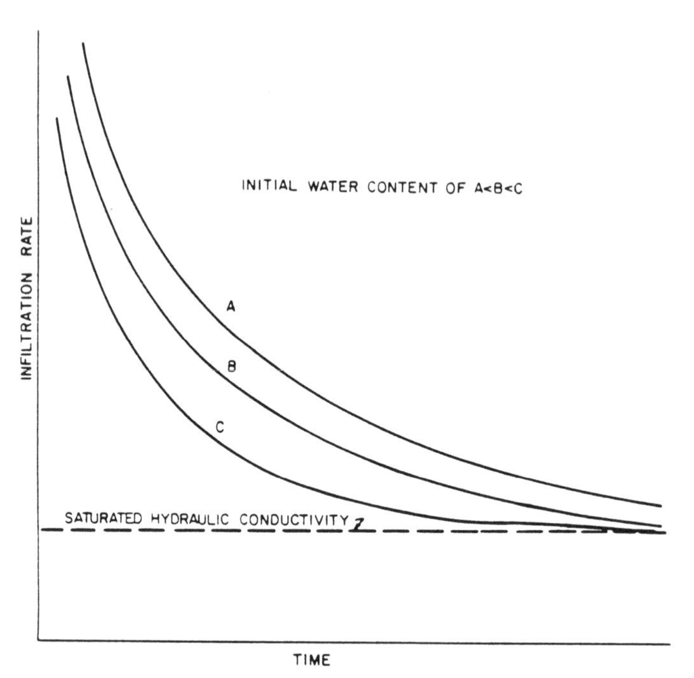

Figure 4-2. The effect of initial water content of a soil on infiltration rates.

Fine sediment entrained in the water may clog the surface pores and effectively create a barrier to the infiltration process. Water movement in the underlying pores becomes unsaturated and flow occurs primarily in the finer pore sequences. Microbial activity may also result in a clogging of soil pores with the byproducts of their metabolic processes, e.g., slimes or gases. Microorganisms may produce a secondary effect by increasing the pH of the overlying water through microbial respiration, resulting in a precipitation of carbonates within the surface pores (Bouwer, personal communication, 1978).

The salinity of the applied water may be a dominant factor in affecting infiltration for soils containing substantial clay fractions (i.e., soils with a tendency to shrink and swell). Thus, it is well known among agriculturists that a high sodium concentration in the applied water relative to calcium and magnesium may disperse the clays, slowing down intake rates. This effect is discussed in detail in Saturated Flow in the Vadose Zone, this section.

FIELD MEASUREMENTS FOR DETERMINING INFILTRATION AT THE LAND SURFACE

Recent reports by the U.S. EPA et al. (1977) and Sidle (1979) present excellent reviews of methods for estimating long-term infiltration rates. For disposal operations involving the surface spreading or ponding of wastes, the two most suitable procedures for determining long-term infiltration are cylinder infiltrometers and test basins. Infiltrometers may be either the single-ring or double-ring type. In the double-ring infiltrometer method, a metal cylinder from 6 to 14 inches in diameter is driven to a depth of 6 inches into the soil at the test site. A larger ring, ranging from 16 to 30 inches in diameter, is placed concentrically around the smaller ring. The areas within inner and outer rings are flooded and the rate of recession of water level in the inner ring is measured via a hook or point gauge.

By keeping the outer region flooded, flow in the inner region is restricted mainly to the vertical direction. Wastewater to be used during the full-scale disposal operation should be used during infiltrometer tests. Because of the spatial variability of soil properties, an inordinate number of ring tests may be required to ensure that results are within a certain percentage of "true" mean values. Even with care, however, experience has shown that double-ring infiltrometers tend to overestimate the true infiltration rate because of the divergence of flow due to shallow impeding layers and unsaturated flow (Bouwer, 1978). Bouwer recommends using a single large cylinder instead of a double-ring infiltrometer to offset the problem of divergence from unsaturated flow.

A superior alternative to infiltrometers is to install a number of small test basins at the site. Test basins should be large enough to permit installation of the desired accessory facilities. For studies of an irrigated field, Nielsen, Biggar, and Erh (1973) used basins of about 400 square feet. Basins can be constructed by digging a narrow trench around the desired rectangular area to a sufficient depth to minimize lateral flow. Side boards, possibly covered with plastic sheeting, are placed in the trench, ensuring that good contact is made with the soil within the plot and at the corners. Sufficient freeboard should be allowed to permit a range of flooding heads. During tests, water is metered into the plot at a discharge rate that maintains a constant head. The discharge rate to maintain this head is the application rate. The infiltration rates at selected times are calculated by dividing the application rate by the surface area.

Alternatively, the head may be allowed to fall a short distance by shutting down the inflow, and the volume, which infiltrates per unit area over a measured period of time, can be converted to infiltration rates.

As with infiltrometers, the same water source to be used during the actual operation should be applied during the tests. The effect of effluent on surface intake rates (e.g., by clogging) can be readily examined. Similarly, the effect of unfavorable exchangeable sodium percentages and electrolyte concentrations on the hydraulic conductivity can be quantified.

Because of the spatial variability of soil properties, even in supposedly homogeneous areas, location of a number of test basins on the proposed waste disposal site is recommended. By arranging the basins in accordance with recommended statistical procedures (e.g., Steele and Torrie, 1960), results amenable to statistical analyses will be generated for infiltration rates and other properties of interest.

A common problem with the test basins and infiltrometers is that shallow impeding layers may promote lateral movement of water in preference to truly vertical flow. Lateral flow rates generally exceed vertical rates. Hence, measured intake rates tend to overestimate the intake rates of a larger disposal area.

If the technique for applying wastewater to the disposal site is sprinkler irrigation, infiltration tests using rainfall simulators are recommended. Such units are described by Bertrand (1965), Sidle (1979), and Hamon (1979).

In conducting infiltration trials, the intake rate should be recorded as a function of time. Plotting the results will produce an exponential relationship between intake and time. The final value of the resultant curve will asymptotically approach that of the saturated hydraulic conductivity.

The water budget method at a pond can also be used to determine infiltration at the land surface. This method entails solving for the seepage component of the water budget equation. This is:

$$\text{Inflow} - \text{outflow} = \pm\Delta S \text{ and } S_L = (I + P) - (D + E) \pm \Delta S$$

where S = storage

S_L = seepage loss

I = inflow from all sources

P = precipitation

D = discharge

E = evaporation.

Field measurements of I, P, D, E, and ΔS are made using flumes, rain gauges, evaporation pans, and staff gauges or water stage recorders. A calibration curve or table of head versus surface area for the pond or impoundment is required. Bouwer (1978) has reviewed the method and gives further details of its implementation. Table 4-1 lists constraints on using the water budget method at ponds or impoundments.

An alternate technique used at ponds or impoundments is the instantaneous rate method. By shutting down all inflows to a pond and all discharges from it, the water level will recede primarily as a result of infiltration. That is, all the components of the water budget equation are set equal to zero except for infiltration, evaporation, and change in storage. Measuring ΔS for

TABLE 4-1. CONSTRAINTS ON THE WATER BUDGET METHOD

Category	Constraint
Geohydrologic	In regions of high water tables, the water table may at times rise above the level of water in the pond. Subsequently, the use of the water budget method to estimate seepage becomes inappropriate.
Chemical	Vapor loss from chemical impoundments may differ from water loss from wastewater ponds. This requires special methods for estimating the evaporation loss term.
Geologic	Geologic conditions may affect seepage rates but not necessarily affect the operation of measuring devices.
Topographic	Elevation changes affect evaporation rates. Must use onsite methods for estimating evaporation.
Physical	1. Installing inflow and outflow measuring devices may be difficult at some sites. 2. Errors occur in all measuring units affecting estimates of seepage. 3. Special techniques for estimating changes in storage will be required where changes in surface level are slow.
Climatic	1. Temperature affects evaporation rates. This requires correcting readings or installing units (e.g., pans) for independent measurement of evaporation. 2. Water in small pans may freeze in winter.

a short time provides a value for infiltration rate (neglecting evaporation). Constraints for the instantaneous method are similar to those found in Table 4-1. One additional constraint is encountered in ponds where seepage rates are minimal and accurate measurement of water level changes requires special techniques, e.g., laser equipment.

Seepage meters can be used to determine infiltration rates at the land surface. This technique has been reviewed by Bouwer and Rice (1963), Kraatz (1977), and Bouwer (1978). Seepage meters are cylinders, capped at one end and open at the other end. The open end of the cylinder is forced into the pond surface and seepage is equated to the outflow from the cylinder when pressure heads inside and outside the cylinder equalize. Several types of meters exist, including the: SCS seepage meter, the USBR seepage meter, and the Bouwer-Rice seepage meter. Table 4-2 lists constraints on the use of seepage meters.

Unsaturated Flow in the Vadose Zone

The two fundamental mathematical expressions describing the flow of water in porous media are Darcy's equation and the equation of continuity. For

TABLE 4-2. CONSTRAINTS ON THE USE OF SEEPAGE METERS

Category	Constraint
Geohydrologic	Local subsurface flow region may affect operation of units.
Chemical	In waste ponds with toxic chemicals, applying this method may be hazardous to personnel.
Geologic	In extensive unlined ponds, different sections of the pond may be underlain by different textured material, causing variable intake rates.
Topographic	Difficult to employ in ponds with steep-sided slopes.
Physical	1. A large number of measurements will be required to account for spatial variability of soil properties (texture, hydraulic conductivity, etc.) 2. Measurements must be obtained on sides of pond as well as bottom. 3. Requires diving to install units in deep ponds.
Climatic	Method cannot be used in frozen ponds.

unsaturated flow, Darcy's equation is written (after Nielsen, Biggar, and Erh, 1973) as:

$$J = -K(\theta)\nabla H \quad (4\text{-}4)$$

where J = specific discharge or flux (volume per unit area per unit time); also called Darcian flux

$K(\theta)$ = hydraulic conductivity, expressed as a function of water content, θ

H = total hydraulic head, the sum of soil-water pressure head, h, and potential head, z

∇H = the hydraulic gradient, expressed in vector form.

The specific discharge has the units of velocity. The negative sign is used by convention because flow occurs in the direction of decreasing hydraulic head.

As expressed by equation 4-4, the specific discharge is a macroscopic quantity since the cross sectional area of flow includes both solids and voids. A more representative depiction of velocity would exclude solids from the flow path and account for the water content of the void spaces. Consequently, an alternative expression for flow in unsaturated media is:

$$\bar{v} = \frac{J}{\theta} = \frac{-K(\theta)}{\theta} \nabla H \qquad (4\text{-}5)$$

where $\bar{v}$ is the average linear velocity (Freeze and Cherry, 1979).

As pointed out by Freeze and Cherry (1979), the average linear velocity is also a macroscopic quantity. The actual or microscopic velocity within pore sequences is greater than v because the tortuous flow paths are greater than the linearized path assumed in defining v. However, v will always be greater than J for a given soil, permitting a more realistic estimate of travel time of nonreactive water-borne pollutants.

The second important flow equation is the equation of continuity, which in one dimension (the z direction) is written as:

$$\frac{\partial \theta}{\partial t} = \frac{\partial}{\partial z}\left(K(\theta)\,\frac{\partial H}{\partial z}\right) \qquad (4\text{-}6)$$

where t is time.

Basically, hydraulic conductivity is a factor representing the ease with which water moves through a soil or aquifer. In equation 4-5, the unsaturated hydraulic conductivity is shown to be a function of water content, θ. However, θ depends on the negative pressure head, h. As a consequence of the h - θ relationship, the unsaturated hydraulic conductivity is also a function of the negative pressure head. Figure 4-3 shows the relationship between the unsaturated hydraulic conductivity and negative pressure head. The figure also shows the effect of hysteresis.

The important feature of Darcy's equation relative to the movement of water into the vadose zone beneath a disposal area is that two factors must be defined to characterize the flow rate. These factors are the hydraulic conductivity and the hydraulic gradient. Thus, the purpose of laboratory and field techniques is to determine $K(\theta)$, or $K(h)$, and the hydraulic gradient.

A number of laboratory techniques are available for estimating the relationship between the unsaturated hydraulic conductivity and pressure head or water content. Because the hydraulic conductivity is dependent on water content (and thus head) in unsaturated soils, measuring $K(\theta)$ and $K(h)$ in the laboratory or field is extremely difficult. Thus, for unsaturated soils, the water content, θ, changes continuously along the flow path and $K(\theta)$ values also vary continuously. That is, a unique value for the hydraulic conductivity cannot be determined.

One laboratory technique for determining $K(h)$ as a function of h is the so-called long-soil column method. Briefly, the method entails introducing water at the soil surface of a long-soil column at a constant rate that produces unsaturated flow in the soil. The soil at the base of a column eventually becomes saturated but the remainder of the soil will be unsaturated. The test is continued until the hydraulic gradient becomes equal to one or, in other words, θ is nearly constant throughout the upper region of the column. Hydraulic heads are measured by tensiometers placed in the column. When the

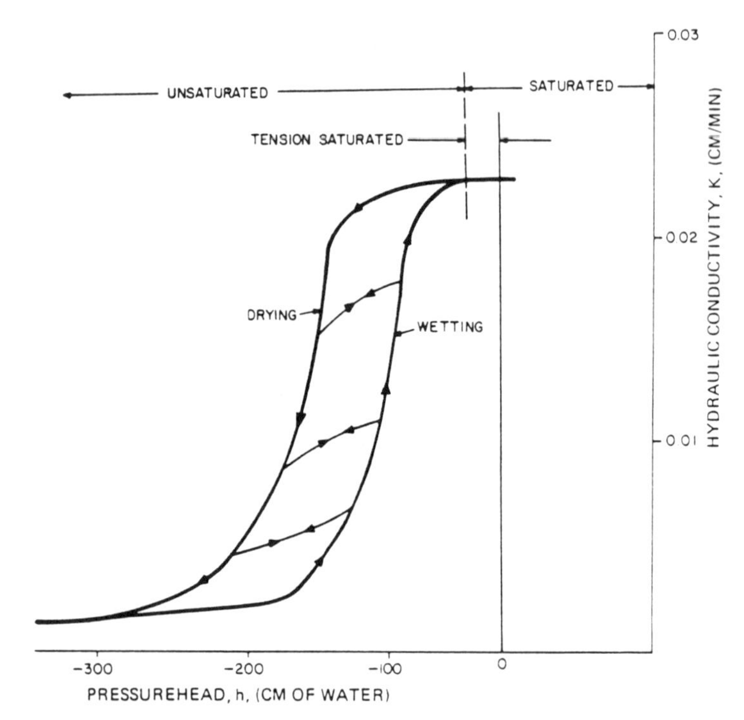

Figure 4-3. Effect of the negative pressure of soil water on the unsaturated hydraulic conductivity and the effect of hysteresis on K (after Freeze and Cherry, 1979).

hydraulic gradient equals one, the unsaturated hydraulic conductivity equals the flow rate divided by the cross sectioned area from Darcy's equation. The resultant hydraulic conductivity is related to the water content in the column at the time of the experiment. The experiment may be repeated at different flow rates and different water content values to provide a range of $K(\theta)$ - θ relationships.

Alternative laboratory methods for determining the unsaturated hydraulic conductivity at successive steady-state flow rates are presented by Klute (1965). These methods are suitable for soil cores. One of the most common laboratory methods, the pressure outflow technique, consists of applying successive air pressures to a moist soil section in an appropriate chamber. Using the air pressures and related water content values, an equation (Kirkham and Powers, 1972, p 274) is solved for corresponding $K(\theta)$ values.

In contrast to such steady-state methods, Alemi, Nielsen, and Biggar (1976) present centrifugal techniques for determining the unsaturated hydraulic conductivity during transient flow conditions.

Indirect methods for estimating $K(\theta)$ have relied on soil-water characteristics (for example, Millington and Quirk, 1959; and Jackson, Reginato, and van Bavel, 1965). Paetzold (1979) indicates that a computer program is available to calculate $K(\theta)$ from H versus θ curves.

Unsaturated Flow Parameters

The principal unsaturated flow parameters that require quantification at a waste disposal site following infiltration include (1) the amount of water moving into the lower vadose zone, (2) the direction of unsaturated water movement, (3) hydraulic gradients, (4) the unsaturated hydraulic conductivity, and (5) flow rates (flux).

Amount of Water Movement Below the Soil Zone--

A soil moisture accounting system may be used to estimate the amount of water draining from the soil zone into the lower vadose zone. Such an accounting system, or water balance, is based on the hydrologic equation for the soil-water system. This equation is often written as:

$$\Delta SM = PPT - PET - RO \quad (4\text{-}7)$$

where ΔSM = the change in soil moisture storage

PPT = precipitation + irrigation

PET = potential evapotranspiration

RO = runoff.

A commonly used water balance method is that developed by Thornthwaite and Mather (1957). This method, essentially a bookkeeping procedure, accounts for annual soil moisture additions and extractions. Basic credit components are the quantities (expressed as depths) of rainfall and irrigation. The principal losses (debits) from the system are evapotranspiration, runoff, and deep percolation. Monthly temperature and precipitation data are required. The potential water storage capacity of the soil must be estimated for the soil depth of interest. Generally, the potential storage capacity can be taken as the difference in water content (on a volumetric basis) of the soil at field capacity and that at a selected drained value.

The analyses of Thornthwaite and Mather (1957) generate values of changes in soil moisture storage. For example, if potential evapotranspiration exceeds rainfall or other water additions to the system, the depth of water in storage is decreased. In contrast, when additions exceed the potential water-holding capacity, the excess either runs off or percolates. By measuring run-off, the amount of percolation is determined by differences. The most difficult component of the analysis to quantify is the consumptive use, or evapotranspiration. A report to the American Society of Civil Engineers (Jensen, 1973) discusses in detail the state of the art for measuring evapotranspiration losses.

Fenn, Hanley, and DeGeare (1975) and Mather and Rodriguez (1978) used variations of the Thornthwaite-Mather method to estimate the potential of infiltrating rainwater to move into a landfill and create leachate. Assumptions were made on the thickness and storage capacities of soil cover and solid waste cells, on packing of solid waste, and on transmission characteristics of water in solid waste. According to Mather and Rodriguez (1978), the advantages of the water budget approach are that it is easy to apply, it provides quantitative values, and it is based on straightforward assumptions. The method should be applicable to landfills in humid regions where precipitation amounts are great enough to produce surplus water in the soil layers. For landfills in arid and semiarid regions protected from sheetflow, surplus water may not be available for deep percolation into solid waste cells and the method may have limited utility.

Direction of Water Movement and Hydraulic Gradients--

Methods for monitoring water movement and hydraulic gradients in saturated regions such as perched groundwater cannot be used to monitor unsaturated flow because of the so-called outflow principle: water will not freely enter a soil cavity unless the soil-water pressure is greater than atmospheric. Consequently, special methods must be used in unsaturated soils. Of the available methods, the three most commonly used to infer unsaturated flow are tensiometers, psychrometers, and neutron moisture logging. Neutron logging and tensiometers have been discussed. A discussion on psychrometers follows.

Psychrometers--The principle of psychrometric measurement of soil-water potential is given by Rawlins and Dalton (1967), who also diagrammed a representative unit (see Figure 4-4). Psychrometers measure soil-water pressure under very dry conditions, where a tensiometer unit is ineffective because of air-entry problems. Basically, the soil-water psychrometer measures the relationship between negative soil-water potential and the relative humidity of soil water. This relationship is given by:

$$\Psi \equiv \frac{RT}{n} \ln \frac{P}{P_o}$$

where Ψ = soil water potential (matric potential plus osmotic potential)

R = ideal gas constant

T = absolute temperature

n = volume of a mole of water

P/P_o = relative humidity.

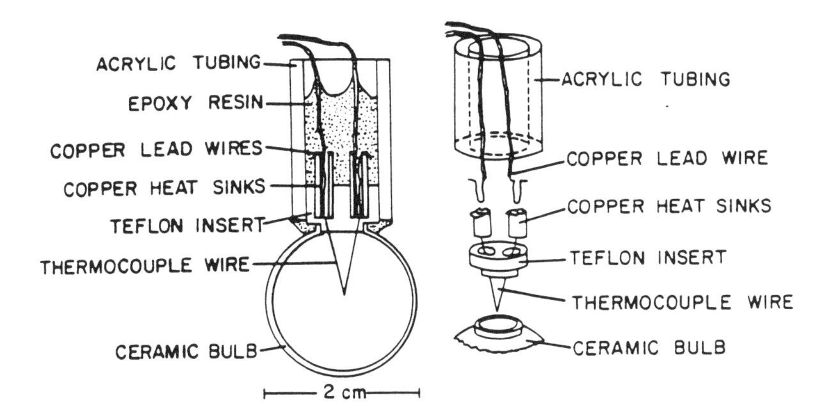

Figure 4-4. Soil psychrometer (after Rawlins and Dalton, 1967).

Operation of the Peltier-effect psychrometer consists of (Merrill and Rawlins, 1972):

> ... (1) measuring the electromotive force (e.m.f.) while the measuring junction is dry (dry e.m.f.); (2) passing a current through the sensor, cooling the measuring junction below the dewpoint, and thereby wetting it; (3) measuring the e.m.f. again after an initial, transient phase of the e.m.f. has subsided, but before the e.m.f. has decayed significantly as a result of depletion of water on the junction (wet e.m.f.); and (4) again passing current through the thermocouple wires in the reverse direction to heat and dry the measuring junction. This mode of operation permits any temperature difference between the measuring junction and the reference junction to be compensated for by subtraction of the dry e.m.f. from the wet e.m.f. (delta e.m.f.).

Psychrometers consist of a porous bulb chamber to sample relative humidity of a soil, a sensitive thermocouple, heat sink, reference electrode, and associated electronic circuitry. Such units use the principle of Peltier cooling to lower the temperature of one junction of the thermocouple below the dewpoint, thereby allowing an evaluation of the relative humidity.

Field implementation--Two types of thermocouple psychrometers (TCP) are available. One unit, installed in access tubes, consists of positioning psychrometers in porous cups at the base of tubing. This unit may be withdrawn for recalibration. The second type of unit, called "sealed-cup" psychrometers by Merrill and Rawlins (1972), has the thermocouple unit permanently sealed into a porous enclosure.

Enfield, Hsieh, and Warrick (1973) used psychrometers at a desert test site located in south central Washington to estimate the direction and rate of water flux in material in the vadose zone with a water table at 311 feet. The test hole was dug using the dry core barrel technique with progressively smaller diameter telescoped casing (12 to 6 inches in diameter) similar to the technique used with a cable tool rig. The reduction in casing diameter minimizes skin friction forces that would be encountered during removal of the casing following instrument implacement. A string of TCPs spaced about 10 feet apart was lowered into the cased hole through a rigid plastic pipe protect the instrumentation during the backfilling procedure. The casing and plastic pipe were gradually removed as the well was backfilled with well cuttings around the instrument cable and the TCPs. The cuttings were replaced at their approximate original stratigraphic position.

Each psychrometer must be calibrated before field installation. Calibration stability of the units was evaluated by Merrill and Rawlins (1972). During their field study, recalibration of the access-tube type instruments was required due to declining sensitivity in many of the units. Sensitivity was restored by immersing the thermocouple wires in hot, alcoholic potassium hydroxide solution (10 percent). The median decline in sensitivity was reported to be 18 percent for 52 units following a mean implacement period of 6 months. Sealed-cup psychrometers were found to be considerably more stable throughout an 8-month study period. The median change for these units was reported to be 5.3 percent with both increases and decreases in sensitivity noted.

Merrill and Rawlins (1972) found the accuracy of the data to vary throughout the -2.5- to -15-bar range tested. Standard deviation of data triplets varied from 0.08 bar for -7.5 to -10.0 bars to a high of 0.23 bar for the -2.5- to -5-bar range. Improved data accuracy was obtained by taking multiple readings using automatic techniques and computer processing instead of manually operating the data collection system.

Range of applications and limitations--Psychrometers have been employed at depths varying from a few inches to over 300 feet and in a wide range of geohydrologic settings. They have been found to be most useful in measuring water potential in dry soils. A primary assumption in their operation is that the ceramic material in which the thermocouple is embedded is in equilibrium with the soil moisture. The degree to which this assumption is met directly affects the accuracy of the measurements.

When installing the psychrometer in coarse material, care must be taken to establish good contact between the cup and the surrounding formation. In tight, well-indurated material or at great depths, this is difficult and nonrepresentative measurements may occur. Steeply sloping terrain increases the difficulty of proper installation and may restrict the use of this monitoring method. Fluctuations in temperature affect reading of the thermocouple psychrometer and can cause errors.

Psychrometer readings are subject to hysteresis in their characteristic curves. Determining whether the test area is drying or being wetted during the measurements is important to accurately interpret the readings. Measurements provide only a point source for water content. The degree to which this measurement reflects regional conditions is dependent on the homogeneity of the material tested. Field studies indicate that psychrometers perform very poorly in wet media (e.g., water pressure greater than 1 atmosphere).

Thermocouple psychrometers are poorly suited for use in acid media due to severe thermocouple wire corrosion problems. These units are suitable for inferring water movement but may not be useful in detecting the movement of volatile organics.

Thermocouple psychrometers are designed to measure negative soil water pressure from relative humidity versus output voltage relationships. In relating water pressure to water content, characteristic curves must be developed for each soil type in which the units are to be installed. This is time-consuming and may not be economically feasible for small monitoring projects. Most of the units are fragile and require great care during installation. Contaminants can cause errors if allowed to enter the sample chamber or thermocouple junction. Response times are generally slow; however, this is not a severe problem in very dry media.

Diurnal fluctuations in apparent water potential can occur in vertically installed units. To avoid this problem, ceramic-type units must be installed horizontally, which is difficult when using psychrometers at depth.

Earlier units were plagued with problems relating to temperature effects on measurement of potential. By designing the instruments to meet certain boundary conditions, these problems were overcome. If the psychrometers were installed near the ground surface, however, the diurnal heat flux would adversely affect the accuracy of the soil-water potential measurements. In addition, Merrill and Rawlins (1972) found that if the cooling period and cooling current for the thermocouple were not kept within a limited range, accuracy would be lost.

Field testing showed that access-tube instruments were susceptible to attack by fungi or bacteria. Merrill and Rawlins (1972) report very fine fungus-like threads on the thermocouple wires after removal from the access tubes in the field. Sealed-cup units protect the thermocouple wires and avoid possible attack by organisms.

Brown (1970) studied the psychrometer response to changing water potential. Bare, ceramic-cup, and unshielded-screen psychrometers were evaluated to determine the relative magnitude of the resistance to transfer of water vapor. Figure 4-5 shows the response lag of the three psychrometers over a 0.3-m KCl solution at 25°C.

Water potential data show that the ceramic cup offers significant resistance to vapor exchange after liquid contact is lost between the soil and the psychrometer during rapid drying. Brown (1970) reports that the ceramic-cup psychrometer lagged behind the screen psychrometer by from 1 to 2 bars; this lag increased to 8 bars at -30 bars. These results indicate that ceramic psychrometers should be used with caution in areas of high flux, e.g., rapid wetting or drying zones near the ground surface.

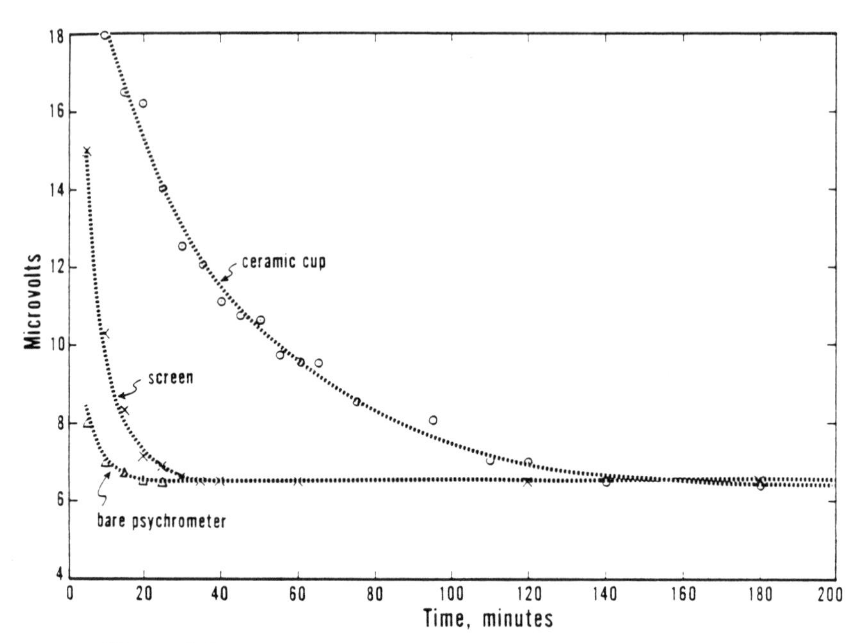

Figure 4-5. Response lag of thermocouple psychrometers to vapor equilibrium (after Brown, 1970).

Flux, Velocity, and Unsaturated Hydraulic Conductivity--

One of the prime goals of vadose zone monitoring is to estimate the flux and velocity of water. Some of the methods for estimating flux and velocity are based directly on the equation of continuity and information on hydraulic conductivity values is not required. However, other methods are based on Darcy's equation and require values of both hydraulic conductivity and hydraulic gradients. In this book, both classes of methods are referred to as "draining-profile" techniques because measurements are made during drainage cycles. Other methods reviewed are: (1) water budget with soil moisture accounting, (2) measurement of hydraulic gradients, (3) methods based on measuring hydraulic conductivity, (4) catalog of hydraulic properties, (5) flowmeters for direct measurement of flux, (6) use of suction cups for estimating velocity, (7) tracer techniques, and (8) an approximate method using water budget data.

By and large, the methods presented may be suitable for characterizing flux and hydraulic conductivity in soils and may have limited applicability to deeper regions of the vadose zone. In many waste disposal operations, however, determining loss in the near-surface region is important. For example, during site evaluation of hazardous waste disposal ponds, a series of tests could be conducted to determine a site with minimal conductivity and flux (K and J) values.

Draining-profile methods--The simplest approach for estimating flux at successive depths in the vadose zone is to obtain a series of water content profiles during drainage. The method is essentially based on the equation of continuity, expressed in terms of flux:

$$\frac{\partial \theta}{\partial t} = \frac{\partial J}{\partial z} \qquad (4\text{-}8)$$

or

$$J = -\int_0^z \frac{\partial \theta}{\partial t}\, dz \qquad (4\text{-}9)$$

To account for all sources and sinks, Bouwer and Jackson (1974) suggest using the following expression for flux:

$$J_1 = R + I - ET - \int_0^z \left(\frac{\partial \theta}{\partial t}\right) dz \qquad (4\text{-}10)$$

where J_1 = combined flux

R = rainfall

I = irrigation

ET = evapotranspiration.

For a draining profile covered to prevent ET, equation 4-10 reduces to equation 4-9.

Figure 4-6 shows an example of hypothetical water content profiles from a field site. Calculation of flux out of a given depth, z, between successive times, t_1 and t_2, requires the solution of equation 4-9. Inasmuch as the right-hand side of the equation is simply the shaded area of the sketch divided by the time difference $t_2 - t_1$, the flux can be found by graphical integration of water content profiles between successive measurement dates. Table 4-3 lists contraints in the uses of the draining-profile method.

Water content profiles can be obtained using neutron moisture logging or tensiometer techniques. An advantage of neutron moisture logging is that profiles can be obtained at considerable depth if access wells are installed carefully to prevent side leakage. The sequence to obtain water content profiles using tensiometers is:

1. Install tensiometers at sequential depth intervals through the soil region of interest

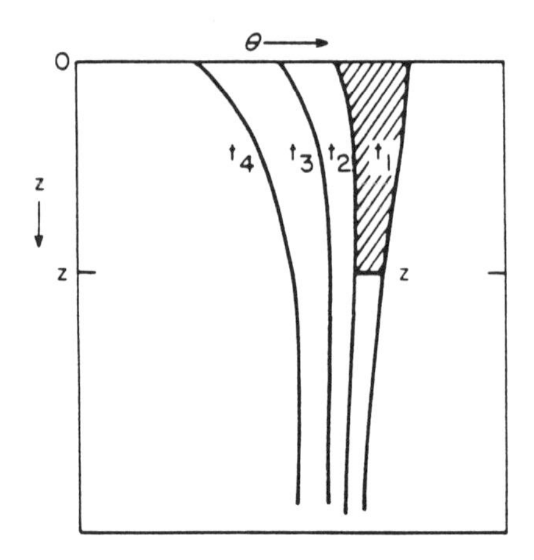

Figure 4-6. Hypothetical water content profiles at several times during drainage of a soil profile (after Bouwer and Jackson, 1974).

TABLE 4-3. CONSTRAINTS ON DRAINING-PROFILE METHOD

Category	Constraint
Geohydrologic	In soils overlying shallow water tables, drainage occurs uniformly and the draining-profile method is applicable. In deeper soils, drainage is not uniform, i.e., upper profile may be draining but lower profile may be wetting. Thus, assuming that flux throughout the vadose zone equals drainage rate in upper profile is not necessarily true for deep profiles.
Chemical	Chemical conditions affecting units used to measure water content changes may introduce errors. For example, chlorine in the soil solution affects neutron logging.
Geologic	1. In well-structured soils, drainage will occur more rapidly than in soil blocks where water content changes are measured. 2. Shallow lenses may impede water flow. 3. Because of the spatial variability of soil-water holding properties in typical vadose zones, a large number of measurements will be required.
Topographic	Steep slopes may preclude the installation of measuring units.
Physical	1. Not suitable for existing landfills and impoundments where measuring units cannot be installed. 2. Physical factors affecting operation of measuring units will affect accuracy of the method.
Climatic	Climatic factors affecting the operation of measuring units will affect the accuracy of the method.

2. Obtain soil cores from the soil at depths bracketing the tensiometer cups

3. Obtain soil-water characteristic curves for each core in order to define the relationship between water content and soil-water pressures; since field data will be obtained during drainage, curves must be obtained for drying cycles

4. Determine pressure values in tensiometers during drainage of the soil profile (see Figure 3-6a)

5. Convert pressure readings to water content values using the appropriate water characteristic curves (see Figure 3-6b)

6. Plot water content versus soil depth (see Figure 3-6c).

An alternative draining-profile method requires two stages: (1) determining the hydraulic properties of soil in a number of plots within a field or disposal area, and (2) using the resultant hydraulic properties and appropriate instrumentation to estimate flux at additional field locations. Basically, the method is an adaptation of a two-step technique presented by LaRue, Nielsen, and Hagan (1968) for estimating soil-water flux below the root zone of an irrigated field.

During step 1, a number of test plots are established at random locations in the area of interest. Each plot is instrumented with a depthwise sequence of tensiometers (installed in triplicate at each depth) and possibly an access tube to permit moisture logging. These units should be installed in the center of the plot to minimize the effects of lateral flow. The periphery of each plot has side walls to facilitate applying a head of water at the soil surface. Figure 4-7 shows a possible arrangement for a test plot. Soil cores obtained during installation of tensiometers are used to construct water characteristic curves for distinct depth intervals in the profile. The following procedure is used:

1. Flood each plot until the reading on the lowermost tensiometer indicates that a maximum degree of saturation has been reached in the surrounding soil

2. Following disappearance of free water, cover the soil surface with plastic and earth to prevent evaporation

3. Allow the soil to drain, taking regular tensiometer readings and logging the access well

4. From water content profile, estimate flux out of various depth intervals using the graphical approach

5. Calculate hydraulic gradients for the selected depth intervals for each sequence of tensiometer readings averaged over the time interval

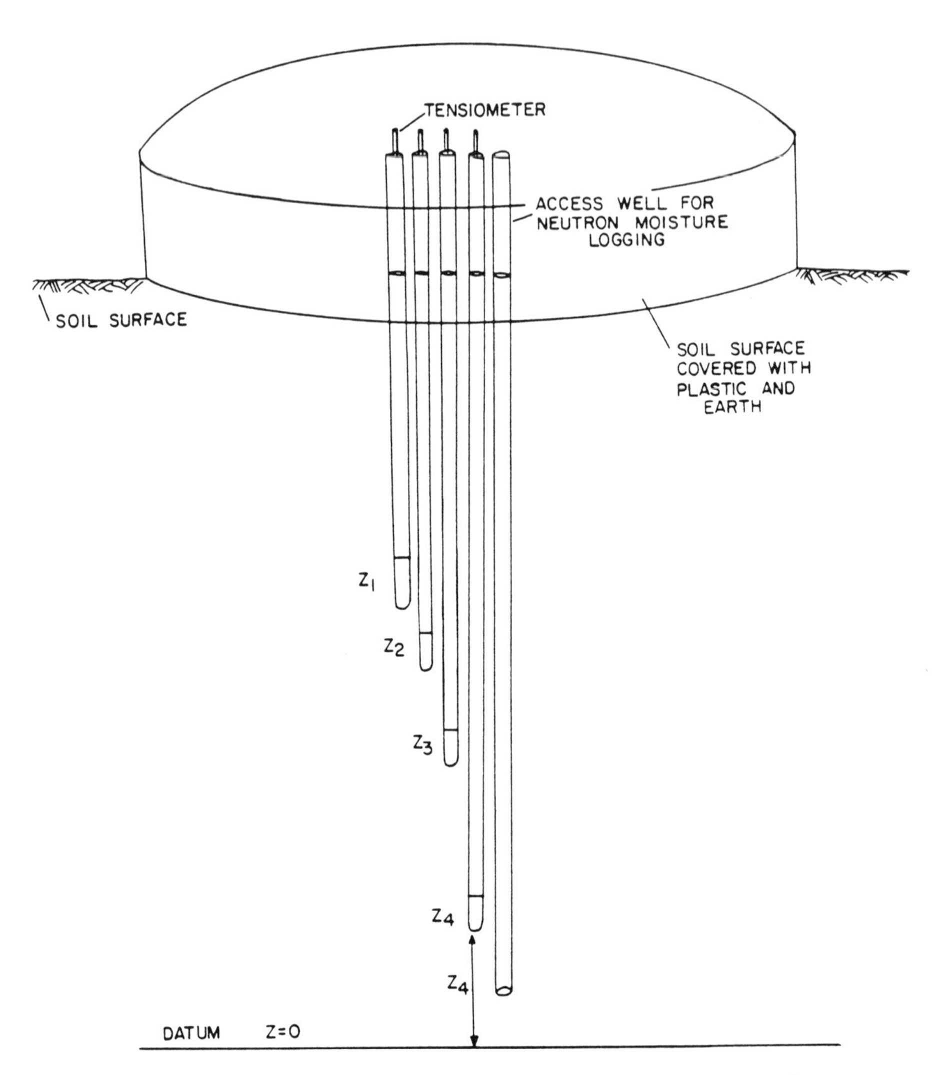

Figure 4-7. Sketch of field plot for determining K and J showing tensiometers and access well.

6. Use the hydraulic gradients and flux values to calculate hydraulic conductivity from Darcy's equation, $J = -K(\theta)\nabla H$.

7. Plot the $K(\theta)$ values versus average water content values (θ) for each depth interval; obtain only one $K(\theta)$ value for each time interval

8. Repeat steps 4 through 7 for additional time steps and obtain K versus θ curves for each depth interval.

An example (courtesy, A.W. Warrick, verbal communication, 1979) will clarify the above procedure for the first step of the LaRue, Nielsen, and Hagan method. Assume that a test plot has been established with tensiometers installed for measuring water pressures at discrete depths to 200 cm. An access well has been installed to obtain neutron-log water content profiles. The soil profile is wetted to 200 cm, following which the surface is covered and water content values and soil-water pressures are measured. Figure 4-8 shows resultant water content profiles for 24 hours and 40 hours after the beginning of drainage.

The area between the curves (shown by cross-hatching) was found by planimetering to be 2 cm of water. Pressure head values measured on tensiometers at the 120-cm and 150-cm depths after 24 hours and 40 hours of drainage were:

	h_1(120 cm)	h_2(150 cm)
t = 24 hours	-36 cm	-29 cm
t = 40 hours	-40 cm	-32 cm .

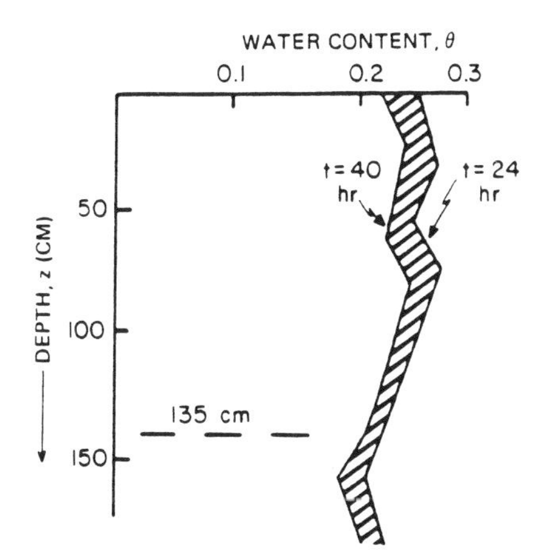

Figure 4-8. Water content profiles in hypothetical soil column at t = 24 hours and t = 40 hours.

The problem is to find a K versus θ relationship for the 120-cm to 150-cm depth interval. Only one value can be found for one time interval; for additional values, profiles for other time intervals are needed. The procedure is:

1. Find the drainage for the time interval t = 24 hours to t = 40 hours. Assume a precise interest in the 135-cm depth (half-way between 120 cm and 150 cm). Since the area in Figure 4-8 corresponds to 2 cm and the time interval is 40 - 24 = 16 hours, then the Darcian flux is:

$$J = 0.125 \text{ cm/hr.}$$

2. Next find the hydraulic gradient. If the reference is taken at the level of the lowermost tensiometer and the positive direction is upward, the hydraulic heads at 120 cm and 150 cm at 24 hours are:

$$H_{150} = (0 - 29) = -29 \text{ cm}$$

$$H_{120} = (30 - 36) = -6 \text{ cm} .$$

The gradient at 24 hours is:

$$\nabla H = [-29 - (-6)]/30 = -0.76 \text{ (24 hours).}$$

Similarly, the hydraulic heads at 120 cm and 150 cm at 40 hours are:

$$H_{150} = (0 - 32) = -32 \text{ cm}$$

$$H_{120} = (30 - 40) = -10 \text{ cm} .$$

The hydraulic gradient at 40 hours is:

$$\nabla H = [-32 - (-10)]/30 = -0.73 .$$

The above values represent averages over the 120- to 150-cm depth interval or, alternatively, at the 135-cm depth.

3. Calculate the hydraulic conductivity for Darcy's law. Using J from step 1 and the hydraulic gradient at 135 cm averaged over the period 24 to 40 hours:

$$J = -K(\theta)\nabla\bar{H}$$

where $\bar{H} = \frac{-0.76 - 0.73}{2} = -0.75,$

$$0.125 = -K(\theta)\ (-0.75)$$

or $K(\theta) = 0.167 \text{ cm/hr}$.

This $K(\theta)$ value corresponds to an average water content value (0.23 cm^3/hr) in the interval 120 cm to 150 cm during the period 24 to 40 hours. The $K(\theta)$ and θ values thus provide one point on a $K(\theta)$ versus θ curve for this depth interval. (Alternatively, K corresponds to an average h value for the four input values; $h_{avg} = (-36 - 40 - 29 - 32)/4 = -34.2$ cm. The corresponding θ value could be found from a soil-water characteristic curve for the depth interval.)

When the hydraulic properties of the field have been estimated, the procedure to follow at additional field sites during step 2 of the LaRue, Nielsen, and Hagan method is:

1. Install tensiometers at depth intervals of interest, corresponding to intervals used on the test plots; alternatively, to minimize expense, two tensiometers could be used to bracket the lowermost interval

2. Measure water pressures during drainage cycles using tensiometers; calculate hydraulic gradients

3. From the water characteristic curve corresponding to the depth interval of interest, convert water pressure value to an average θ value

4. Using the K versus θ curve for the interval, convert θ values to K values

5. Use the K and hydraulic gradient values with Darcy's equation to calculate flux.

Nielsen, Biggar, and Erh (1973) evaluated a number of simplified techniques for measuring flux and hydraulic conductivity values in the field. One of the promising methods is based on the following relationship:

$$J_L = K_o \left(1 + a\, K_o t L^{-1}\right)^{-1} \qquad (4\text{-}11)$$

where J_L = flux at depth L

K_o = steady-state hydraulic conductivity

a = an empirical factor relating K and θ

t = time of observation.

According to Warrick and Amoozegar-Fard (1980), this equation is based on the following assumptions:

1. The soil is initially at a uniform water content θ_o

2. The unsaturated conductivity is given by the equation:

$$K = K_o \exp\left[a(\theta - \theta_o)\right] \qquad (4\text{-}12)$$

3. A unit hydraulic gradient exists at all times

4. Evaporation and transpiration are negligible.

With the hydraulic gradient equal to unity, the equation of continuity, expressed by equation 4-6, may be written as:

$$\int_0^L \frac{\partial \theta}{\partial t}\, dz = K(\theta)\Big|_{z=L} \tag{4-13}$$

or

$$L \frac{d\bar{\theta}}{dt} = K(\theta)\Big|_{z=L} \tag{4-14}$$

where θ is average soil-water content in profile above a level L. The hydraulic conductivity is evaluated at depth $z = L$.

In practice, hydraulic conductivity values are determined from test plots (e.g., Figure 4-7) by (1) measuring θ with a neutron moisture meter, or (2) converting pressure readings from tensiometers to θ values using appropriate water characteristic curves, and (3) solving equation for $K(\theta)$ at depths of interest. If a graph of $K(\theta)$ versus θ produces a straight line, equation 4-12 is assumed to be valid and equation 4-11 can be used to calculate J_L. The graph is also used to determine the factor a, which is simply the slope of the straight line.

Nielsen and Biggar (1973) present a plot of ln $K(\theta)$ versus θ for test plot data for Yolo loam, an unusually uniform soil. The scatter of points is minimal. For a more heterogeneous soil profile, values of $K(\theta)$ ranged over four orders of magnitude for a given water content. However, the flux of water at a given depth was still adequately described using an average straight line from the ln $K(\theta)$ versus θ plot. Problems arising from the spatial variability in soil properties are described at the end of this section.

Equation 4-14 provides yet another alternative method for estimating flux. That is, the left side of the equation is an alternative form for expressing the flux. Thus:

$$L \frac{d\bar{\theta}}{dt} = J_L = K(\theta)\Big|_{z=L}$$

In other words, if the hydraulic gradient is equal to unity in the flow system, the flux at a depth equals the corresponding hydraulic conductivity value.

Flux values determined from the draining-profile techniques are converted to values of average linear velocity by dividing by the appropriate water content values. For example, if the calculated flux of a soil is 0.125 cm/hr and the average water content is 0.23, the average linear velocity is:

$$\bar{v} = J/\theta \tag{4-15}$$

$$= 0.125/0.23 \text{ cm/hr}$$

$$= 0.54 \text{ cm/hr}.$$

Water budget with soil moisture accounting--The water budget method of Thornthwaite and Mather (1957) can be used to a given soil depth (e.g., root zone of an irrigated field; final soil cover on a landfill, etc.) to determine flux in the vadose zone. Inflow components include rainfall and/or irrigation. Outflow components include runoff, evapotranspiration, drainage, and deep percolation (flux). The change in storage equals water content change at the depth of interest. Flux is then calculated by equating it to known inflow and outflow components and changes in storage (Δ). Additional references that describe the method in full include Willmott (1977); Mather and Rodriquez (1978); and Fenn, Hanley, and DeGeare (1975). Of the parameters that require measurement, evapotranspiration may be the most difficult component to measure. Jensen (1973) covers alternate methods of measuring evapotranspiration. Table 4-4 lists constraints on using the water budget method to determine flux in the vadose zone.

Methods utilizing hydraulic gradients--Two methods have been developed to determine flux in the vadose zone using hydraulic gradients. The first method is based on solving Darcy's equation for unsaturated flow; $J = K(\theta)i$ where $K(\theta)$ designates that the hydraulic conductivity is a function of water content θ and i is the hydraulic gradient. Hydraulic gradients are measured by installing tensiometers, moisture blocks, or psychrometers. Calibration curves are used to relate negative pressure measurements to water content. Water content is then related to unsaturated hydraulic conductivity, although separate curves are required for each textural change encountered.

This method is very precise; however, it is more complex than methods using water content values, generally restricted to shallow depths in the vadose zone, and results are subject to hysteresis in the calibration curves. Additional constraints in the use of this method are listed in Table 4-5. For further descriptions of the method, see LaRue, Nielsen, and Hagan (1968); Bouwer and Jackson (1974); and Wilson (1980).

A second method is based on assumptions that hydraulic gradients are unity. Darcy's equation is solved for unsaturated flow with unit hydraulic gradient assumed so that $J = K(\theta)$. With this simplification, only one pressure measuring unit is required at each depth of interest to permit estimating θ from a pressure versus water content curve. $K(\theta)$ is then taken from a separate curve. A somewhat more complex version of this method is described by Nielsen, Biggar, and Erh (1973). An alternative approach is to use the relationship $J = K(\Psi m)$. This method requires a curve showing the changes in hydraulic conductivity with matric potential, Ψm. Bouma, Baker, and Veneman (1974) describe the so-called "crust test" for preparing a $K(\theta)$ versus Ψm curve. This field procedure is carried out on cylindrical columns constructed in a test pit. Each column is instrumented with a tensiometer, a ring infiltrometer, and gypsum-sand crusts. A series of crusts is used during different

TABLE 4-4. CONSTRAINTS ON THE WATER BUDGET METHOD

Category	Constraint
Geohydrologic	High water tables will affect application of the method.
Chemical	Chemical reactions in the soil solution may affect structural properties of soils, possibly changing water transmission and water holding properties.
Geologic	1. In highly-structured soils, flow may occur primarily in cracks and not within soil blocks. Thus, changes in soil-water content may not be related to flux below the soil zone. 2. Spatial variations in water-holding properties of soils occur in most fields. Assuming water-holding capacity for a field based on one soil-mapping unit may not represent "average" conditions.
Topographic	In poorly levelled fields, water may pond in low spots and run off rapidly in other areas. Consequently, the true flux may be different from the "average" flux calculated by assuming uniform water application.
Physical	Method relies on accurately measuring inflows and outflows. Errors affect estimates of flux.
Climatic	1. Method requires accurate measurement of climatological factors required to estimate actual evapotranspiration (ET) potential. Errors affect accuracy of flux method. 2. Requires accurate measurement of rainfall and ambient temperature. 3. Freezing and thawing in soils may affect water-holding capacity by changing texture.

TABLE 4-5. CONSTRAINTS ON THE MEASURED HYDRAULIC GRADIENT METHOD

Category	Constraint
Geohydrologic	1. Requires obtaining calibration curves for the following relationships: θ versus Ψm, Ψm versus $K(\Psi m)$, θ versus $K(\theta)$. 2. Requires installation of units for measuring Ψm and/or θ.
Chemical	Chemical properties of wastes may affect the operation of units used to measure hydraulic gradients.
Geologic	1. Requires installing units in depthwise increments throughout the vadose zone. Gradients across layers may suggest vertical flow when actually horizontal flow is predominant. 2. Calibration curves (Ψm versus θ, Ψm versus $K(\Psi m)$, θ versus $K(\theta)$) required for each change in texture. 3. May not be suitable at sites underlain by fractured media.
Topographic	Topographical constraints may affect the installation of measuring units.
Physical	1. Factors affecting the installation and operation of mea-measuring units will affect results. 2. The spatial variability of vadose zone properties requires a large number of measurements. 3. The Ψm versus θ, Ψm versus $K(\Psi m)$, and the θ versus $K(\theta)$ relationships are subject to hysteresis. 4. May not be possible to install units at existing impoundments and landfills.
Climatic	1. Climatic factors may affect the operation of units. 2. Evaporation must be prevented in test site.

runs to impose varying resistances to flow. During each run, infiltration rates and tensiometer values are monitored.

Methods based on the assumption of unit hydraulic gradients are simpler and less expensive than methods that require gradients. However, the assumption of unit hydraulic gradients may fail, particularly in layered media. Table 4-6 lists other constraints on the method.

Methods utilizing measured hydraulic conductivity--The premise of these methods is that if hydraulic conductivity (K) values are available, the flux can be estimated by assuming hydraulic gradients are unity and that Darcy's law is valid. Several methods have been developed for measuring K. These can be generally divided into laboratory and field methods. Laboratory methods include (1) permeameters, (2) grain-size and hydraulic conductivity relationships, and (3) catalog of hydraulic properties. Field methods include (1)

TABLE 4-6. CONSTRAINTS ON THE UNIT HYDRAULIC GRADIENT METHOD

Category	Constraint
Geohydrologic	1. Assumption of unit hydraulic gradient may fail in layered media. 2. Requires installing a single measuring unit at depths of interest. 3. Calibration of relationships of K(θ), θ, and Ψm are required.
Chemical	1. Chemical properties of wastes may affect the operation of measuring units. 2. Calibration relationships may not hold for certain pollutant types, e.g., volatile organics.
Geologic	1. Calibration relationships are required for change in texture. 2. Assumption of unit hydraulic gradient may fail in layered media. 3. Because of spatial variability in soil properties, a large number of measuring sites may be required.
Topographic	Topographical constraints may affect the installation of measuring units.
Physical	1. If measuring units are not installed properly, results may not be valid (e.g., poor contact between tensiometer cup and soil). 2. May not be possible to install measuring units at existing landfills or impoundments.
Climatic	Climatic factors may affect the operation of measuring units.

methods for measuring saturated K in the absence of a water table, (2) instantaneous profile method, (3) USBR single-well method, (4) USBR multiple-well method, (5) Stephens-Neuman single-well method, and (6) air permeability method.

Catalog of hydraulic properties--Mualem (1976) prepared a catalog of hydraulic properties of soils. If these soils are similar to those in the vadose zone, both saturated and unsaturated K values can be obtained.

The method is simple, quick, and may be used to estimate relative variations in K caused by stratification. It is inexpensive if grain-size data are available. Unfortunately, because of errors in measuring K(θ), values for a particular soil type may not be transferable to similar soil types. To obtain a closer estimate, K(θ) must be evaluated for each soil or horizon of interest (Evans and Warrick, 1970).

Direct Measurement of Flux

Attempts have been made in recent years to develop equipment to measure soil-water flux in the unsaturated state, which does not require information on the hydraulic conductivity.

Two types of flowmeters reported by Cary (1973) involve (1) direct flow measurement, and (2) the displacement of a thermal field by water in motion. The direct flow unit measures the flow of soil-water intercepted by a porous tube containing a sensitive flow transducer. The second unit entails measuring very accurately the transfer of a heat pulse in water moving in a porous cup buried in the soil. Because of the intimate contact between the soil and porous cup, the water moving in the cup forms a continuum with soil-water. Laboratory calibration curves are prepared to relate output of a sensitive millivolt recorder during imposition of heat pulses to empirically measured flow rates.

For field installations, the porous discs containing either flow transducers or heat sources are mounted in cylinders buried in the soil. A limitation, therefore, is that flow is measured in disturbed soils.

According to Dirksen (1974a), the accuracy of the above flowmeters can be improved by:

1. Minimizing convergence or divergence

2. Extending the range of water fluxes and soil types

3. Reducing or preferably eliminating very tedious calibration procedures

4. Minimizing soil disturbance during installation and its effect on original flow pattern.

Dirksen (1974b) discusses the design of a fluxmeter he developed. A feature of his unit that circumvents some of the foregoing difficulties is that, as soil hydraulic conductivity changes, the resistance of the meter is adjusted so that the head loss across the meter matches the head loss in the soil as measured by nearby tensiometers. The fluxes through the meter and soil then are equal. Dirksen also presents a method for installing his fluxmeter to minimize soil disturbance.

To date, no data are available on the use of such flowmeters in the lower vadose zone of deep profiles. Table 4-7 lists other constraints on the use of flowmeters.

Indirect Estimates of Velocity Using Suction Cups

An indirect estimate of vertical velocity in the vadose zone beneath a surface source entails observing the response of suction cups (see Suction Samplers, Section 5) as the wetting front moves into the profile. Basically, the technique involves periodic attempts to sample from an array of suction samplers within the vadose zone. Generally, when the surrounding pore-water system is unsaturated, very little, if any, sample will be obtained. However, when the wetting front reaches a particular cup, samples are more readily obtainable. By observing the response of depthwise units, an apparent vertical velocity may be inferred.

TABLE 4-7. CONSTRAINTS ON FLOWMETERS

Category	Constraint
Geohydrologic	1. Intercepting-type meters require fairly wet soils to perform effectively. 2. Intercepting-type meters may be difficult to install in layered media without affecting flow lines. 3. Fluxmeters featuring variable hydraulic resistance control require measurement of negative pressure gradients, e.g., by tensiometers.
Chemical	Physical properties of some chemical waste fluids may be dissimilar to those of water, causing units to give erroneous results. For example, units based on relating heat flux to water flux may be affected by the difference in heat-conducting properties of water and chemical fluids.
Geologic	1. May not be applicable in fractured media. 2. May not have good contact between units and measurement volume in very coarse materials. 3. Meters obtain only point measurements. Because of the spatial variability of soil hydraulic properties, a large number of units should be installed to determine "average" flux.
Topographic	Difficult to install units in very steep terrain.
Physical	1. Calibration relationships are required for most units. For multilayered media, a large number of curves are required. 2. Intercepting type units are restricted to shallow depths. 3. Hydraulic resistance units are restricted to shallow depths because of the need to construct a pit from which a horizontal cavity is dug for installing the meter. 4. Soil disturbance during installation may affect flow paths through meters. 5. It may not be possible to install meters beneath existing waste disposal facilities.
Climatic	1. Frost heaving may shift units and reduce contact between meter and surrounding media. 2. Diurnal temperature fluctuations affect the operation of hydraulic resistance type units.

Meyer (personal communication with L.G. Wilson, 1978) used this method to follow the wetting front during deep percolation of irrigation water in the San Joaquin Valley. Signor (personal communication, 1979) used a similar approach during recharge studies in Texas.

Tracers

A direct method for measuring the average linear velocity of water movement in the vadose zone is to introduce tracers at the soil surface. These include physical tracers such as temperature, ionic constituents such as chloride, organic tracers such as fluorescent dyes and fluorocarbons, and radioactive tracers such as tritium. Evans, Sammis, and Warrick (1976) evaluated the relative merits of two tracers, temperature and tritium, for determining flux beneath an irrigated field. Results were compared with flux values found by using Darcy's equation. In general, results using temperature and tritium compared poorly with calculated values. They point out that problems with tritium include a short half-life (12.26 years) and possible reactions with clay particles. Table 4-8 lists other constraints on the use of tracers.

Frissel et al. (1974) compared ^{36}Cl, ^{60}Co, and tritium for tracing moisture movement in laboratory soil columns. From observed breakthrough curves for the tracers, it was determined that for clay and sand soils, tritium adsorption was negligible but that the use of ^{36}Cl and ^{60}Co ions would lead to errors because of anion exclusion.

A basic problem with fluorescent tracers is sorption on clays, resulting in errors in estimating flow velocities. Fluorocarbons may have promise, although sorption occurs on organic constituents of a soil (Brown, personal communication, 1980).

Calculating Velocity Using Flux Values

Flux values obtained by using a suitable tracer (e.g., tritium, iodide, bromide, fluorocarbons), together with estimated or measured water content values, are substituted in the following relationship:

$$v = J/\theta$$

where v = velocity

J = flux

θ = water content.

This method of determining velocity in the vadose zone assumes that the hydraulic gradients are unity, an average water content can be determined, flow is essentially vertical, and homogeneous media are being evaluated.

The method provides a quick approximation of the travel time of pollutants in the vadose zone. It is simple and inexpensive when coupled with other methods. Velocities determined for structured media will tend to be

TABLE 4-8. CONSTRAINTS ON TRACERS

Category	Constraint
Geohydrologic	1. If flow velocities are slow, a long time interval may be required to obtain measurements. 2. Operation of sampling units may affect the flow field, leading to incorrect results.
Chemical	1. "Average" velocity of water-borne tracers may not be the same as "average" velocity of chemical liquids. 2. Some tracers (e.g., organic type) may require elaborate analyses. 3. Nonreactive type tracers must be used.
Geologic	1. In structured or fractured media, flow may occur primarily in large cracks. "Average" velocity for these conditions may be difficult to determine. 2. A large number of test sites is required to measure an "average" velocity because of the spatial variability of soil properties.
Physical	1. Installation of sampling units beneath existing disposal facilities may not be possible. 2. If soil becomes very dry, the air entry value of suction cup samplers may be exceeded and samples cannot be obtained.
Climatic	1. Freezing and thawing may alter soil structure/texture, affecting flow paths and velocity. 2. Freezing and soil heaving may dislocate suction cups, changing their sampling depths and altering contact with soils. 3. Heavy rainfall or flooding at site may dilute tracer.

higher than those calculated by this method. Since the method assumes vertical flow, perching layers that cause lateral flow would preclude its use. Table 4-9 lists additional constraints on the use of the method. For further details regarding the method see Bouwer (1980) and Wilson (1980).

Velocity Calculation Using Long-Term Infiltration Data

This method assumes that long-term infiltration rates, I, at a site are equal to the steady-state flux, J, in the vadose zone. Consequently,

$$v = I/\theta = J/\theta.$$

This relationship assumes that the hydraulic gradients are unity, that average water content for the media tested equals θ, and that flow is vertical in a homogeneous media. If these assumptions are valid, then velocity can be easily obtained.

TABLE 4-9. CONSTRAINTS ON VELOCITY CALCULATIONS USING FLUX VALUES

Category	Constraint
Geohydrologic	1. Hydraulic gradient must be unity. 2. Flow in the vadose zone should be primarily in the vertical direction.
Chemical	1. Method assumes that I, the long-term intake rate, equals the saturated hydraulic conductivity (I = Ki, where i = hydraulic gradient = unity). For liquid chemicals, the long-term intake rate may not equal the "hydraulic" conductivity. 2. May be difficult to determine an equivalent θ value for liquid wastes.
Geologic	1. In highly layered media, flow may be primarily in the horizontal direction and vertical velocity will be overestimated. 2. In structured or fractured media, intake rates will be affected by flow in cracks, e.g., flow within peds will be slower than estimated "average."
Physical	1. Requires information on the average water content, θ, of the vadose zone. 2. Requires accurate measurement of intake rate, I. 3. May not be possible to measure long-term intake rates of leachate beneath solid waste facilities.
Climatic	Freezing and heaving may affect the surface texture/structure, modifying the long-term intake rate.

The method is simple and inexpensive and probably satisfactory as a first approximation of true velocities in the vadose zone. In multilayered media, an average θ and v may be difficult to obtain. Similar to velocities calculated using flux values, structured media calculations will tend to overstate velocities (Bouwer, 1980, and Warrick, personal communication, 1981).

Simplified Method for Estimating Velocity Using Water Budget Data

For many disposal operations, it may not be possible to use the refined flux and velocity calculation methods for estimating the velocity of waterborne pollutants in the vadose zone. An approximate procedure is required to estimate whether pollutants have reached the water table. The following is suggested as one such approach. The method is based on a variation of equation 3-1, expressed as:

$$d_{vz} = \frac{d_w}{\theta} \tag{4-16}$$

where d_{vz} = depth of penetration of water below a given soil depth

d_w = depth of water applied

θ = volumetric water content at field capacity.

The method assumes vertical piston movement of water beneath the disposal site with water content values equal to field capacity.

In practice, an estimate of seepage rate is determined using a water budget method such as that of Thornthwaite and Mather (1957), or other simplified methods. The rate is converted into an equivalent depth of water per unit time. Average field capacity values for materials in the vadose zone are estimated from drillers' logs.

As an example, assume that the depth of water seeping beneath a disposal facility for 1 year is 5 feet. If the volumetric water content corresponding to field capacity is 15 percent, the depth of penetration in the vadose zone is:

$$d_{vz} = \frac{5.0}{0.15}$$

$$= 33 \text{ feet.}$$

Thus, if the vadose zone is 100 feet thick, seepage water would reach the water table in about 3 years.

Saturated Flow in the Vadose Zone

For saturated flow, Darcy's equation is written in the form:

$$v = -K\nabla H \tag{4-17}$$

where v = specific discharge

K = saturated hydraulic conductivity

∇H = hydraulic gradient.

To account for flow only within the pore space, Darcy's equation is also written as:

$$\bar{v} = -\frac{K}{S_t}\nabla H \tag{4-18}$$

where $\bar{v}$ = average linear velocity

S_t = porosity.

As with unsaturated flow, the average linear velocity is a macroscopic quantity of lesser magnitude than the actual microscopic velocity in the pore sequences.

In the context of this book, saturated flow is considered only in perched groundwater regions of the vadose zone. The complex flow patterns within groundwater mounds immediately above a water table are not considered.

A common assumption is that perched groundwater is always created by a relatively impermeable layer that impedes vertical water movement. Although this concept is true, perched groundwater may also develop at the interface between two layered regions that individually may be quite permeable.

Figure 4-9, reproduced from Bear, Zaslavsky, and Irmay (1968), illustrates the requisite condition for perched groundwater formation when a region of higher permeability overlies a region of lesser permeability in the vadose zone. For the first case, the flow rate, J, is less than K_2, the hydraulic

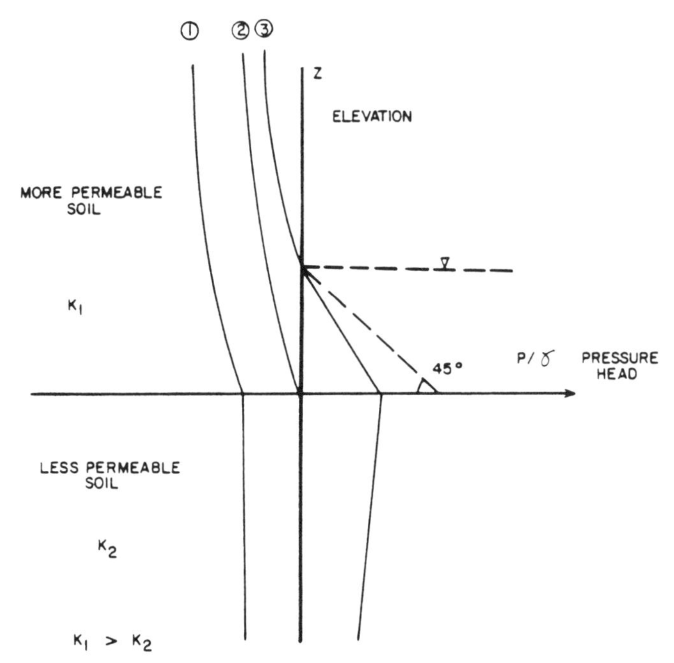

Figure 4-9. Pressure head diagram in a transition from a more permeable to a less permeable layer (after Bear, Zaslavsky, and Irmay, 1968). For 1, the flow rate $J < K_2$; for 2, $J = K_2$; and for 3, $J > K_2$.

conductivity in the lower strata. Flow is unsaturated and water could not be extracted from the system except by means of a suction cup. In the second case, J equals K_2 and the soil-water pressure is atmospheric. Again, it may not be possible to obtain a water sample except by means of a suction cup. In the third case, J is greater than K_2 and a region of positive water pressure is present at the interface. Water samples could be obtained from a well or open cavity.

For saturated media, the hydraulic conductivity in Darcy's equation is not generally expressed as a function of the water content. However, it is necessary to account for possible variations in the hydraulic conductivity depending on the direction of flow. For example, in layered alluvial deposits, the horizontal hydraulic conductivity is generally greater than in the vertical direction. The condition of variable K values depending on direction is called anisotropy. In anisotropic media, the hydraulic conductivity is a second rank tensor (Bear, 1972).

Because the hydraulic conductivity, K, of a soil is a measure of the ease of water movement, it follows that K depends greatly on soil texture and structure. Thus, a sandy soil will have greater hydraulic conductivity than a clay soil, even though the latter has greater porosity. Similarly, a well-structured soil will permit rapid flow through the cracks. Such a soil will have greater hydraulic conductivity than a poorly structured soil, even though the soil type may be the same. Ranges of hydraulic conductivity values for representative water-bearing materials in California are listed in Table 4-10.

Sodium Adsorption Ratio--

Although the saturated hydraulic conductivity of a soil is not dependent on the water content, K is not necessarily a constant. For example, a decrease in temperature may dissolve air bubbles entrapped in the soil pores, causing an increase in the hydraulic conductivity. The hydraulic conductivity of soils containing clays will also be affected by the cationic constituents and solute concentrations of applied water. Of particular concern with respect to the cationic composition of the applied water is the relationship of monovalent sodium to divalent calcium and magnesium. This relationship is commonly represented by the sodium adsorption ratio (SAR):

$$SAR = \frac{Na^{+}}{\sqrt{\frac{Ca^{++} + Mg^{++}}{2}}} \quad (4\text{-}19)$$

where Na^{+}, Ca^{++}, and Mg^{++} are expressed in milliequivalents per liter.

SAR values greater than 6 and 9 in an irrigation source are expected to reduce the hydraulic conductivity of shrinking-swelling soils (Ayers and Westcot, 1976), thus decreasing the long-term infiltration rate. However, the ratio of carbonate and bicarbonate levels in irrigation water must be considered. As stated by Ayers and Westcot: "When drying of the soil occurs between irrigations, a part of the CO_3 and HCO_3 precipitates as Ca-$MgCO_3$, thus removing Ca and Mg from the soil water and increasing the relative

TABLE 4-10. TYPICAL HYDRAULIC CONDUCTIVITY VALUES IN CALIFORNIA (after Cooley, Harsh, and Lewis, 1972)

Material	Hydraulic Conductivity (gal/d per square foot at 60°F)
Granite	0.0000009 to 0.000005
Slate	0.000001 to 0.000003
Dolomite	0.00009 to 0.0002
Hematite	0.000002 to 0.009
Limestone	0.00001 to 0.002
Gneiss	0.0005 to 0.05
Basalt	0.00004 to 1
Tuff	0.0003 to 10
Sandstone	0.003 to 30
Till	0.003 to 0.5
Loess	1 to 30
Beach sand	100 to 400
Dune sand	200 to 600
Clay	0.001 to 1
Silt	1 to 10
Very fine sand	10 to 100
Fine sand	100 to 1,000
Medium sand	1,000 to 4,500
Coarse sand	4,500 to 6,500
Very coarse sand	6,500 to 8,000
Very fine gravel	8,000 to 11,000
Fine gravel	11,000 to 16,000
Medium gravel	16,000 to 22,000
Coarse gravel	22,000 to 30,000
Very coarse gravel	30,000 to 40,000
Cobbles	Over 40,000

proportion of Na which would increase the sodium hazard." The older concept of SAR has been adjusted to accommodate the effect of lime dissolution or deposition by means of the adjusted SAR:

$$\text{adj SAR} = \text{SAR}\,[1 + (8.4 - \text{pHc})] \qquad (4\text{-}20)$$

where Na, Ca, and Mg are obtained from water analyses. The pHc is the theoretical pH of water equilibrated with $CaCO_3$. In practice, pHc is calculated using equations given by Ayers and Westcot.

The relationship of exchangeable sodium percentage (ESP), the fraction of the exchange complex occupied by sodium, to the total salt concentration of the soil solution also affects the hydraulic conductivity of soils with clays. According to McNeal (1974), a combination of high ESP and low total salt concentration leads to swelling and dispersion of the soil minerals. Swelling reduces the hydraulic conductivity of the soil. Dispersion may also cause a migration of clay particles into lower regions of the soil profile, possibly forming a layer that restricts free water movement.

Quantitative interrelationships of ESP, total salt content of the soil solution, and hydraulic conductivity were demonstrated on laboratory soil columns by Quirk and Schofield (1955). In particular, they observed that for a given ESP, the hydraulic conductivity, K, increased with an increase in total salt content. Figure 4-10 shows the relationships between K, ESP, and salt content for Pachappa sandy loam (McNeal, 1974). This figure also illustrates that a greater and greater total salt content is required to offset the effect of sodium on K as ESP increases.

Quirk and Schofield (1955) coined the expression "threshold salt concentration" to represent the level at which the salt content would need to be reduced to produce a 10- to 15-percent reduction in K at a given value of ESP.

As pointed out by McNeal (1974), these concepts have applicability in land disposal operations as well as in irrigated agriculture, particularly if a sizable fraction of the effluent contains softened water.

Permeameters--

Laboratory devices for measuring the hydraulic conductivity of disturbed or undisturbed soils are generally classified as permeameters. Klute (1965) describes the construction and operation of common permeameters. Soil cores taken in the field may be used directly in the permeameter. For research purposes, columns are filled with dried and sieved soils packed to the desired bulk density. Klute (1965) recommends use of water to be applied to the field soil in conducting the tests.

Two general types of permeameters that are used are the constant head type and the falling head type. Generally, the permeameters are arranged so that flow is vertically downward through the soils. For the constant head type of permeameter, water is ponded continuously at a constant distance above the surface of the soil core. The rate of water movement through the core is measured via a graduated cylinder and stopwatch. The hydraulic conductivity is calculated from Darcy's equation. In the falling head method, the soil core is connected to a standpipe. After saturating the soil, the water level in the standpipe is allowed to fall. The time for the head to drop through successive distances is measured and Darcy's equation is used to calculate K.

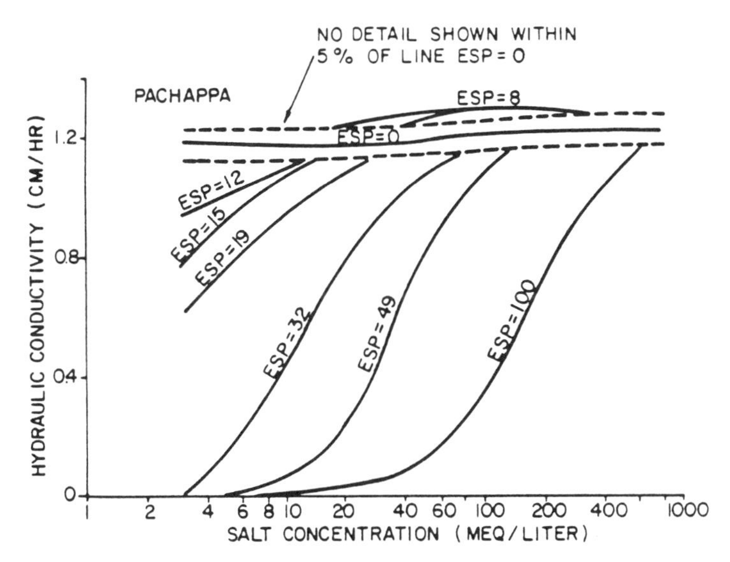

Figure 4-10. Relationships among hydraulic conductivity, salt concentration, and ESP for Pachappa sandy loam (after McNeal, 1974).

These methods are simple and may be used to determine variations in K values caused by stratification. With a large number of samples, however, the method becomes quite expensive and the accuracy is questionable because of wall effect. Table 4-11 lists further constraints on the method. The methods are described in more detail by Bouwer (1978) and Freeze and Cherry (1979).

As Klute (1965) points out, the results of the permeameter methods are of questionable value when applied to a large area. Thus, a large number of samples would be required to account for the spatial variability in field hydraulic conductivity of soils. When samples are also obtained from deeper regions of the vadose zone for permeameter tests, the problem of spatial variability assumes an even more three-dimensional character.

Grain-Size Distribution--

Saturated K values can be determined from hydraulic conductivity and grain-size distribution curves obtained from vadose zone material. The hydraulic conductivity is then calculated from equations that account for the representative grain-size diameter or from the spread in the gradation curve. These relationships are discussed by Freeze and Cherry (1979) and other references given therein.

TABLE 4-11. CONSTRAINTS ON PERMEAMETERS

Category	Constraint
Geohydrologic	1. Assuming saturated K overestimates the flux. 2. Because sample is confined by cylinder walls, only unidirectional flow (and K) is measured. That is, method does not account for anisotropic flow under field conditions.
Chemical	Wastewater from disposal site should be used for tests to account for chemical interactions in the soil pores (e.g., precipitation of salts decreases the intrinsic permeability; dissolution of medium increases the intrinsic permeability).
Geologic	1. Method cannot be used to estimate K in fractured media. 2. For disturbed core samples, results may be affected by root channels, cavities, and cracks.
Physical	1. Air bubbles in the sample affect test results. 2. Disturbed samples will probably not give results that are representative of in situ values. 3. Unless samples are very large, test results apply only to points in the vadose zone. May not represent "average" K values.
Climatic	Change in temperature during test will affect results.

This method provides a first approximation of K when other data are unavailable. It can also be used to estimate relative variations in K due to stratification. Unfortunately, the accuracy is questionable and expensive if grain-size distribution data are unavailable. Table 4-12 lists additional constraints on the use of the method.

Saturated K can also be measured in the absence of a water table by bringing a portion of the soil zone to saturation with saturated K values estimated for the flow system thus created. Appropriate measurements and equations are used to solve for K. Alternative methods for creating the flow system include (1) pump-in method, (2) air-entry permeameters, (3) infiltration gradient method, and (4) double tube method. The advantages and disadvantages of each method are discussed by Bouwer and Jackson (1974). Table 4-13 lists constraints on the method.

TABLE 4-12. CONSTRAINTS ON GRAIN-SIZE K RELATIONSHIPS

Category	Constraint
Geohydrologic	1. Assumption of saturated K overestimates the flux in unsaturated regions of the vadose zone. 2. Method estimates vertical K, whereas in the field, horizontal K is more important.
Chemical	The "hydraulic" conductivity values estimated by this method may not apply to other liquids, e.g., hydrocarbons.
Geologic	1. Cannot be used to estimate K in fractured media. 2. Different curves required for each layer. 3. Method essentially provides point estimates of K. Because of the spatial variability of hydraulic properties of vadose zone sediment, the values estimated by this method may be differ vastly from "average" values over a wide area.
Physical	1. Preparation of samples for determinations of grain-size distribution destroys the original texture and structure. Thus, if the in situ flow occurs between peds, actual K will be greater than that calculated by this method. 2. Drill cuttings may not be truly representative of in situ texture because of wall sloughing, etc. Cores probably give better results.

TABLE 4-13. CONSTRAINTS ON MEASURING K IN THE ABSENCE OF A WATER TABLE

Category	Constraint
Geohydrologic	1. Shallow pump-in method measures K mainly in horizontal direction. 2. Cylinder permeameter measures K in vertical direction. 3. Infiltration gradient method measures K in vertical direction. 4. Air entry permeameter measures K in vertical direction. 5. Double tube affected by K in vertical and horizontal directions. 6. Must combine double tube and infiltration gradient in anisotropic soils. 7. K values overestimate flux in unsaturated soils.
Chemical	1. Methods are for water. May not be applicable for chemical waste liquids. 2. For wastewater disposal sites, should use actual wastewater during tests to account for effects of chemical interactions on K.
Geologic	1. Air permeability method, infiltration gradient method, and double tube method cannot be used for gravelly soils. 2. Shallow flow-impeding layers affect results.
Topographic	Difficult to apply methods on steep-sided terrain.
Physical	1. Restricted to shallow depths of the vadose zone. 2. Generally, a large amount of water or wastewater is required. 3. For uncased methods, smearing of walls tends to clog surface pores, resulting in an underestimation of K.
Climatic	Temperature changes during the tests will affect results.

The instantaneous profile method can be used to determine K values. The basis of this method is the Richards (1965) equation, rewritten as follows:

$$K(\theta) = \frac{\partial\theta/\partial t}{\partial\Psi/\partial z}\, z$$

where z is depth. In practice, a soil plot in the region of interest is instrumented with a battery of tensiometers, with individual units terminating at measurement depths of interest. A neutron probe access tube for moisture logging is installed with the battery of tensiometers. The soil is wetted to saturation throughout the study depth. Wetting is stopped and the surface is covered to prevent evaporation. Water pressure and water content measurements are obtained during the drainage period. Curves of Ψ versus z and θ versus t are prepared. Slopes of the curves at the depths of interest are used to solve for K(θ). Values of K(θ) at varying times can be used to prepare K(θ) versus θ and K(Ψ) versus Ψ curves. Bouma, Baker, and Veneman (1974) contains a detailed description of the method and step by step procedures.

The instantaneous profile method can be used in stratified soils, is simple, and reasonable accuracy can be obtained at each measuring site. It provides hydraulic conductivity values for draining profiles. Because of hysteresis, however, these values are not representative of the wetting cycle. The method is time-consuming and relatively expensive. Table 4-14 lists additional constraints on the use of the instantaneous profile method.

TABLE 4-14. CONSTRAINTS ON THE INSTANTANEOUS RATE METHOD

Category	Constraint
Geohydrologic	1. A draining-profile method. Consequently, because of hysteresis, results cannot be used for wetting conditions. 2. Primarily used to estimate vertical hydraulic conductivity. In normal soils, horizontal K may be greater than vertical K.
Chemical	1. Method is suitable for water and wastewater but not necessarily for chemical liquid wastes. 2. Interaction between wastewater and solids may affect results (e.g., dispersion of clays).
Geologic	1. Because of the spatial variability of soil hydraulic properties, a large number of measurements is required to obtain "representative" values. 2. Cannot be used in fractured media. 3. Results may be affected by shallow flow-impeding layers.
Topographic	Cannot be used too effectively on steep-sided slopes.
Physical	1. Restricted to shallow regions of the vadose zone. 2. Not applicable to existing ponds or landfills.
Climatic	Temperature affects results through viscosity and surface tension affects.

Test Well Methods--

Three similar well test methods have been developed to determine K values. These include: (1) U.S. Bureau of Reclamation (USBR) single-well method, (2) USBR double-well method, and (3) Stephens-Neuman single-well method. Each method requires pumping water into a borehole at a steady rate to maintain a uniform water level in a basal test section. Saturated K is estimated from appropriate curves and equations, using dimensions of the hole and inlet pipes, length in contact with formation, height of water above base of borehole, depth to water table, and steady-state intake rate. The multiple-well method is used to estimate K in the vicinity of widespread lenses of slowly permeable material. For this method, an intake well and series of piezometers are installed in the area or horizon of interest. Water is pumped into the well at a steady rate and water levels are measured in the piezometers. Two types of single-well tests and the multiple-well method have been described (USBR, 1977) including (1) open-end casing tests in which water flows only out of the end of the casing, and (2) open-hole tests in which water flows out of the sides and bottom of the test hole.

These methods can be used for measuring K at great depths in the vadose zone as well as for obtaining a profile of K values through multiple testing. K is underestimated, however, because the single-well method assumes that the flow region is entirely saturated (free surface theory), which is not the case. This problem has been addressed in the Stephens-Neuman empirical formula based on numerical simulations using the unsaturated characteristics of four soils, thereby accounting for the unsaturated flow within the test setup (Stephens and Neuman, 1980).

Advantages and disadvantages for the USBR methods are given in the cited reference. Each of these methods requires trained personnel, is expensive, and is time-consuming. The Stephens-Neuman formula can be used to estimate the saturated hydraulic conductivity of an unsaturated soil with improved accuracy. In addition, waiting for steady-state conditions during field operations is unnecessary because the final flow rate can be estimated from data developed during the transient stage. Additional field testing of the Stephens-Neuman method is needed and Table 4-15 lists other general constraints that apply to all the methods.

Air Permeability Method--

The air permeability method measures air pressure changes in specially constructed piezometers during barometric changes at the land surface. Pressure response data are coupled with information on air-filled porosity to solve equations leading to air permeability. If the Klinkenberg effect is small (Weeks, 1978), air permeability is converted to hydraulic conductivity.

The method can be used to estimate hydraulic conductivity values of layered materials in the vadose zone. This indirect method requires trained personnel and is time-consuming and expensive. The presence of excessive water limits its utility. Table 4-16 lists additional constraints.

Saturated flow occurs in the vadose zone primarily within perched groundwater bodies that develop at the interface of regions of varying hydraulic conductivity (see Figure 4-9). The principal saturated flow characteristics that should be quantified for the vadose zone are the following: (1) hydraulic gradients, (2) direction of flow, (3) hydraulic conductivity values, and (4) flow rates.

TABLE 4-15. CONSTRAINTS ON THE USBR AND STEPHENS-NEUMAN WELL METHODS

Category	Constraint
Geohydrologic	1. For USBR methods, it is assumed that the flow region is entirely saturated. Because this is not true, calculated K values are less than the actual values. In addition, the governing equations may be questionable. 2. Although a profile of K values at a given site is obtained, different profiles will be obtained at different sites because of spatial variability of hydraulic properties.
Chemical	1. Formulas are developed for water and may not be applicable to chemical wastes. 2. Should use wastewater from disposal site during tests. Note, however, that particulate matter will clog borehole surfaces and affect results. Therefore, test water should be strained.
Geologic	1. Impeding layers beneath end of borehole will affect results. 2. Not suitable for fractured media (alternative "pressure-permeability" tests for stable rock are available, however).
Topographic	Cannot be used in steeply sloping terrain.
Physical	1. The USBR methods depend on reaching a constant head in the borehole before taking readings. Time to reach constant head may be excessively long, particularly in dry media. 2. Drilling test holes and installing casing and appurtenances requires the services of trained drillers. 3. Particulate matter in test water will affect results. 4. Boreholes should be developed and cleared prior to testing.
Climatic	1. Temperature of test water may affect results. 2. As with most field methods, climatic conditions may cause testing to be physically difficult on personnel.

TABLE 4-16. CONSTRAINTS ON THE AIR PERMEABILITY METHOD

Category	Constraint
Geohydrologic	1. An indirect method in that air permeability is used to estimate hydraulic conductivity. 2. The presence of pore water may affect results by decreasing the air permeability. For best results, sediment should be at field capacity.
Chemical	1. Results are related to "hydraulic" conductivity and may not be valid for chemical fluids, which may have different intrinsic permeabilities. 2. Chemically induced structural changes during flow of wastewater will cause hydraulic conductivity values to be different from those calculated from air permeability values in dry media.
Geologic	1. May not be applicable to fractured media. 2. In fine-grained material, the permeability to air is greater than hydraulic permeability because of the Klinkenberg effect.
Topographic	Difficult to install test wells in steep terrain.
Physical	If swelling-shrinking clays are present in vadose zone, air permeability values obtained during dry conditions will be greater than hydraulic conductivity values during wetting conditions, after clays swell.

Hydraulic Gradients and Flow Direction in Perched Groundwater--

Two basic tools for measuring hydraulic gradients in perched groundwater are the piezometer and observation well. As will be shown later, these devices are also useful for obtaining water samples for analyses.

Piezometers--Piezometers consist of small-diameter pipes drilled into a saturated zone or a zone in which saturation is expected. Reeve (1965) discusses in detail common techniques for installing and cleaning new piezometers. In general, a tight fit between the outer wall of the piezometer and the surrounding media is essential. For shallow units, piezometers may be installed by augering and driving with a sledge hammer. Deeper units will require jetting or use of standard drilling equipment. It may be necessary to fill the cavity between pipes and boreholes with grout to ensure tightness of fit. As with regular wells, piezometers should be developed by pumping or bailing to clean and open up the material at the base of the unit. In some cases, it may be necessary to install piezometers with screened well points to prevent the upward movement of saturated sediment into the unit.

In many field situations, it is desirable to install a number of piezometers at different depths within a perched groundwater body. Two methods of construction are possible: installing a battery of separate piezometer wells, or installing a cluster of wells within a common borehole. Clustered wells are constructed by installing the individual units in the borehole at desired depths and surrounding the well points with sand. The region of the borehole between units is backfilled with bentonite. Fenn et al. (1977) presents additional details on methods for installing piezometer clusters together with their advantages and disadvantages.

In operation, the piezometer provides a measure of the hydraulic head at the terminus of the tube. The total head consists of the sum of gravity and pressure heads. The gravity head is referenced to an arbitrary datum, say, the ground surface. For a nest of piezometers, it is convenient to use the elevation of the lowermost unit as the datum. The pressure head is the height of water above the bottom opening of the unit.

Depth to water in piezometer units is measured by chalked tape, electric sounders, or air lines. By referencing piezometric values to a common datum, mapping of the piezometric surface is facilitated. Such mapping will provide clues on the direction of water movement.

Although piezometers are generally used in the soil zone, they can also be installed into the lower vadose zone. For example, Wilson and DeCook (1968) describe units placed into perched groundwater within the 80-foot-thick vadose zone of a recharge site near Tucson, Arizona. The depths of individual units were based on moisture logs in a network of access wells. Each piezometer contained a screened well point, and the outside of the casing was grouted in the drill hole to inhibit side leakage.

An array of piezometers with the ends of individual units terminating at different depths may be used to monitor the vertical movement of water in perched water tables. As an example, Figure 4-11 indicates three hypothetical flow situations in a soil containing three piezometer units beneath a water table. For purposes of calculation, the datum is selected as base of the deepest piezometer. For the first case, that of perched groundwater above an impermeable layer, vertical flow is not occurring. This may be verified by calculation. The total hydraulic head, H, for the water levels in all wells is the same and corresponds with the water table. For unit No. 1, the gravitational head, z_1, is zero and the pressure head is h_1. Similarly, for unit No. 2, the positional head is z_2 and the pressure head is h_2. But $h_2 + z_2 = h_1$. Consequently, $H_1 = H_2$ and no flow occurs between units No. 1 and No. 2. Similar analysis would show that flow does not occur between units No. 2 and No. 3.

For the second case, water applied to the surface moves downward in the profiles. The perched water body is underlain by an unconfined gravel aquifer. In this case, flow occurs vertically downward. Water levels in the individual piezometers are at the base of the unit; that is, pressure heads are all zero. For unit No. 1, the total hydraulic head is zero because $z_1 = 0$ and $h_1 = 0$. For unit No. 2, the positional head equals z_2 and the pressure head =

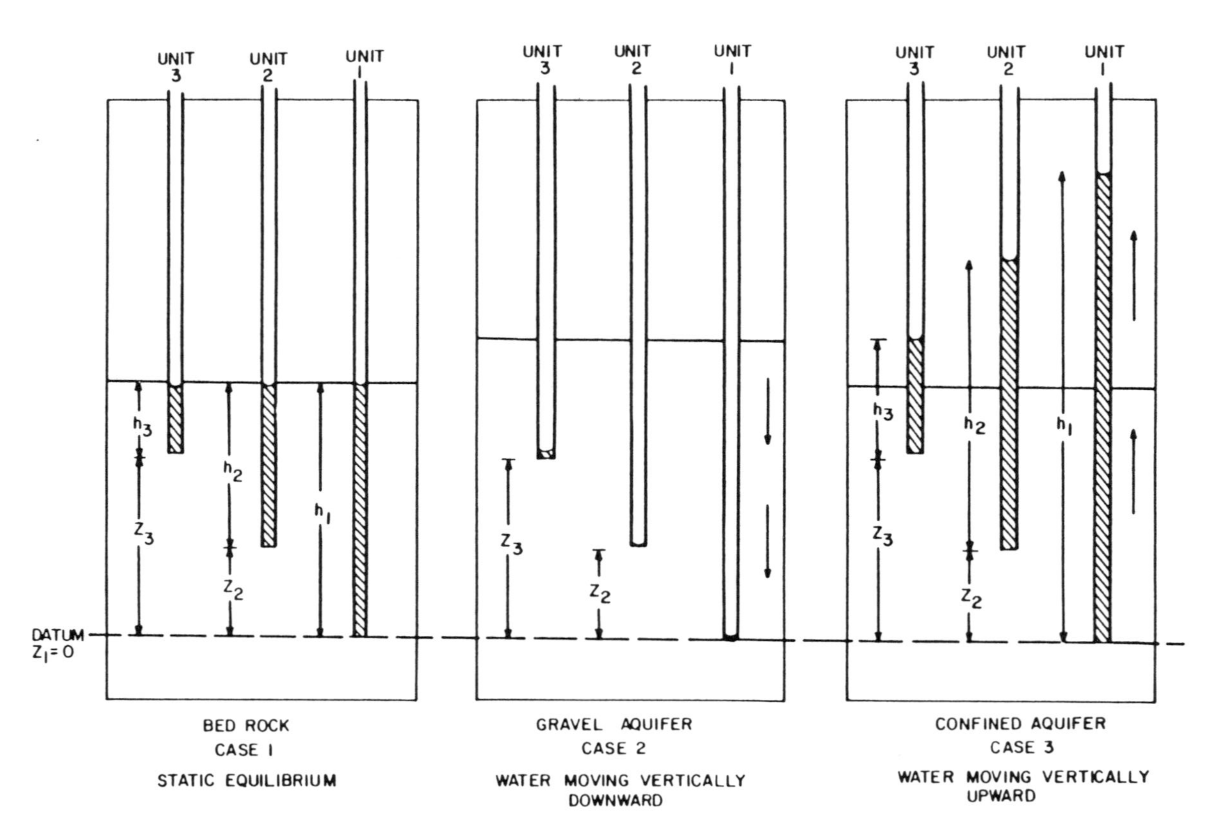

Figure 4-11. Hydraulic head distribution in three piezometers during three hypothetical flow cases (after Reeve, 1965).

0. The total hydraulic head at unit No. 2 is z_2. Inasmuch as $H_2 > H_1$, flow occurs between units No. 2 and No. 1. Also, because the vertical distance between the lower ends of units No. 1 and No. 2 is z_2, the hydraulic gradient is unity, i.e.:

$$\frac{z_2 - 0}{z_2} = 1 \ .$$

A similar analysis also shows that flow occurs between units No. 3 and No. 2 under a unit hydraulic gradient.

The third case represents a water table region underlain by a confined aquifer, with evaporation occurring at the land surface. The depth to water is greater in the shallower units. That is, the pressure head increases with depth in the profile. An analysis using hydraulic principles would show that the total hydraulic head decreases with distance above the datum so that water is moving vertically upward in the system below the water table. The last two cases demonstrate an important principle: when vertical flow occurs, it may be difficult to locate the actual depth to a water table in a moving system using only the data from piezometers.

The cases shown in Figure 4-11 do not include the anisotropic effects of lateral flow superimposed on vertical flow. In many field soils, the hydraulic conductivity in the horizontal direction may be much greater than in the vertical direction. For example, Bouwer and Jackson (1974) observed the ratio of horizontal to vertical hydraulic conductivity to be 7:1 for a finely stratified, loamy sand. For such soils, it may be advisable to install more than one array of piezometers in a lateral direction away from a flooding operation. The basal openings of corresponding units in sequential arrays should terminate at the same elevation. Such an arrangement permits observations of lateral head differences resulting from horizontal flow.

A basic problem with piezometers is that a lag may occur in the response of the units to a rapidly changing system and when flow occurs in stratified material. Bouwer and Jackson (1974) present a time lag equation for piezometers:

$$T_L = \pi r^2 \ Y_o/AK \ Y_o = \pi r^2/AK \qquad (4\text{-}21)$$

where T_L = time lag constant

r = radius of tube

Y_o = distance of sudden rise of water level

A = geometry factor

K = hydraulic conductivity.

The value of A is determined from suitable equations, graphs, or tables. It can be shown that the basic time lag manifests the time for which 63 percent of the sudden change in actual water level position has been registered by the water level in the measuring device (Bouwer and Jackson, 1974).

Since T_L is directly proportional to r, more rapid response may be obtained by reducing the area of the piezometer opening. Lissey (1967) presents a piezometer design based on this principle.

Observation wells--An observation well consists of an uncased borehole or perforated pipe extending from ground surface into a perched water table. For shallow water tables, observation wells can be installed by hand augers or simple drilling equipment. For deeper wells (e.g., those extending below the main water table), standard drilling equipment is necessary. Development of cased wells by surging or pumping may be necessary to ensure free flow of water into and out of perforations.

A principal function of observation wells is to permit observation of water table fluctuations. During water-spreading operations, for example, water level changes in observation wells may manifest the arrival of downward-flowing effluent. Water levels are measured by means of chalked tape, electric sounders, or air lines. Automatic water-stage recorders can be used for continuous records of water table fluctuations. Water table readings should be referenced to a datum, e.g., the top of the casing, so that levels from a network of wells can be used to plot isopiestic maps.

Under steady flow conditions, an observation well may perform adequately in providing data on the location of the water table. However, during transient vertical flow, reliable estimates are difficult to obtain for reasons stated by Kirkham (1947):

> If there is downward seepage, the bottom of the perforated pipe acts as a sink and the water table is accordingly depressed about the pipe; if there is upward seepage, the bottom of the perforated pipe acts as a source, the water table then mounding up about the pipe.

Saturated Hydraulic Conductivity--

Alternative techniques have been developed for estimating the saturated hydraulic conductivity of vadose zone materials either in the presence or absence of a water table. Detailed reviews of these methods are presented by Bouwer and Jackson (1974) and Bouwer (1978).

Estimates of K within perched groundwater bodies of the vadose zone may be obtained by pumped hole techniques such as the auger hole method, the piezometer method, tube method, well point method, and multiple-well methods (Bouwer and Jackson, 1974). The auger hole, piezometer, and well point methods entail installing a single cavity (cased or uncased) below the water table, lowering the water level in the cavity by pumping or bailing, and measuring the rate of recovery of the water level. Knowing the geometry of the cavity and the known or assumed depth to an impermeable layer, suitable equations, curves, or nomographs may be used to calculate K. In the multiple-well method, water is pumped from one well or set of wells into other wells until the difference in water levels of the wells becomes stabilized. Knowing the geometry of the cavities and flow system, appropriate equations or curves are used to determine K (Bouwer and Jackson, 1974). All of the above methods, except the tube method, determine K primarily in a horizontal direction.

If the perched groundwater region is extensive and contains productive wells, it may be possible to use standard pumping test methods to determine the transmissivity, T, of the formation. Such methods are reviewed in detail by Lohman (1972). Knowing the transmissivity of the system and the thickness of the layer, m, the perching layer, the hydraulic conductivity is estimated by the relationship $T/m = K$. A problem may exist in using this method if leakage beneath the perching layers is substantial.

In planning waste disposal operations, estimates of the saturated hydraulic conductivity in the vadose zone are useful even though a water table is absent. For soils, the saturated K value indicates the potential intake rate. Thus, comparison of K values of soils at a number of sites using the methods discussed subsequently will facilitate locating a disposal operation to minimize deep seepage. Similarly, a range of K values in deeper strata of the vadose zone will indicate the location of possible perching layers. In addition, if hydraulic gradients are assumed to equal unity, the K values represent an upper limit of flux through the vadose zone.

A number of field techniques have been developed for estimating hydraulic conductivity, K, in shallow soils in the absence of a water table. Determining K in the absence of a water table generally involves techniques that bring the soil at the measuring point to saturation or near-saturation. Bouwer and Jackson (1974) reviewed five possible methods: (1) the shallow well pump-in method, (2) the cylinder permeameter method, (3) the infiltration gradient technique, (4) the air-entry permeameter technique, and (5) the double-tube method. The shallow well pump-in method measures K mainly in the horizontal direction and is suitable for stoney soils. The cylinder permeameter and infiltration gradient techniques measure K in the vertical direction and are not suitable for stoney soils. The air-entry permeameter measures K in a vertical direction and with care may be used in stoney soils. The double-tube method produces a value of the hydraulic conductivity in a direction oriented between the horizontal and vertical flow. It is a time-consuming technique requiring a 24-hour period to characterize a volume about the size of a 4-inch core. It is not suitable for stoney soils. An important consideration in using these methods is to minimize air entrapment in the pores of the media. Obviously, occluded air will decrease the measured K value below the true saturation value. Additional details on these methods, including associated equations and techniques for determining K, are included in the review of Bouwer and Jackson (1974).

Both laboratory and field methods are available for measuring the hydraulic conductivity in deeper unsaturated regions of the vadose zone. Laboratory-related techniques that have been used to estimate K values from drill cuttings include permeameter tests and empirical relations between the grain-size distribution and K. Permeameter tests are not particularly meaningful because of the disturbance caused by the drilling process. In contrast, certain empirical relationships might exist between grain-size analyses and hydraulic conductivity. Davis and De Wiest (1966, p 375) present a table relating hydraulic conductivity to the dominant size of selected sediments.

The USBR (1977) has developed two field techniques for estimating the saturated hydraulic conductivity, K, of unconsolidated unsaturated sediments of the vadose zone. One approach entails pumping water into a borehole at a steady rate to establish a uniform water level within a basal test section. Knowing the dimensions of the hole and inlet pipes, the depth of water, and the constant inflow rate, appropriate equations and curves are consulted to calculate K. For depths less than 40 feet, the hole is cased to the desired depth. A water inlet pipe and a separate water level measurement pipe are placed inside the casing. Gravel is then poured inside the casing to ensure a gravel pack throughout the test section. Subsequently, the casing is pulled back and a test is initiated. For depths greater than 40 feet, a preperforated casing is necessary to isolate the test section. The total open area of the perforations must be known. An observation pipe is placed in the casing on a 6-inch bed of gravel. The preperforated casing permits driving the casing in depth-wise increments to obtain a profile of K values.

The second method for estimating K is used in the vicinity of a widespread lens of slowly permeable material. The method entails installing an intake well and a series of observation wells. Water is pumped into the well at a steady rate and the water level response in the observation wells is recorded. Appropriate equations and curves are consulted to calculate K.

Weeks (1978) presents a method for measuring vertical air permeability values of layered materials in the vadose zone. Basically, the method measures air-pressure changes in specially constructed piezometers during barometric pressure changes at the land surface. By coupling pressure-response data with auxiliary information on air-filled porosity and numerical solutions of the one-dimensional flow equation, an estimate of the air permeability is obtained. If the material is well-drained and permeable enough that the Klinkenberg effect (see Glossary) is minimal, air permeability values may be converted to the corresponding hydraulic conductivity values. Weeks (1978) contains additional details.

Flow Rates in Perched Groundwater--

The velocity of water movement in extensive perched groundwater and the total discharge are frequently of interest. These characteristics are estimated from the following forms of Darcy's equation:

$$v = -K\left(\frac{H_2 - H_1}{L}\right) \tag{4-22}$$

and

$$Q = -KA\left(\frac{H_2 - H_1}{L}\right) \tag{4-23}$$

where v = specific discharge

K = hydraulic conductivity

H_2 = total head in downstream well

H_1 = total head in upstream well

L = distance between wells

A = cross sectioned area of aquifer normal to flow direction

Q = total discharge.

The hydraulic conductivity, K, of the perched system can be estimated using methods previously discussed. An example of the use of these equations is in Bouwer (1978, pp 46 and 47).

Combination of Methods for Observing Saturated and Unsaturated Flow--

A combination of methods is necessary for monitoring the flux of wastewater from ground surface into a perched groundwater table. Such a system is shown in Figure 4-12. The tensiometers and electrical resistance blocks can be used to monitor flux, hydraulic gradients, and the direction of water movement in the unsaturated region. The observation well is useful in estimating the response of the water table to recharge. The piezometer nest is used to detect the vertical direction of flow beneath the water table.

Again, observation wells and piezometers will not accurately detect the location of the water table in a fluctuating system. However, by installing an access well and monitoring water content, positional changes in the water table may be monitored via a neutron logger. Although piezometers are more effective than observation wells in monitoring water-level response, the neutron logger measures water content changes and not energy.

Effect of Spatial Variability on Soil Properties

When using field data to estimate such properties as flux and hydraulic conductivity, variations in these parameters may exist because of soil heterogeneity. For example, large variations in soil texture and structure, even with a given mapping unit, will affect values of related properties such as water content, porosity, and hydraulic conductivity.

The problem of such spatial variabilities was examined in detail by Nielsen and Biggar (1973), Coelho (1974), and Guma'a (1978). For their study area, Nielsen, Biggar, and Erh (1973) found that variations in water content were normally distributed with depth and horizontal distance in the field, whereas hydraulic conductivity values were log-normally distributed. The following conclusion is particularly noteworthy:

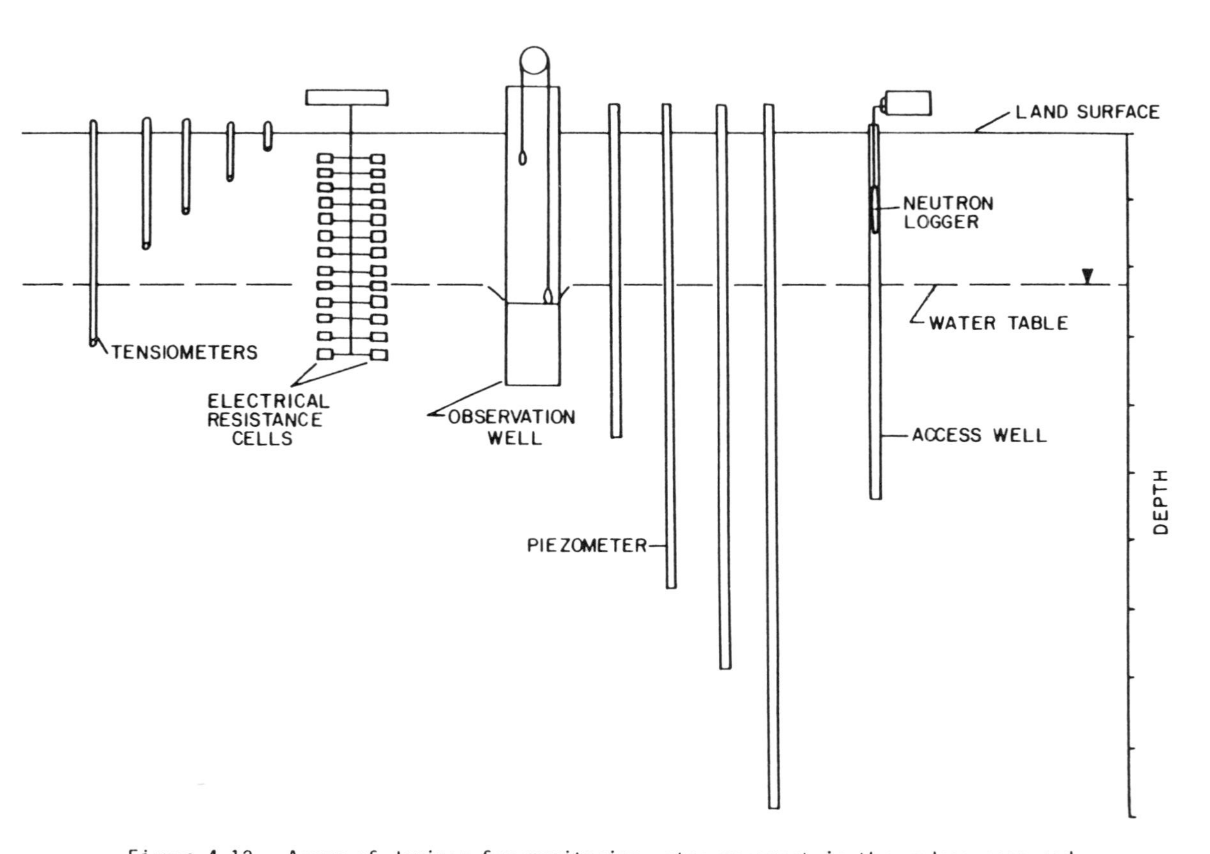

Figure 4-12. Array of devices for monitoring water movement in the vadose zone and perched groundwater (after Freeze and Cherry, 1979).

> Even seemingly uniform land areas manifest large variations in hydraulic conductivity values. Variations in texture, bulk density, and water content are much less. For a given location, methods for measuring water content, hydraulic conductivity, and hydraulic gradients will yield values that are much more accurate than required to characterize an entire field because of the heterogeneity of the soil. Thus, our ability to make predictions over a large area from a single plot can range from good to unsatisfactory, depending on the particular prediction parameter of interest.

Based primarily on the results of studies by Nielsen, Biggar, and Erh (1973) and Guma'a (1978), the variabilities in various soil parameters were ranked as follows by Warrick and Amoozegar-Fard (1980):

Low variability: (Coefficient of variability less than 20 percent)

- Bulk density
- Water content at a zero tension

Medium variability: (Coefficient of variation 20 to 75 percent)

- Textures (sand, silt, or clay)
- Field water content
- Water content at specified tension between 0.1 and 15 bars

High variability: (Coefficient of variation greater than 100 percent)

- Saturated hydraulic conductivity
- Unsaturated hydraulic conductivity
- Apparent diffusion coefficient
- Pore water velocity.

Problems arising from the spatial variability of hydraulic properties of the vadose zone will increase with depth in the profile. For example, even small lenses may affect soil-water pressures, hydraulic conductivity, and flux. Consequently, estimating flux out of the soil zone of a waste disposal facility may be the best that can be expected.

5. Active and Abandoned Site Monitoring Methods

In active and abandoned site monitoring programs, the spatial and temporal variations of specific water-borne pollutants will receive major emphasis. In the ensuing discussion, such pollutants are categorized as major inorganic chemical constituents, trace chemical constituents, organic chemical constituents, microbial constitutents, and radionuclides.

Particular constituents within each of these categories that should be monitored in the vadose zone are site-specific, i.e., they depend upon the waste disposal operation. For example, when monitoring agricultural return flows, specific pollutants may include the nitrogen series, particularly NO_3-N, phosphate, total dissolved salts (TDS), and pesticides. For land treatment sites, monitoring may emphasize bacteria, viruses, heavy metals, NO_3-N, phosphate, TDS, total organic carbon (TOC), biochemical oxygen demand (BOD), and chemical oxygen demand (COD). For sanitary landfills, a range of pollutants in all five categories should be monitored in leachate, with particular emphasis on heavy metals. Finally, hazardous waste disposal operations on monitoring should place great emphasis on the fate of specific organic toxins during flow in the vadose zone.

In the following discussion, reactions affecting the movement of pollutants in the vadose zone are briefly described and current methods for vadose zone monitoring are reviewed.

TECHNICAL REVIEW

As with water movement, the present state of knowledge on pollutant movement in the vadose zone is derived from the efforts of specialists working in soils and deeper geological regions. For example, soil chemists and soil microbiologists have examined chemical and microbiological interactions, respectively, in soils. Similarly, geochemists have studied low-temperature chemical reactions in groundwater. Among the references that may be consulted for greater detail on soils and geochemistry relative to pollutant movement are the following: (1) Bohn, McNeal, and O'Connor (1979), a review of the modern concepts of soils chemistry; (2) Alexander (1961), a text on soil microbiology; (3) Hem (1970), a detailed review of water chemistry, including a discussion of chemical constituents in groundwater; and (4) texts by Davis and De Wiest (1966), Bouwer (1978), and Freeze and Cherry (1979) containing chapters on water quality.

Several review papers have also been published in recent years dealing with interactions among water, pollutants, and vadose zone materials. Such papers include those of Ellis (1973), Rhoades and Bernstein (1971), Murrmann and Koutz (1972), McNeal (1974), Pratt et al. (1978), and Chang and Page (1979). In his review paper, Runnells (1976) itemizes 11 physical-chemical processes that may operate in the subsurface to purify liquid wastes. These processes are: dilution, buffering of pH, precipitation by reaction of wastes with indigenous water or solids, precipitation by hydrolysis, removal due to oxidation and reduction, mechanical filtration volatilization and loss as a gas, biological assimilation or degradation, radioactive decay, membrane filtration, and sorption. Runnells (1976) briefly discusses each of these factors, and his paper should be consulted for details. Fuller (1977) also presents a review paper dealing with the movement of selected heavy metals in soils. Among the factors that Fuller (1977) examines relative to the movement of these metals are: soil pH, oxidation-reduction (eH), surface area of soils, pore-size distribution, organic matter, concentration of ions or salts, and presence of hydrous oxides. Pratt et al. (1978) discusses the removal of biological and chemical contaminants by soil systems. The removal process includes filtration, adsorption, decomposition, ion exchange, oxidation-reduction, chemical complex formation, chemical precipitation, and other chemical reactions.

In recent years, a number of soil column experiments have been conducted in the laboratory to identify the factors influencing the mobility of pollutants. The column studies of Fuller (1978) and his associates at the University of Arizona on the mobility trace contaminants from landfill leachate are an example of such laboratory studies. Based on statistical analyses of these laboratory studies, Korte et al. (1976) found that the following factors are dominant in affecting the movement of trace contaminants in soils: soil texture and surface area, percentage of free oxides, and pH.

Field studies have also been conducted to determine the factors affecting the movement of pollutants in vadose zones. Apgar and Langmuir (1971) obtained water samples to a depth of about 55 feet from the vadose zone beneath a landfill in Pennsylvania. (The technique used to obtain these samples was suction cup lysimeters; see Suction Samplers, this section. Based on their field observations, Apgar and Langmuir (1971) concluded that the following factors are important in attenuating pollutants: dilution and dispersion, oxidation, chemical precipitation, cation exchange, and anion exchange. During field studies to determine chemical changes in water during artificial recharge at a site in Texas, Wood and Signor (1975) examined the following mechanisms: cation exchange, anion exchange, mineral solution, adsorption, desorption, and sulfate reduction. They concluded that at their site ion exchange and desorption are the major mechanisms affecting water quality in the vadose zone.

Ion Exchange, Cation Exchange Capacity, and Sorption/Desorption Effects in Soils

Soil scientists generally recognize that among the factors affecting the mobility of the chemical constituents of water in soils, ion exchange and sorption/desorption are of prime importance. A brief review of these factors is presented in this subsection. For details, the text of Bohn, McNeal, and O'Connor (1979) should be consulted.

Ion exchange is an important soil-water interaction for altering the composition of the soil solution. Ion exchange is a function of the cation exchange capacity (CEC) of a soil, which in turn is related to colloidal clay minerals, soil organic matter, iron, and aluminum sesquioxides and hydrous oxides.

Clays commonly produce an overall net negative charge on their exchange complex. Murrmann and Koutz (1972) point out that the cation exchange property arises from the need to balance the negative charge on clay micelles to maintain neutrality. To do this, positive ions in the soil solution become associated with the negative charge on the exchange complex. These ions are mobile and readily exchange with other cations in the soil solution to maintain chemical equilibrium. This process represents cation exchange. Representative CEC values range from 100 milliequivalents per 100 g (meq/100 g) for montmorillonite clays to 5 meq/100 g for kaolonite-type clays.

For most soils, the level of exchangeable cations on the exchange complex is greater than the amounts in the soil solution. Thus, a plot of ionic concentration versus distance from the clay surface (i.e., from the micelle surface) shows a marked drop-off in cation concentration away from the surface, approaching asymptotically to the level in the soil solution. In contrast, the concentration of anions in the vicinity of the micelle is lower than the concentration in the soil solution. In fact, because of this repulsion of anions, the anion concentration in the solution extracted from clay may be greater than in the solution originally added (Rose, 1966), a phenomenon known as negative adsorption.

The ease with which a given ion on the exchange complex exchanges with ions in solution is, among other things, a function of the valence of the ion--the smaller the valence of the ion, the more readily it is exchanged. The ease of replacement of common soil cations is (Rose, 1966): Li > Na > K > Mg > Ca > Ba > Al.

The organic matter fraction of a soil may contribute substantially to the CEC. In fact, additions of organic matter through sludge application may improve the CEC of soils with naturally low CEC (Broadbent, 1973). In addition to contributing a negative charge to the soil exchange complex, organic matter together with the sesquioxides may contribute a positive charge. Apparently, this phenomenon is highly pH dependent.

A complex relationship exists among soluble salts in the soil solution, those on the exchange complex, and salts in the solid (precipitated) phase. The complexity is increased when it is necessary to account for the effects of soil water. Aspects of this problem are reviewed in this section under the discussion of methods to estimate CEC and soluble salts.

In the opinion of Murrmann and Koutz (1972), adsorption of metals onto the surface of soils is the most important process for removing chemicals from wastewater. In contrast to ion exchange, in which ions retain their mobility, in adsorption reactions ions are held so tightly that they become essentially immobile. Exact mechanisms for adsorption are not clear, although covalent bonding appears to be important (Chang and Page, 1979). Methods to quantify the extent of adsorption of ions onto solids have been developed. In particular, adsorption isotherms, such as the Freundlich isotherm and the Langmuir adsorption isotherm, are commonly used. Ellis (1973) discusses the use of these isotherms to quantify the adsorption of phosphate, sulfate, boron, and heavy metals by soils.

Rhoades and Bernstein (1971) indicate that the tremendous sorptive area of soils, coupled with high cation exchange capacities, influences the fate of pesticides, radionuclides, nutrient elements, and other inorganic solutes during disposal in soils.

In their artificial recharge studies in Texas, Wood and Signor (1975) concluded that silica was the only constituent affected by desorption. The source of the mobilized silica apparently was not dissolution of quartz or amorphous silica, but desorption of monomeric silica from the aquifer matrix. They concluded that, "The amount of mobilized silica was very large, and the process of desorption should be considered in a predictive model of water quality in areas similar to the one desorbed."

A commonly used measure of the partitioning of ionic species between solid and liquid phases as a consequence of processes such as adsorption is the "distribution coefficient," K_d. This coefficient is defined (Wheeler, 1976) as:

$$K_d = \frac{\text{concentration of ionic species on solid phase (g/g)}}{\text{concentration of ionic species on soil water (g/ml)}} . \quad (5\text{-}1)$$

According to Freeze and Cherry (1979), the distribution coefficient "... is a valid representation of partitioning between liquid and solids only if the reactions that cause the partitioning are fast and reversible and only if the isotherm is linear." Wheeler (1976) points out that K_d varies from zero for nonreactive species such as chloride to 10^5 for some actinides (the actinides include uranium, plutonium, and americium).

Wheeler presents the following expression to describe the rate of flow of an ionic species relative to water during partitioning:

$$\frac{v_i}{v_w} = \frac{1}{1 + \frac{D_b}{\theta} K_d} \quad (5\text{-}2)$$

where v_i = pore velocity of adsorbed species

v_w = pore velocity of migrating water

θ = volumetric water content

D_b = bulk density of soil.

This equation applies to both saturated and unsaturated media. Factors affecting the use of equation 5-2 are: (1) the assumption that reactions are fast and reversible may not be satisfied, (2) the adsorption isotherm may not be linear, (3) the distribution coefficient may not be constant with varying soil-water contents, and (4) concentrations of other ionic species may affect the magnitude of K_d (Wheeler, 1976).

Laboratory Methods: Soluble Salts, CEC, Exchangeable Ions, Sorptive Capacity, Column Studies, and Batch Tests

Soluble Salts--

As defined by Bower and Wilcox (1965), soluble salts refer to inorganic soil constituents that are appreciably soluble in water. The following steps are usually taken in an analysis for soluble salts: preparing a sample by bringing the soil-water content to some prescribed value, extracting the sample, and measuring or analyzing the salt content of the extract.

The soil-water content for extracting soluble salts is of concern. Saturation extracts, for example, are commonly used by agriculturists to relate soluble salts to field moisture range. A problem with diluting the sample when obtaining saturation extracts is that while the concentration of some ions increases on dilution the concentration of others may decrease (Reitemeier, 1946). For example, as water is added to the soil, precipitated calcium carbonate or gypsum may gradually dissolve, releasing calcium and possibly magnesium ions into solution. Concurrently, the additional concentrations of calcium and magnesium may displace sodium on the exchange complex, increasing the sodium level in the soil solution. Overall, calcium, magnesium, and sodium will increase in the soil solution. In contrast, chloride and nitrate will decrease in concentration because of dilution and negative adsorption. Such trends have been experimentally observed during laboratory studies (Reitemeier, 1946). Reitemeier recommends that for arid soils dissolved ions be determined at or near the water content at which the results are to be applied. Consequently, the soil solutes of samples from deeper horizons of the vadose zone should be extracted at the prevailing soil-water pressure.

Pressure membrane apparatus can be used to extract soil solution samples in the dry range (Richards, 1954). Thus, if a soil-water characteristic is available for the sample, together with gravimetric water content data, the equivalent soil-water pressure can be determined. This pressure is then applied to the soil sample in the membrane. A problem with this technique is that because the solution is extracted mainly through the larger pores, the chemical composition may differ from that for solution from the smaller sequences (Rhoades and Bernstein, 1971).

For surface soils, other factors should be examined in the determination of soluble salts. For irrigated soils or on land disposal operations subject to wetting and drying, the water content of soils may range over large values because of evaporation and conversion to the field water content value may not be meaningful (Pratt, Jones, and Hunsaker, 1972). In this case, the saturation extract technique is recommended (Rhoades and Bernstein, 1971). Water content by this method represents about twice that at field capacity. Therefore, the salt content extracted from the saturated sample is about one-half the concentration at field capacity. Bower and Wilcox (1965) explain in detail the procedure to obtain a saturated extract. Briefly, deionized water is mixed into a weighed oven-dry sample of soil until the soil glistens and no free water has collected on the soil surface. If this endpoint is sustained for an hour or more, the sample is placed in a buchner funnel. Vacuum is applied and the filtrate collected for analyses. Total salt concentration of the extract may be estimated by measuring the specific electrical conductance and using the relationship (Bower and Wilcox, 1965): salt concentration (mg/l) = 640 x electrical conductivity, $mmho/cm^3$ (Ayers and Westcot, 1976).

If detailed information is required on specific ionic constituents in extracts, chemical analyses using procedures in Methods of Soil Analyses (Black, 1965) may be used. According to Bower and Wilcox (1965), the principal ions of importance in soils studies are: Ca^{++}, Mg^{++}, K^+, Na^+, $CO_3^=$, $SO_4^=$, CL^-, and boron, as well as the nitrogen series.

Cation Exchange Capacity and Exchangeable Ions--

The sum of individual exchangeable basis of a soil sample is equal to the cation exchange capacity. Alternatively, the CEC may be obtained directly using methods detailed by Chapman (1965). Briefly, the exchangeable cations in a soil sample are replaced by either ammonium acetate or sodium acetate, and the amounts of ammonium and sodium ions adsorbed are determined. A problem may develop with the use of ammonium ions because this ion becomes strongly adsorbed on some clays. Cation exchange capacity is expressed as milliequivalents per 100 g (meq/100 g) of sample.

Sorptive Capacity: Specific Surface--

The cation exchange capacity and specific surface together govern the sorptive characteristics of a soil. Mortland and Kemper (1965) discuss the principles of adsorption relating to the specific surface of clays and review a number of adsorption isotherms. A method for determining specific surface based on sorption of ethylene glycol is presented.

Column Studies and Batch Tests--

An approximate idea of the attenuating properties of vadose zone materials for specific pollutants can be obtained from laboratory column studies. In practice, cylindrical columns are packed with vadose zone samples (core samples can also be used) and subsequently flooded with wastewater from the disposal site. Samples of column effluent are collected and analyzed and the breakthrough of particular constituents is determined. Advantages of laboratory column tests include: (1) the method is simple, (2) results are related to a specific mass of material, and (3) results may generally be obtained in a short time. Disadvantages include: (1) if disturbed samples are used, water movement in the column may differ from in-place flow, (2) flow along the column walls may occur, and (3) flow rates may differ from in-place rates.

Batch testing is an alternative method for estimating the attenuation of water-borne pollutants as a result of interactions in the vadose zone. Batch tests consist of placing fragmented samples of vadose zone material in flasks together with a measured aliquot of wastewater. The flasks are shaken via a mechanical shaker for a given period of time. The fluid is subsequently drained from the flasks and analyzed for constituents of interest. Advantages of batch tests include: (1) the method is simple and inexpensive, (2) results are related to a specific mass of material, (3) results are obtained in a short time, and (4) results may be used to prepare adsorption isotherms or selectivity coefficients in ion exchange reactions (Freeze and Cherry, 1979).

Disadvantages of batch tests include: (1) results may be affected by sample disturbance, (2) flow conditions differ from those in place, and (3) samples are exposed to the air, i.e., oxidized, and their adsorptive capacity may differ from reduced material, affecting the transferability of results to field cases (Freeze and Cherry, 1979).

Attenuation of Specific Constituents in the Vadose Zone

Major Chemical Constituents--

Certain wastewaters such as landfill leachate may contain excessive concentrations of major chemical constituents, including calcium, magnesium, sodium, nitrate, chloride, sulfate, phosphate, and bicarbonates. The fate of both calcium and magnesium is dependent upon precipitation and affected by sorption on the cation exchange complex of clays. Generally, calcium and magnesium will precipitate during reactions with bicarbonate and sulfate. McNeal (1974) discusses precipitation reactions in detail. Precipitation of carbonates occurs as a result of the concentration of the soil solution, although the solubility is affected by the partial pressure of CO_2 in the gas phase and the salt concentrations of the solution. McNeal (1974) illustrates the use of the "Langelier saturation index" to predict the approximate amounts of calcium carbonate precipitating from waters.

Next to precipitation in the carbonate form, precipitation of calcium in the form of $CaSO_4$ (gypsum) is important as a process for removing calcium from solution. Precipitation of gypsum can be predicted using Debye-Huckel Theory.

The solubility of $MgCO_3$ is also influenced by the presence of CO_2, and free movement within the vadose zone will occur until the pH elevates. Sodium and potassium salts are soluble and mobile unless concentrations are increased to several thousand parts per million.

In addition to precipitation, calcium and magnesium mobilities are limited by exchange reactions in clays. For this case, calcium and magnesium compete with sodium for exchange sites. The relative degree of sodium adsorption is expressed by the adjusted sodium adsorption ratio, adj SAR, defined by Ayers and Westcot (1976) as:

$$\text{adj SAR} = Na^{+}/((Ca^{++} + Mg^{++})/2)^{1/2}\ (1 + (8.4 - pHc)). \qquad (5\text{-}3)$$

Terms of this equation are defined in Section 4.

The mobility of nitrogen compounds in vadose zone materials is primarily related to the oxidation-reduction potential. Thus, if the system is aerobic, nitrification of organic-N sources occurs readily, producing NO_3-N as an end product. Nitrate-N is highly mobile, moving readily with the soil solution into the lower vadose zone and ultimately into groundwater. Under anaerobic conditions, nitrification is inhibited and the NH_4-N form predominates. Ammonium-N is attenuated in soils with clays or organic materials by two mechanisms: sorption of the positively charged ammonium ion on the clay-organic exchange complex, and fixation of the ammonium ion within the crystal lattice of clay minerals. The latter mechanism is particularly pronounced in clays with a 2:1 silica-alumina ratio (Nommik, 1965). Ammonia sorption on the exchange complex is affected by the presence of other cations such as calcium and magnesium, which compete for available sites.

In soil systems in which wastewater is applied intermittently, nitrification of sorbed NH_4-N occurs during drying cycles (Bouwer, 1978). During subsequent flooding cycles, NO_3-N may be leached into the lower vadose zone.

Under anaerobic conditions, nitrogen may be lost through denitrification. In particular, denitrification occurs quite readily as a result of the activity of heterotrophic bacteria, which convert NO_3-N to volatile gases such as N_2O and N_2 via a number of intermediate compounds. An organic substrata is necessary as an energy source for denitrifying microorganisms. During controlled field studies, Rolston and Broadbent (1977) noted that denitrification was also affected by soil-water content, soil temperature, and by the presence or absence of a growing crop in the soil.

The mobilities of sulfate and bicarbonate are linked to reactions with calcium and magnesium. In addition, sulfate may be sorbed to a minor extent in the aquifer matrix and be retained by iron hydrous oxides (Keeney and Wildung, 1977). The latter reaction, however, requires low pH. The presence of organic acids in leachate migrating into the vadose zone may result in reducing sulfate mobility. Reduction of sulfate under anaerobic conditions may lead to the formation of H_2S and eventually insoluble sulfides such as ZnS and FeS_2. An increase in bicarbonate often accompanies an increase in sulfate (Wood and Signor, 1975).

Factors influencing the solubility of iron include pH, the reduction-oxidation (redox) potential (Eh), and the dissolved CO_2 and sulfur species (Hem, 1970). The reduced form, or ferrous ion, is soluble and mobile. The oxidized form, or ferric ion, forms relatively insoluble precipitates with sulfur and carbonate species. Hem (1970) presents a pH-Eh diagram that demonstrates the conditions under which iron solubility is very low.

Phosphate retention and mobility are discussed by Keeney and Wildung (1977). Under acid conditions, phosphorus is sorbed on the surface of iron- and aluminum-containing minerals. Organic acids may have a local effect on lowering pH values, promoting the above effect. For alkaline conditions, the sorption of phosphorus on $CaCO_3$ or formation of calcium phosphate minerals may occur (Keeney and Wildung, 1977). Phosphate retention on clays and hydrous oxides may also be important.

Trace Contaminants--

Adsorption appears of prime importance among the general mechanisms associated with the removal of trace metals. In reviewing the results of recent laboratory studies on metal attenuation, Chang and Page (1979) indicates:

> ... strong adsorption of trace metal ions occurred at the surface of amorphous iron and manganese oxides and aluminum minerals.... Other soil properties such as texture and cation exchange capacity did not appear to significantly influence soil adsorption characteristics. Unlike the electrostatic cation exchange reactions, covalent bonding-induced adsorption generally is more specific and the reactions are not easily revised by the presence of other cations.... Even trace elements which usually form anions in aqueous solution may be effectively adsorbed or rapidly converted into insoluble forms.

Murrmann and Koutz (1972) also compared the relative effects of cation exchange and adsorption for removal of heavy metals in wastewaters. They indicate that in contrast to cation exchange, "... soil has a capacity to retain heavy metals so tightly that they can be replaced only with difficulty."

Based on results of laboratory studies, Korte et al. (1976) developed figures on the relative mobility of 11 trace contaminants in the 10 most prominent soil orders of the United States. These figures are reproduced in Figures 5-1 and 5-2.

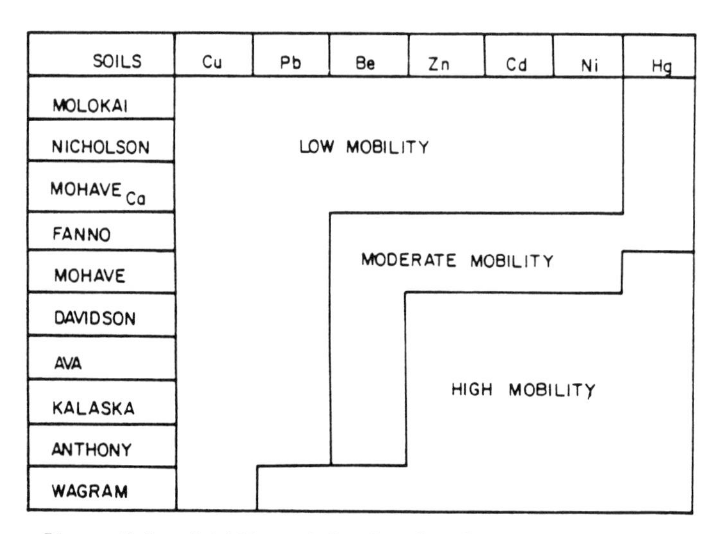

Figure 5-1. Mobility of Cu, Pb, Be, Zn, Cd, Ni, and Hg in 10 soils (after Korte et al., 1976).

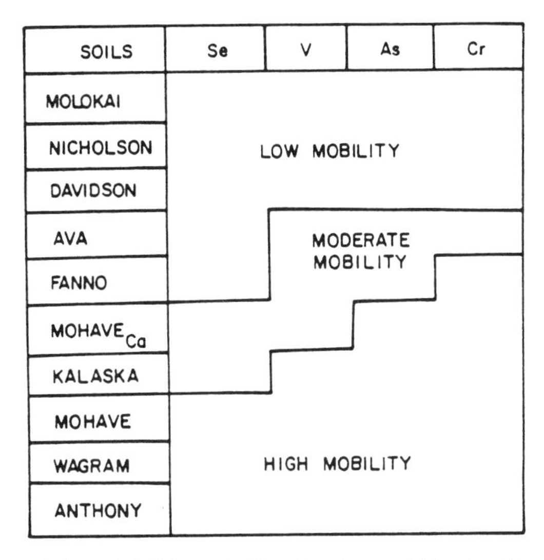

Figure 5-2. Mobility of Se, V, As, and Cr in 10 soils (after Korte et al., 1976).

Fuller (1977) published a comprehensive report on the movement through soils of the following elements: arsenic, beryllium, cadmium, chromium, copper, cyanide, lead, mercury, selenium, and zinc. According to Fuller, the principal mechanism for attenuating arsenic is adsorption by soil colloids. However, if waterlogging should occur during migration of leachate, reducing conditions will favor the mobilization of arsenic. Finally, according to Fuller, "At the low concentrations usually found in waste waters, landfill leachates, and other aqueous waste streams, As probably will not precipitate in soils except possibly as an impurity in phosphorus compounds formed over a long period of time."

The complex chemistry of selenium is described by Fuller. The behavior of selenium is closely related to that of sulfur in acid formation and other properties. Based on experimental studies at the University of Arizona, Fuller concluded that other factors being equal, selenium is less mobile in acidic than in neutral or alkaline soils.

Regarding the mobility of zinc, Fuller (1977) indicates that Zn^{2+} forms slowly soluble precipitates with carbonate, sulfide, silicate, and phosphate ions. Sulfide values may increase in leachate if waterlogging occurs, promoting precipitation of zinc (and other cationic heavy metals). Zinc is also strongly sorbed on the exchange complex of soil.

Unlike other members of the halogen group, fluoride compounds tend to be rather insoluble (Hem, 1970). The mobility of fluoride in leachate may be limited by the formation of fluorite (CaF_2) with a solubility product of $10^{-10.57}$ (Hem, 1970). High concentrations of calcium in leachate would tend to favor fluorite formation in spite of common ion effects.

Reactions of strontium in water are similar to those of calcium. Strontianite, formed by the reaction of strontium and bicarbonate, is slightly less soluble than calcite (Hem, 1970). In addition, relatively insoluble strontium sulfate may form in sulfate rich waters (Davis and De Wiest, 1966). Both reactions may occur, limiting the mobility of strontium.

Copper appears to be strongly complexed to organic matter (Fuller, 1977); consequently, soluble and mobile copper-organic chelates may form in leachate. The formation of hydrous oxides of manganese and iron provides the main control in the immobilization of copper. Hem (1970) reports that copper solubility is generally lower in reducing systems than in oxidizing systems, particularly if reduced sulfur species are present. Reducing or anaerobic conditions can exist if saturation develops. Reduction of sulfate, present in leachate, would then lead to precipitation of copper (as well as iron, zinc, cadmium, lead, and mercury (Fuller, 1977)).

Both nickel and cobalt are strongly adsorbed by iron and manganese oxides (Hem, 1970). The low solubility of $CaCO_3$ may be an important factor in limiting cobalt concentrations in solutions in which HCO_3^- concentrations are high. According to Hem (1970), there are no effective solubility controls over molybdenum concentrations in water. Consequently, the mobility of the anionic form, molybdate, will probably be high.

Jurinak and Santillan-Medrano (1974) examined the transport of lead and cadmium in soils. Processes considered important in governing cadmium and lead movement in soils include: precipitation and dissolution, ion-pair formation, pH flux, cation exchange, and adsorption. Jurinak and Santillan-Medrano (1974) concluded that the principal mechanism regulating lead solubility in noncalcareous soils was precipitation of the forms $Pb(OH)_2$ and $Pb_5(PO_4)_3OH$. In calcareous soils, $PbCO_3$ also precipitates: "Calcareous soils appear to be an excellent sink for Pb^{2+} ions."

Jurinak and Santillan-Medrano found that the solubility of cadmium in soils is about 100 times that of lead in the pH range 5 to 9: "The data suggest that particularly at low concentrations, the adsorption of Cd by the soils is a more important mechanism in retention than in the case of Pb where precipitation of slightly soluble compounds regulates solubility."

Pratt et al. (1978) indicate that in solution, mercury exists in inorganic form as a monovalent or bivalent ion and as complex ions or ion pairs. Regarding mobility of mercury, these authors state, "The concentrations of Hg in soil solutions are governed by ionic adsorption by organic and inorganic materials and by the low solubilities of Hg as phosphate, carbonate, and sulfide."

Alesii (1976) studied the mobility of simple and complex forms of cyanide in soil columns. He observes, "Cyanide as $Fe(CN)_6^{3-}$ and CN^- in water were both found to be very mobile in soils. Soil properties, such as low pH, percentage free iron oxide and Kaolin, chloride, and gibbsite type clay (high positive charges) tended to increase the mobility of the cyanide forms." When an abundant energy source is available, microorganisms volatilize cyanide into less harmful forms of nitrogen. As noted by Fuller (1977), however, anaerobic conditions inhibit the microbial degradation of a cyanide. In a survey of municipal wastewaters in California, Pratt et al. (1978) found that the maximum cyanide level was 0.16 mg/l. They conclude that during recharge, cyanide levels would be reduced well below the recommended limit of 0.2 mg/l.

Organic Constituents--

A problem in specifying the mobility of specific organic pollutants is that quantitative studies have only recently been reported. One problem is that analytical procedures to identify organic constituents are still being developed. "Methods of sampling and analyzing for most organics are in their relative infancy compared to those used for other pollutants" (U.S. EPA, 1979d).

Two methods that may have applicability in characterizing organic pollutants during flow in the vadose zone have been proposed by the U.S. EPA (1979d). One method is quantitative, employing GC/MS (gas chromatography/mass spectrometry). The other method provides for "screening" to detect the presence or absence of organic toxins by GC/MS, followed by extra GC or LC (liquid chromotography) quantification of identified pollutants.

Leenheer and Huffman (1976) describe the development of the dissolved organic carbon (DOC) technique for fractionating organics into hydrophobic and hydrophilic components using macroreticular resins. The technique was applied to several natural waters. It has advantages over other methods for concentrating organics, such as activated carbon. For example, Robertson, Toussaint, and Jorque (1974) report that only 10 percent of organics present in groundwater beneath a landfill in Oklahoma were identified using carbon adsorption followed by carbon chloroform and carbon alcohol extraction.

The principal mechanisms for attenuating organic pollutants are adsorption and decomposition (Pratt et al., 1978). As an example of the importance of adsorption, Robertson, Toussaint, and Jorque (1974) observe that polychlorinated biphenyls (PCBs) tend to be strongly adsorbed in soils. Leenheer and Huffman (1976) indicate that both hydrophobic and hydrophilic organics may be sorbed by sediment. Chang and Page (1979) discuss adsorption of organic substances on soils in some detail:

> The adsorption of dissolved organic substances usually is complicated by the chemical characteristics (e.g., molecular weight, structure and the presence of various functional groups) of each organic compound.... Since many organic substances tend to form negatively charged colloids in aqueous solution, the most likely sites for adsorption to take place would be the positively charged edges of clay minerals. These organic substances also form organometallic complexes which may be adsorbed as neutral or positively charged molecules.

Biodegradable organics are decomposed by microorganisms rendering potentially harmful constituents into gases. Davis (1956) reports that microbiologists have observed the utilization of hydrocarbons by certain bacteria, actinomycetes, filamentous fungi, and yeasts. In general, however, hydrocarbons are not as readily decomposed as carbohydrates, proteins, or fats. Furthermore, the cyclic hydrocarbons are less susceptible to microbial decomposition than are the aliphatic hydrocarbons. Chlorination of wastewaters before recharge may destroy both pathogenic organisms and microorganisms responsible for decomposition (Pratt et al., 1978).

pH may also be a factor in the mobility of organics. For example, Leenheer and Huffman (1976) note the formation of an organic precipitate upon acidification of a groundwater sample from an oil shale area near Rock Springs, Wyoming. Raising the pH dissolved the precipitate. The pH of leachate during aerobic decomposition within a sanitary landfill may be low enough to cause the flocculation of certain organics. In the underlying media, flocs would then be filtered out. However, in alkaline soils, the pH would eventually elevate to a point in the flow system at which organic flocs would dissolve.

In studies on leaching of spent oil shale, Schmidt-Collerus (1974) notes that the solubilization of polycondensed organic matter was enhanced by the presence of seepage water high in TDS. Possibly, the flux of organics in wastewater would be accelerated in a similar fashion if the initial TDS were high.

Microorganisms--

According to Gilbert et al. (1976), bacteria are removed at the soil surface by filtration, sedimentation, and adsorption. Virus removal occurs mainly by adsorption, which increases with decreasing pH. Other factors listed by Gilbert et al. as important in attenuating bacteria and virus are: salt concentration, pH, organic matter, soil composition, infiltration rates, and climatic conditions. Survival and movement of microorganisms within a soil relate to soil moisture content, temperature, pH, nutrient availability, and antagonisms. The presence of excessive salt levels in wastewater, coupled with the presence of high levels of toxic substances (e.g., trace contaminants, pesticides), may limit the migration of bacteria and virus to a small segment of the vadose zone beneath a landfill.

Table 5-1 summarizes specific factors affecting the movement of virus in soils. Chang and Page (1979) caution that immobilization of viruses by soil should not be equated with virus inactivation. "Many adsorbed virus particles have been demonstrated to be infectious for significant periods of time. Viruses immobilized by soil adsorption may also become desorbed when the chemical composition of the percolating waste water is changed."

Pesticides--

Factors influencing the fate and behavior of pesticides in soil systems include (1) chemical decomposition, (2) photochemical decomposition, (3) microbial decomposition, (4) volatilization, (5) plant or organism uptake, and

TABLE 5-1. FACTORS THAT INFLUENCE THE MOVEMENT OF VIRUSES IN SOIL (after U.S. EPA et al., 1977)

Factor	Remarks
Rainfall	Viruses retained near the soil surface may be eluted after a heavy rainfall because of the establishment of ionic gradients within the soil column.
pH	Low pH favors virus adsorption; high pH results in elution of adsorbed virus.
Soil composition	Viruses are readily adsorbed to clays under appropriate conditions and the higher the clay content of the soil, the greater the expected removal of virus. Sandy loam soils and other soils containing organic matter also are favorable for virus removal. Soils with a low surface area do not achieve good virus removal.
Flow rate	As the flow rate increases, virus removal declines, but flow rates as high as 32 ft/d can result in 99.9-percent virus removal after travel through 8.2 ft of sandy loam soil.
Soluble organics	Soluble organic matter competes with viruses for adsorption sites on the soil particles, resulting in decreased virus adsorption or even elution of already adsorbed viruses. Definitive information is still lacking for soil systems.
Cations	The presence of cations usually enhances the retention of viruses by soil.

(6) adsorption-desorption (Bailey and White, 1970). The last factor, adsorption-desorption, directly or indirectly influences the magnitude of the other five factors and is considered to be the prime factor governing the interactions between pesticides and soil colloids. Leonard, Bailey, and Swank (1976) classify organic pesticides as ionic or nonionic. In turn, ionic pesticides are subclassified as cationic (paraquat, disquat), basic (s-triazones), and acidic (benzoic acids, phenols, picolinic acid). Nonionic pesticides include chlorinated hydrocarbons and organophosphates. Cationic pesticides are retained tenaciously on the exchange complex. Changes in soil pH have a profound but complex effect on pesticide forms. For example, a decrease in pH increases the molecular form of an acidic pesticide but increases the conjugate acid form of the base (Leonard, Bailey, and Swank, 1976). These changes will modify adsorption-desorption characteristics and the mobility of pesticides. Leonard, Bailey, and Swank also review the effect of clay on sorption-desorption properties and point out that organic matter greatly increases the sorptive tendency of soils.

Temperature affects adsorption-desorption of pesticides. For example, an increase in temperature decreases adsorption and promotes desorption (Leonard,

Bailey, and Swank, 1976). Temperature also affects sorption solubility and vapor pressure.

Regarding soil moisture effects, Leonard, Bailey, and Swank indicate, "A decrease in soil moisture (a) causes an increase in concentration per unit volume, ... and may increase surface acidity, and, thus increases adsorption; and (b) causes a decrease in competition with water for adsorption sites, which should increase adsorption." Furthermore, when the pesticide concentrates to the solubility product, crystallization will result.

Pesticide degradation will occur in soils. Photodecomposition will reduce pesticide levels near the soil surface but may be inconsequential with depth. Chemical degradation is a complex phenomenon, related to pH, redox potential, and surface acidity.

Microbial decomposition of pesticides depends on such factors as "microbial population ecology, soil moisture and temperature, organic matter content, pH, redox potential, pesticide concentration, availability for degradation, and nutrient concentration and availability" (Leonard, Bailey, and Swank, 1976). In addition to decomposition by microbial activity, pesticide compounds may be taken up by plants.

Volatilization of pesticides will reduce their concentration in the soil. Volatilization depends on such factors as temperature and soil moisture, as well as on the vapor pressure of specific compounds.

Bailey and White (1970) list the following factors as being most significant in governing the leaching and movement of pesticides in soils: adsorption, physical properties of the soil, and climatic conditions. The effects of adsorption are reviewed above. Regarding soil physical properties, Bailey and White indicate that pesticides are leached to a greater degree in light-textured soils than in heavier-textured soils. The porosity of soils affect diffusion rates of volatile pesticides. Air diffusion plays a prominent role in the eventual loss of pesticides from soil due to volatilization. Pore-size distribution of soils affects the rate at which the water infiltrates and moves through a soil.

For a given soil type, leaching of pesticides is increased with an increase in the amount and frequency of rainfall. The same relationship holds for irrigation. Evapotranspiration will tend to increase the concentrations of pesticides at the soil surface.

Radioactive Wastes

The principal mechanisms relied upon for the attenuation of radionuclides in the vadose zone are ion exchange and adsorption. Consequently, the presence of clays, organic matter, and hydrous oxides at disposal sites is of paramount importance in reducing the potential for groundwater pollution for radioactive wastes. The "distribution coefficient," K_d, is a commonly used measure of the ability of a solid matrix to retard the movement of radionuclides (and other solutes) via sorption effects.

Borg et al. (1976) reviewed the chemical and physical factors affecting the measured values of the distribution coefficient for radionuclides. They state:

> Among other variables, mineralogy, particle size, nature of solution, and chemical nature of radioactive species are important.... In general, a decrease in particle size results in an increase in K_d. Fresh silicate rocks have lower K_d's than their altered counterparts. For the same reasons, old fractures absorb more than fresh fractures in a given rock. The sorption of Cs and Sr is greater than that of Ru and Sb and is probably related to the fact that the latter two elements form anionic complexes which do not readily take part in ion exchange processes. Cs in clay-rich rocks often is sorbed more than Sr because of lattice shrinkage that traps the larger ion. Pu forms a positively charged polymer that is highly sorbed in the pH range of 2 to 8.

Regarding ions in the soil solution competing with radionuclides for sorptive sites, Borg et al. (1976) cite the work of Nelson (1959). In particular, Nelson (1959) conducted batch and column experiments on the sorption of strontium on Hanford soil in the presence of cesium, sodium, barium, calcium, magnesium, and aluminum. He found that water of hydration was a significant factor because cations with the combination of highest valence and lowest hydrated radii provide the greatest competition with strontium for sites. The work of Kokotov, Popova, and Urbanyuk (1961) is also cited by Borg et al. (1976). Kokotov, Popova, and Urbanyuk (1961) measured K_d values for strontium and cesium in Russian soils. For strontium, the order of effectiveness in lowering the K_d (i.e., increasing the mobility) is $Sr^{2+} > Ca^{2+} > Mg^{2+} > K^{2+} > NH_2^+ > Na^+$. For cesium, the order of effectiveness in lowering K_d values is $Cs^+ > Rb^+ > NH_4 > K^+ > H^+$.

The effect of soil pH on distribution coefficients of various radionuclides has been examined by several investigators. The basis for their studies was that changes in pH affect cation exchange properties of soils and concomitantly K_d values. In particular, increasing the soil pH changes cation selectivity by increasing both the cation exchange capacity and the preference of the exchange complex for polyvalent over monovalent ions (Bohn, McNeal, and O'Connor, 1979). Rhodes (1957b) evaluated the effect of pH on the uptake of the following polyvalent radionuclides: plutonium, cerium, zirconium, yttrium, and ruthenium, as well as monovalent cesium and bivalent strontium. The soil used in his tests was from Hanford, Washington. Uptake of the polyvalent species exhibited a maximum between about pH 4 and pH 8. Above pH 8, a region of reduced uptake was observed, persisting up to pH 11. Rhodes (1957b) notes that the effect of increasing pH on uptake of polyvalent radioisotopes is complicated by two possible consequences of changing pH: (1) changing ion species, and (2) formation of polymers or colloids with changing pH.

For example, in an earlier paper, Rhodes (1957a) concludes that plutonium forms a positively charged polymer between pH 2 and pH 8 which is highly sorbed. Apparently, because of a change in the nature of the polymer, sorption of plutonium decreases above pH 8, reaching a minimum at about pH 12.

In his experiments using monovalent cesium, Rhodes (1957b) observes that pH has a negligible effect on the uptake of this ion when HCl is used to adjust the pH of the solution. However, when NaOH is used to adjust the solution pH, a marked reduction in the distribution coefficient occurs. Apparently, sodium ions compete with cesium ions for exchange sites. Experimental results with strontium indicated that when the pH was adjusted with either NaOH or HCl, the distribution coefficient of strontium increased from about pH 4 to about pH 10. Above pH 10, uptake was reduced by the presence of large sodium ion concentrations.

Alternative combined forms of radionuclides may have an effect on distribution coefficients and associated relative velocities. For example, Jakubick (1976) determined that the velocity of PuO_2 in soils is about 100 times faster than the velocity of $Pu(NO_3)_4$.

FIELD METHODS FOR MONITORING POLLUTANT MOVEMENT IN THE VADOSE ZONE

Source monitoring, a primary activity of a monitoring program, is not covered in this section. A comprehensive methodology for source monitoring, however, has been developed by Huibregtse and Moser (1976). For convenience, field monitoring in the vadose zone is categorized as follows: (1) indirect methods, (2) direct methods for solids sampling, (3) direct methods for solution sampling in unsaturated media, and (4) direct methods for sampling saturated regions.

Indirect Methods

Two properties of a wastewater that can be used to estimate pollutant mobility in the vadose zone are temperature and electrical conductivity. Temperature can serve as an indicator of wastewater movement if the source is at either an elevated or lowered temperature, and if the disposal site is underlain by shallow perched groundwater. Insertion of sensitive thermistors into the soil and groundwater system could possibly monitor the spread of the source. The limitation of this method is that the temperature wave would be dampened by contact with soil and groundwater. That is, the extent of the plume could be poorly defined.

Resistivity Methods for Soil Salinity--

An indirect measure that has been extensively used to characterize soil salinity and to delineate the areal distribution of shallow pollution plumes is electrical resistivity, or its inverse property, conductivity. As pointed out by Rhoades and Halvorson (1977), most soil minerals are insulators. Consequently, electrical conductivity in saline soils is mainly through pore water containing dissolved electrolytes. Exchangeable cations do not contribute extensively to electrical conductivity in nonsodic soils because they are not present in abundance and they are less mobile than the soluble electrolytes. According to Rhoades (1979a), the electrical conductivity of a saline soil, EC_a, depends primarily on the electrical conductivity of the liquid, EC_w; on the volumetric water content, θ; on the tortuosity, T; and on the extent of surface conductance, EC_s. For a given soil, the specific conductance of a saturation extract, EC_e, is uniquely related to EC_w. For simplicity, a relationship between EC_w and EC_a is obtained at a uniform water content value, θ, in order to standardize θ and T effects. In practice, the standard water content is taken to be that at field capacity. The relationship between EC_a and EC_e then becomes:

$$EC_a = A\ EC_e + B \qquad (5\text{-}4)$$

where $B = EC_s$

A = slope of the EC_a versus EC_e line.

The four-electrode method--Rhoades and Halvorson (1977) present three methods for determining apparent bulk soil conductivity, EC_a, and three methods for establishing conductivity of saturated extract, EC_e, versus EC_a calibrations. One method for measuring soil electrical conductivity in situ uses the Wenner four-probe array (see Figure 5-3). This method is a common surface resistivity technique used by geophysicists (Zohdy, Eaton, and Mabey, 1974).

Field implementation--The method entails placing four electrodes into the soil surface at a constant distance "a" apart. An electrical current is applied to the outer two electrodes and the resultant electrical resistance is measured as a potential drop across the inner two electrodes. Resistance is measured via a resistance meter. EC_a is calculated by the following equation:

$$EC_a = \left(\frac{5.222}{a}\right) \frac{f_t}{R_t} \qquad (5\text{-}5)$$

where R_t = measured resistance in ohms at temperature t

a = spacing of electrodes

f_t = factor to adjust the reading to a reference temperature of 25°C.

EC_a is reported in mmhos/cm.

In the absence of layering, the depth of penetration of the electrical current is about one-third the outer electrode spacing, y (Rhoades, 1979a). Similarly, the volume of soil measured by a given "a" spacing in a single EC_a determination is about $\pi y^3/9$. To increase the volume of soil scanned during a salinity survey, it is thus a matter of increasing the "a" spacing of the electrodes. The limitation on "a" spacing, and thus depth, is the degree of uniformity of texture within the subsoil. A depth of 4 feet appears to be a reasonable lower limit. Figure 5-3 depicts the theoretical mass of the soil scanned by the method.

Rhoades and Halvorson (1977) successfully used the four-probe method to determine the average bulk soil salinity in soils of the Northern Great

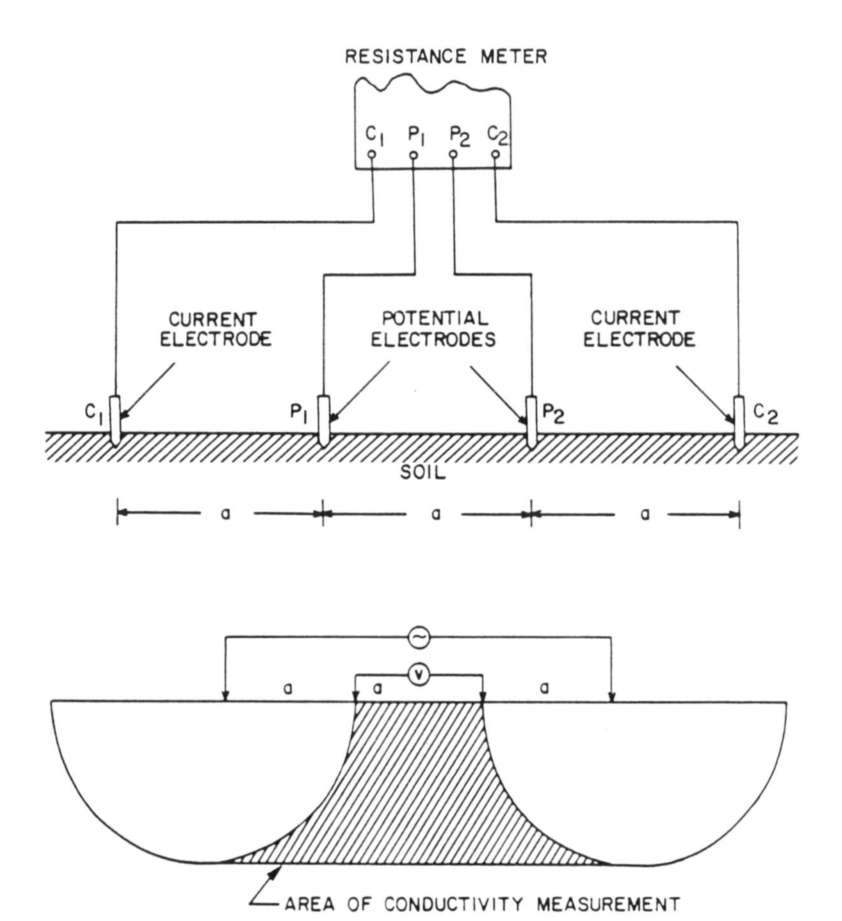

Figure 5-3. Wenner four-probe array and associated area of conductivity measurement (after Rhoades and Halvorson, 1977).

Plains. The four-probe method was also used by Rhoades to detect the presence of a saline shallow water table by varying the "a" spacing. In two soils examined by this method, water tables were detected at depths of about 1 meter and 3 meters. The conductance of the soil solution at discrete depth intervals can also be obtained by varying the "a" spacing and using a difference equation presented by Rhoades and Halvorson (1977). They used this method to delineate salt-affected regions in incremental soil depths in Montana.

According to Rhoades (1979a), one of the major advantages of the four-probe method is that because of the high degree of correlation between EC_e and EC_a, the method facilitates obtaining representative values of the salinity of typically heterogeneous soils without excessive expenditures of time and money. Other advantages cited by Rhoades (1979b) are: larger volumes of the salinity of soil are scanned than when soil samples, suction cups, or salinity sensors are used; and because the method is fast and simple, it is particularly well-suited for routine salinity monitoring and mapping. A disadvantage is that the accuracy of the method decreases if soil layering is pronounced.

Methods for obtaining EC_e versus EC_a calibrations--Field measurements of EC_a are converted into corresponding EC_e values via calibration relationships. Rhoades and Halvorson (1977) discuss three calibration methods in detail. In particular, measurements of conductivity are obtained using the Wenner four-electrode method, the EC-probe, or the four-electrode cell. Soil samples are then obtained within the volume of soil measured by the selected method and laboratory measurements are made to determine the conductivity of the saturated extract, EC_e. A number of soil samples of differing salinity must be measured in order to obtain a range of EC_e versus EC_a values. In addition, the field EC_a measurements should be obtained at a uniform water content, such as that at field capacity.

Range of applications--The accuracy of the four-probe electrode method decreases as the complexity of soil layering increases. The resistivity associated with varied soil textures can confound the resistivity readings when several layers are monitored at once.

The installation of the electrodes in fractured or consolidated media can be quite difficult. In addition, it would be difficult to obtain a calibration for the equipment under these conditions.

The four-probe electrode method requires the presence of some soil moisture in order to get a resistance reading. Therefore, this method is not appropriate for extremely dry soils.

The four-probe electrode method is based on the ability of the soil and soil water to conduct electrical current. Therefore, if the liquid waste is a poor conductor, unreliable readings may result. To detect movement of pollutants, there must be a contrast between the conductivity of the waste and the background conductivity of the soil solution. Also, this method does not provide specific information regarding the chemical characteristics of the waste being monitored.

EC-probe--Rhoades and van Schilfgaarde (1976) developed an "EC-probe" to detect changes in soil salinity in discrete depth intervals within stratified soils. Figure 5-4 shows construction features of the unit. The probe utilizes the Wenner array of four electrodes with a fixed "a" spacing of 2.6 cm. Salinity can thus be assessed in 15-cm depth intervals. The corresponding soil volume scanned with this electrode setting is about 90 cm^3. The electrodes in the unit designed by Rhoades and van Schilfgaarde (1976) were constructed of brass embedded within a lucite housing. The probe is attached to a shaft with a handle. Wire leads from the electrodes are brought to the surface through the hollow shaft and handle. The outer two electrodes are attached to a current source and the inner two electrodes are attached to a resistance meter.

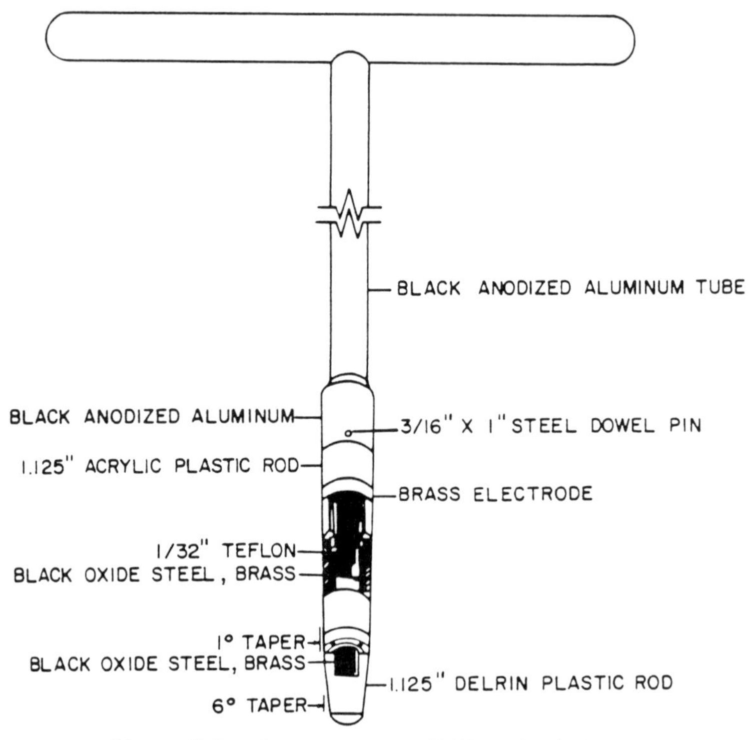

Figure 5-4. Cross section of EC-probe (after Rhoades and Halvorson, 1977).

Field implementation--In operation, a soil cavity is augered to the depth of interest using a soil auger or sampler of about the same diameter as the probe. The probe is placed within the cavity. An electric current is applied to the outer electrodes and the potential drop between the inner electrodes is measured via a resistance meter. The resistance value is converted to conductance by means of the cell constant for the probe. The cell constant is determined by submerging the unit in a container filled with solutions of known conductivity. Individual EC_a versus EC_e calibrations must be obtained for each of the soil strata.

Frequently during salinity studies, it is advantageous to obtain periodic measurements at the same location. For such cases, Rhoades (1979c) designed an inexpensive probe that can be left in place within a cavity. "... Implanted probes offer certain advantages, like avoiding the need for making new access holes, remaining in the same position with time, minimizing the complications arising from several access holes being made in the sampling area...." Ostensibly, a number of probes could be clustered within a common hole.

Range of applications--Soils normally have considerable spatial variability in soil properties and, therefore, a large number of EC readings would be necessary in order to detect movement of pollutants. When layered soils are being monitored, a separate calibration must be established for each strata. Use of the EC-probe could be difficult in rocky, consolidated, or fractured materials.

For probes left within the soil profile, care must be taken not to influence the flow of soil solution preferentially into the probe installation cavity. During operation of the probe in the field, it is important to measure salinity at a uniform water content, such as at field capacity.

The readings from the EC-probe may be erroneous if the pollutant is a poor conductor. Similarly, if the pollutant and the soil water do not contrast, the movement of pollutant will not be detectable. As with other resistivity methods, the EC-probe does not provide information regarding the chemical composition of the pollutant.

The EC-probe method cannot be used beneath waste disposal ponds nor in deep installations such as in a landfill. When using the stationary installation, care should be taken to avoid displacement of the unit in expanding soils or during freeze thaw cycles.

The soil salinity probe shown in Figure 5-4 is available from Micron Engineering and Manufacturing Inc., Riverside, California. Several resistivity meters are available to complement the salinity probe. One such meter is the Bison 2350.

Four-electrode conductivity cell--The four-electrode conductivity cell is described in detail by Rhoades et al. (1977). Undisturbed soil cores are obtained using a soil-core sampler with lucite column sections as core inserts. The lucite sections are removed from the sampler and segmented to form individual cells. Electrodes are inserted into threaded holes in the cell walls. The EC_a of the soil is subsequently measured using a resistivity meter.

In the unit described by Rhoades et al. (1977), light stainless steel electrodes are inserted into the soil surface at 45-degree intervals around a circumference to a fixed depth of 3 mm. Consequently, any four neighboring electrodes could be regarded as a Wenner array, the outer two electrodes being current electrodes and the inner two electrodes being potential electrodes. By rotating the connectors, eight independent EC_a measurements can be made.

A cell constant must be obtained to permit converting resistance values to conductivity values. In practice, cell constants are obtained by filling the units with solutions of known conductance and measuring the cell resistance.

Earth resistivity measurements--Howe (1982) provides a good technical description of the earth resistivity method:

> The earth-resistivity method indirectly measures the earth's resistance to an induced current. The current originates from the source (usually a type of dry cell battery) and travels to a current electrode (see Figure 5-5), through the ground to the the other current electrode and back to the battery through the surface lines. The voltage drop is measured between the two inside potential electrodes. The configuration of the electrodes and the distance between them dictates the depth and volume of the earth material measured.

The earth resistivity technique is based on the resistivity of solute in the pore space of the vadose zone being less than that of dry rocks. By the same token, a highly saline groundwater plume offers less resistivity to an applied current than less saline native groundwater. Consequently, by obtaining a series of resistivity-depth profiles in a grid within a waste disposal site, the lateral extent, and possibly the depth, of the plume could be delineated (Fenn et al., 1977).

Field implementation--The basic approach used for earth resistivity surveys is the same as that for soil salinity surveys. That is, the Wenner array is used with increasing "a" spacings in order to observe resistivity changes with depth. According to Fenn et al. (1977), it is more appropriate to refer to "apparent resistivity" rather than "resistivity" alone. This term accounts for subsurface heterogeneity and is defined as "... the weighted average of the actual resistivities of the individual subsurface materials or strata within the depth of penetration at the resistivity measurement" (Fenn et al., 1977). Howe (1982) presents a formula for converting direct measurements of current, I, and voltage, V, to apparent resistivity, pa:

$$pa = K(V/I) .$$

The value for K depends on the electrode configuration and may be determined from the equation $K = 2\pi a$ (where a is the electrode spacing) when using the Wenner configuration. The conversion of voltage and current field readings to apparent resistivity eliminates the need to consider changes in resistivity relative to changes in the volume of material measured (Howe, 1982).

Howe (1982) presents a series of steps that are helpful in setting up and conducting a successful earth resistivity monitoring program. He suggests beginning with a survey of all available data for the site, including geophysical or drillers' logs from wells in the area, information on the chemical composition of native waters and the contaminant, and maps or air photos that might reveal changes in vegetation or topography (drainage patterns, etc.) that might suggest the direction of movement for the pollutant. This information may also reveal that the site is too complex to be monitored by the earth resistivity method.

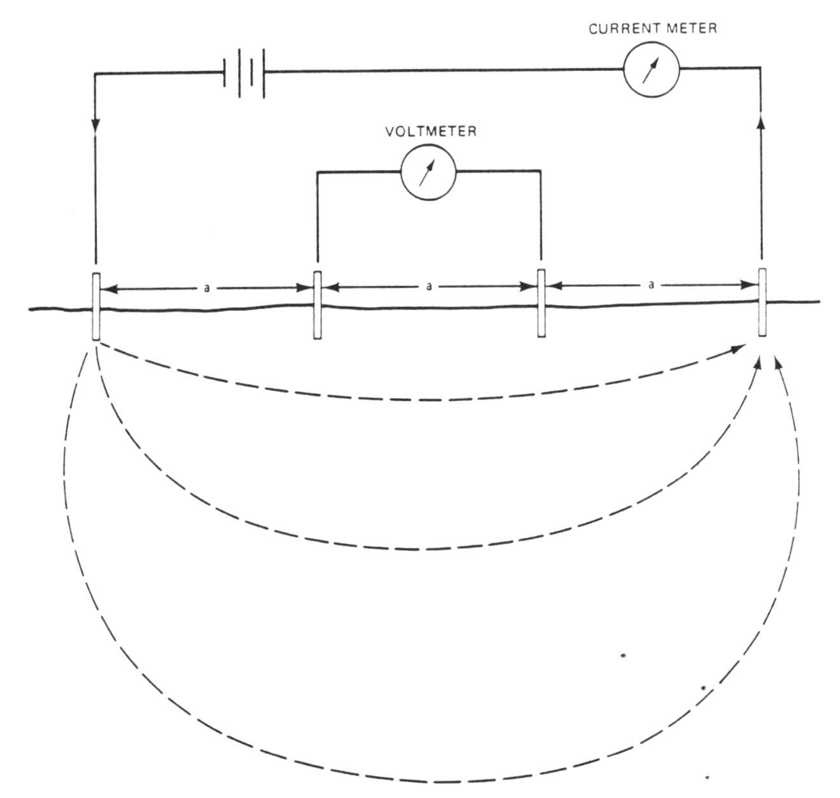

Figure 5-5. Wenner electrode configuration with constant "a" spacing. Broken lines show current flow (after Howe, 1982).

The second step recommended by Howe (1982) is to develop a map with the pollutant source, probable direction of movement, and potential discharge areas delineated.

A reconnaissance-level earth resistivity survey should then be made covering the entire survey area with a low level of data collection. "Vertical soundings should be made at each point on the survey or until the lower limit of the contamination plume is defined." Once the lower limit and/or horizontal extent of the contamination plume has been identified, a slightly greater electrode spacing can be applied to the remainder of the site in order to better define the contaminated area. Comparison of resistivity data with water quality from existing wells in the area should be made whenever possible to field verify the resistivity information. The data collected during the reconnaissance-level investigation should show a contrast in apparent resistivity both horizontally and vertically. Howe (1982) suggests plotting the rough location of the contamination plume on the previously developed map and using the new information to plan a detailed survey to better define the extent of the pollutant plume and to monitor its advancement. The detailed survey should include the installation of monitoring wells inside and outside of the pollutant plume to assure the accuracy of the resistivity measurements.

Range of applications and limitations--Use of earth resistivity methods to delineate the lateral and vertical extent of pollutant plumes has been successful in several studies. Cartwright and his associates observed pollution plumes emanating from landfills in Illinois using earth resistivity surveys (Cartwright and McComas, 1968; Cartwright and Sherman, 1972). Warner (1969) discussed the use of the method as a preliminary approach for delineating zones of polluted groundwater. Stollar and Roux (1975) demonstrated the successful delineation of pollutant plumes from industrial waste sites and landfills when the depths to polluted groundwater ranged from 5 to 50 feet below the land surface.

When used under the proper circumstances, the method can be an effective monitoring tool that can reduce costs when compared to monitoring programs relying solely on monitoring wells. The earth resistivity technique has some limitation, however, based on varied geologic, hydrologic, chemical, and physical settings.

Changes in rock type can confound the apparent resistivity measurements from a given site to the point that readings bear no resemblance to the actual pollutant plume. Under more complex geologic environments, two methods are used to interpret the field resistivity data--theoretical or empirical (Fenn et al., 1977).

For the theoretical approach, the field resistivity values are plotted. "... The resulting curve [is] compared with sets of master curves developed for numbers of resistivity layers with definite ratios of resistivity and thickness. By this method, the value of resistivity for each geologic unit as well as its thickness and depth can be determined."

The empirical method relates the field resistivity data to geological information, e.g., the properties of layered sediments. The success of the empirical approach thus depends greatly on the availability of geological information, such as texture and the vertical and areal distribution of the layered sediments, and the chemical quality of perched groundwater. Under conditions where significant stratigraphic changes occur laterally, the earth resistivity method is not applicable. Also, buried electrical lines or other manmade conductors will interfere with the resistivity measurements. Conversely, under relatively simple geologic settings, the earth resistivity method can be quite accurate in delineating pollutant plumes.

The depth to the top of the contaminated groundwater body and its associated thickness can affect the success of the earth resistivity monitoring. The case studies by Stollar and Roux (1975) were successful under conditions where the contaminated groundwater was as deep as 60 feet.

In order for the earth resistivity method to be an effective monitoring tool, a significant contrast between the salinity of the pollutant and native groundwater must be present. Most pollutants tend to remain intact rather than mix with the host groundwater system. Thus, when a difference in resistivity (salinity) occurs between the two fluids, a distinct boundary can be detected.

Earth resistivity meters such as the ER-1 or ER-2 models are available from Geophysical Specialties Division, Minnetech Laboratories Inc., Minneapolis, Minnesota.

Salinity Sensors--

Another indirect method for in situ evaluation of soil salinity is the so-called "salinity sensor." The basis of these devices is the relationship between specific electrical conductance, EC_e, of soil solution and the total concentration of salts in solution.

As described by Richards (1966), the basic idea of the salinity sensor is that electrodes embedded in porous ceramic forming a hydraulic continuum with soil water can be used to directly measure the specific conductance of the soil solution. From suitable calibration relations, therefore, specific conductance values can be directly related to the total salt content. The unit described by Richards comprises a plate about 1 mm thick with platinum electrodes fired in-place on opposing faces. An important feature of this sensor is that the unit is spring-loaded to ensure good contact with soil.

Because of the strong dependency of specific conductance on temperature, it is important to accurately measure the temperature of the soil solution. Richards used a thermistor to provide temperature compensation in his unit.

Oster and Willardson (1971) reviewed problems arising from calibration of sensors. They also reported on field studies using the unit. Of particular importance is their observation that sensors should not be used at soil-water pressures less than -2 atmospheres. Also, they indicate that when sensors are placed in trenches at field sites, the permeability of the materials in the backfilled trench tends to be greater than in indigenous soil. During leaching trials, therefore, the salinity measured with the sensors tended to be lower than in adjacent soil. Differences were attributed to greater leaching in the trench. However, structural differences were also present, e.g., the pore sizes in the trench soil were probably of a different range than indigenous soil.

Rhoades (1979b) indicates that a principal advantage of using salinity sensors compared to soil sampling for assessing salinity is that readings are taken in the same location. Thus, by installing a network of sensors, a chronological record of in situ salinity changes can be obtained from a large volume of soil. Rhoades also indicates that salinity sensors "... are simple, easily read, and sufficiently accurate for salinity monitoring purposes."

The U.S. Salinity Laboratory staff (1977) discusses the use of a data acquisition system that automatically obtains electrical conductivity values from a number of salinity sensors. The units were installed during field studies on salt movement in irrigation return flow. By means of a computer program, the system activated stepping switches, read the sensors in sequence, adjusted the readings for soil temperature, and printed out the calculated EC values. The resultant data were transmitted by means of a teletype to Riverside, California.

Rhoades (1979b) compares the use of salinity sensors and four-probe units for assessing soil salinity. He concluded that four-probe units have six major advantages over the salinity sensor: (1) they are less subject to calibration change, (2) they are more durable and less costly, (3) they do not suffer from time lags in the response to changing salinity, (4) they can be sized to measure within different soil volumes, (5) they are more versatile, and (6) they are well suited for mapping and diagnosis, as well as monitoring.

Solids Sampling

Soil sampling methods used by soil scientists and irrigation and drainage engineers to evaluate the physical properties of soils are also suitable for sampling inorganic chemical constituents. Particular hand tools that have been used include the following types of hand augers: screw-type augers, post-hole augers, barrel augers, and dutch augers (Donnan, 1957). Another common type of sampler is the split-spoon sampler, which is a barrel-type auger, one side of which pivots on a hinge. Tube-type samplers have been used extensively to obtain soil cores. The Veihmeyer sampler is an example of such a unit. The end of the sampler is bevelled and sharpened to facilitate insertion into the soil. A drive hammer is used to force the tube to the desired depth. An intact core is removed from the tube. Such a core may show the distinct breaks in soil layering.

A problem associated with using augers and tube-type samplers is that in very dry soil the sample tends to fall out of the unit when it is being withdrawn. "Core catchers" could be used to overcome this problem (Thomas, personal communication, 1979). Another problem is that if the soil is wet, chemical interactions may occur between the soil water and the metal parts of the sampler. Thus, trace metals could be introduced into the sample.

The problem of contamination between a soil sample and sampling device is of particular importance when sampling for organic chemicals and microorganisms. Bordner, Winter, and Scarpino (1978) recommend the following procedure for obtaining soil samples for microbial analysis:

> ... scrape the top one inch of soil from a square foot area using a sterile scoop or spoon. If a subsurface sample is desired, use a sterile scoop or spatula to remove the top surface of one inch or more from a one-foot square area. Use a second scoop or spoon to take the sample. Place samplings in a sterile one-quart screw-cap bottle until it is full. Depending on the amount of moisture, a one-quart bottle holds 300-800 grams of soil. Label and tag the bottle carefully and store at 4°C until analyzed.

Techniques for sampling organic and microbial constituents have been examined by the Groundwater Research Branch of the Robert S. Kerr Environmental Research Laboratory, as reported by Dunlap et al. (1977). One method used with success comprises an auger and dry-tube coring procedure. The method entails augering a hole to the top of the desired sampling depth and forcing a "dry-tube" core sampler into the sampling region. The core sampler used by Dunlap et al. (1977) was a steel tube about 18 inches long and 3 inches in diameter. The core barrel is fitted with a steel drive shoe of slightly smaller inside diameter than the diameter of the barrel to facilitate removing the core. As pointed out by Dunlap et al. (1977), contamination problems are minimized by the auger/dry-tube coring technique primarily because drilling fluid is not used. It would appear that the core barrel could be modified to include a lucite plastic insert to minimize contamination of soil samples obtained for trace metal analysis.

The core sampling method of Dunlap et al. (1977) has been used to obtain solids samples for analyses of organic chemicals and microorganisms in depths to about 25 feet. In addition to dry samples, solids have also been obtained from saturated regions. Problems developed, however, when sampling loose saturated material. An alternative approach to overcome such problems that was considered by Dunlap et al. was to use a hollow-stem auger instead of a regular auger. Dunlap et al. also describes a piston-type sampler for collecting solids for organic analyses when fluid is used in the drilling process. To date, the sampler has not been extensively field tested.

The methods discussed above for obtaining soil samples could, with difficulty, be used to sample from the lower vadose zone. For example, Fenn et al. (1977) indicate that post-hole augers could be used to sample to depths down to 80 feet in certain material by using a tripod and pulley. A more convenient faster approach is to employ engine-driven augers and drilling equipment. Among the power-driven sampling units available are continuous-flight spiral augers (which could be hollow-stemmed), core samplers, bucket augers, cable-tool drill rigs, and rotary drill rigs. The spiral-type units are not particularly suitable for sampling because it is not always possible to distinguish the sampling depth. By using the hollow-stem auger, however, it is possible to insert a core sampler inside the auger and sample from discrete depths. Trailer-mounted engine-driven soil sampling mechines are available commercially for obtaining either auger samples or cone samples by attaching appropriate tools. Alternatively, sampling machines may be purchased for mounting on pick-up trucks and driven by power take-offs. Hydraulic controls facilitate drilling or coring operations. Travelling Kelley bars permit sampling at depths greater than 25 feet (courtesy, Giddings Machine Company).

A bucket auger consists of a large-diameter bucket fitted with cutting blades (Fenn et al., 1977). The bucket is rotated in the hole until filled. When brought to the surface, the bucket is dumped and a sample can be easily obtained. Alternatively, a core is obtained before dumping. Rible et al. (1976) used this method to sample for nitrate distribution in the vadose zone underlying irrigated fields in California.

Drill rigs can be used for sampling purposes while test wells and monitor wells are being installed at the waste disposal site. There are some problems, however. The cable-tool, or percussion, technique entails alternately raising and dropping a heavy string of tools in the cavity. The drill bit crushes large-diameter material. Water is commonly added to the borehole and, after removing the tool string, a bailer is inserted. The fine cuttings are removed by the bailer. Casing is driven in as drilling progresses. An obvious disadvantage of this method is that the water added during drilling changes the chemistry of the native pore water. This problem can be obviated somewhat by analyzing the water and adjusting the analyses obtained from solids samples.

The hydraulic rotary method utilizes a rotating bit to excavate a borehole and a continuous flow of water to remove drill cuttings. Drilling mud is added to the circulating water during drilling to hold open the hole. Obviously, solids samples are badly contaminated by the mud and water mixture used with this method. Also, the precise depths at which cuttings are obtained are impossible to define.

In the air rotary method, air is used instead of water to bring the drill cuttings to the surface. Thus, the mud and water contamination encountered with hydraulic rotary drilling is not a factor. Defining the sampling depth and sample composition precisely may be difficult, however; e.g., particle-size segregation may occur because of density differences.

Field Implementation--

Field implementation of solids sampling methods for undisturbed areas utilizes standard drilling, coring, augering, and boring techniques to obtain the sample material. At existing waste disposal sites, however, these methods can cause short circuiting of pollutants to greater depths. Kaufmann et al. (1981) discuss a driven pile method for well completion in hazardous waste. This method would also be useful in collecting solid samples at existing waste sites.

Whenever possible, pore water should be extracted from solid samples in the field and analyzed immediately for unstable constituents such as temperature, pH, Eh, and EC (Mooij and Rovers, 1976). When the soils are unsaturated, special techniques are required to obtain samples of the pore water. Fenn et al. (1977) reviewed a number of alternative methods for extracting pore water from core samples, including a method by Lusczynski (1961) that uses CO_2 under pressure to displace the fluid. Another method described by Fenn et al. (1977) is a "hydraulic squeezer," which uses a pressurized piston to force pore water into a syringe.

Instead of using pore water directly, it is often more convenient to prepare saturated pastes in the field and extract water from them for specific conductance measurements. Rhoades and Halvorson (1977) describe a convenient extraction unit for field use. Saturated paste is transferred into a small-diameter filter funnel made of plastic with a filter pipe in place. Vacuum is provided by a hand pump and the sample is collected in a reservoir. The EC of the sample is determined via a portable EC bridge.

Solids samples collected for chemical analyses of pore water in the laboratory should be stored in paper bags unless they are wet, when plastic bags will be necessary. Depending on the intended analyses, the samples are air-dried in the laboratory. If the analyses are intended for the nitrogen series, the handling room atmosphere should be free of ammonia.

Samples collected for determining microorganisms and organic pollutants using the "dry-tube" core sampler require special handling. Dunlap et al. (1977) describe the procedure in detail. Briefly, the core is pressed from the barrel of the sampler and the first 5 to 8 cm of solids are removed for analysis of physical/chemical parameters. In order to obviate microbial contamination of the core sample by contact with the walls of the sampler, a subsample is removed from the center of the core (see Figure 5-6) by inserting a sterile stainless steel tube into the center of the core. The subsample is extracted using a sterile rod and placed in containers. Subsamples for analyses of adenoside triphosphate (ATP) are also obtained by subsampling from the center of the core.

Two 5-cm-long subsamples for organic analyses are obtained after removing the central cores. The subsamples are dropped into aluminum baking pans and covered with aluminum foil. The pan and subsamples are then placed in an insulated polystyrene box with liquid nitrogen for quick-freezing.

Range of Applications and Limitations--

An important aspect of a monitoring program for the vadose zone underlying a waste disposal operation is to obtain samples of the solid materials throughout this region. Solids are defined as soils and underlying geologic materials. Dunlap et al. (1977) state the rationale for solids sampling in the vadose zone:

> Only by analysis of earth solids from the unsaturated zone underlying pollutant-releasing activities can those pollutants which are moving very slowly toward the water table because of sorption and/or physical impediment be detected and their rates of movement and degradation measured. Such pollutants, which probably include a major proportion of organics and microorganisms, are not likely to be detected in groundwater until the activities releasing them have been in operation for protracted periods. Because of their potential for long-term pollution of groundwater, it is imperative that the behavior of these pollutants in the subsurface be established at the earliest practicable time.

Dunlap et al. (1977) also list reasons for sampling from the groundwater zone ("zone of saturation"). These reasons apply equally well for sampling from perched groundwater:

> Analyses of organic pollutants in solid samples from the zone of saturation are needed for a realistic evaluation of the total extent and probable longevity of organic pollution in an aquifer. Such analyses provide a measure of the quantity of

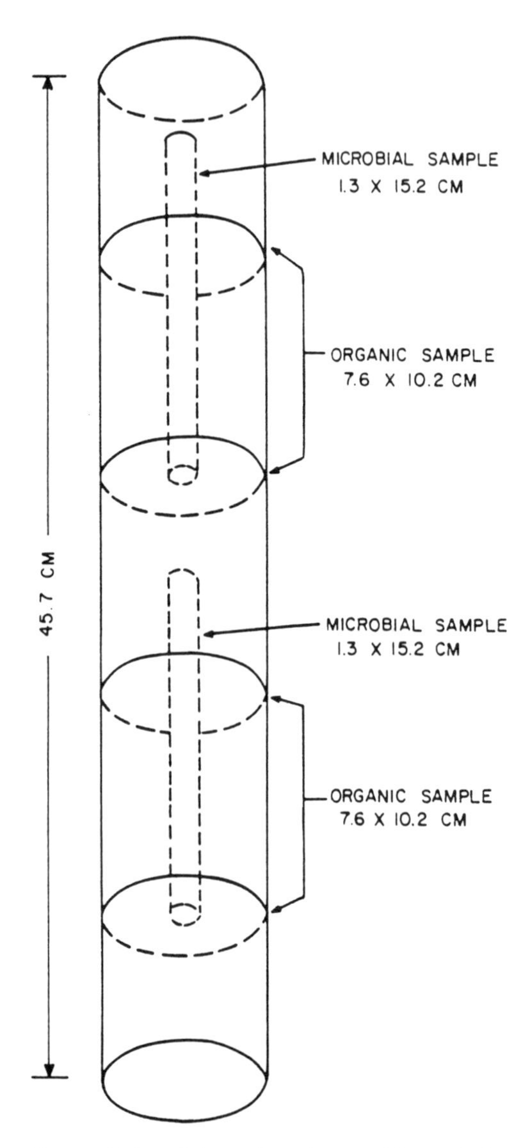

Figure 5-6. Subsamples from a "dry tube" core sampler (after Dunlap et al., 1977).

pollutants which are sorbed on aquifer solids and which are in equilibrium with, and in essence serve as a reservoir for, pollutants in solution in the adjacent groundwater.

Microbial populations which may be involved in the biological alteration of pollutants in subsurface formations are likely to be in such close association with subsurface solids that they will not be present in well waters in numbers which are quantitatively indicative of their presence in the formations; hence, analyses of subsurface solids are needed for accurate evaluation of such populations.

Even when the best well construction and groundwater sampling procedures are used, it is difficult to completely eliminate the possibility that contaminating surface microbes may be present in groundwater samples. Solids taken from the interior of cores carefully obtained from the zone of saturation probably provide the most authentic samples of aquifer microorganisms that can be obtained.

The rationale and need for obtaining solid samples has been clearly established by previous studies. Specific geologic, hydrogeologic, chemical, and physical site characteristics, however, can limit the representativeness of the samples obtained. These should be incorporated into the interpretation of the resulting field data.

Problems With Solids Sampling--

Drilling a large number of holes to obtain representative samples may alter infiltration patterns and cause short-circuiting of pollutants to greater depths. Chemical composition of pollutants in pore spaces of the solids samples may change after exposure to air. Volatile organics are especially prone to loss after sampling. These are reasons why immediate field analysis of pore water extract from the solids samples is recommended.

Obtaining samples at sites with toxic wastes will be dangerous for the drilling and sampling personnel. Solids sampling is a destructive process. The same location cannot be resampled to monitor temporal changes in pollutants. Shallow hand methods of sampling may not be possible in freezing conditions. Steep terrain may preclude solids sampling from areas adjacent to the waste disposal site.

An important limitation of field sampling of solids is that an inordinate number of samples may be required to ensure that the measured mean value of a particular parameter falls within a given range of a true value. A principal reason for the problem is that variations arise because of the natural spatial variability of soil properties. Such variations were noted by Rible et al. (1976); Pratt, Warneke, and Nash (1976); and Lund and Pratt (1977) during studies on nitrate leaching in irrigated fields in California. Pratt, Warneke, and Nash (1976), for example, showed that 5 to 21 sampling sites were needed (eight samples per site) from four fields ranging from 62 to 223 acres to obtain measured mean nitrate concentrations that were within 30 percent of the true mean at the 95-percent confidence limit. They also estimated that sampling 23 to 161 sites per field would have been necessary to be within 10 percent of the true mean.

In summarizing the utility of solids sampling for nitrate leaching from irrigated fields, Lund and Pratt (1977) states:

> Monitoring nitrate leaching by soil sampling does not appear to be feasible for basin-wide studies. As a research tool to study the components of the nitrogen cycle it has proved useful, but we would not recommend it as a monitoring method.

Rhoades (1979b) describes another problem with solids sampling: "Some changes in soil-water composition occur as soil is removed from its natural condition, dried, ground, sieved, extracted, etc., hence only relative changes in composition, not absolute, can be determined from soil sampling."

Suction Samplers

Wells and open cavities cannot be used to collect solution flowing in the vadose zone under suction (negative pressures). The sampling devices for such unsaturated media are thus called suction samplers. Three types are (1) ceramic-type samplers, (2) hollow fiber samplers, and (3) membrane filter samplers.

Ceramic-Type Samplers--

Two types of samplers are constructed from ceramic materials: the suction cup and the filter candle. Both operate in the same manner. Basically, ceramic-type samplers (also called suction "lysimeters") comprise the same type of ceramic cups used in tensiometers. When placed in the soil, the pores in these cups become an extension of the pore space of the soil. Consequently, the water content of the soil and cup become equilibrated at the existing soil-water pressure. By applying a vacuum to the interior of the cup such that the pressure is slightly less inside the cups than in the soil solution, flow occurs into the cup. The sample is pumped to the surface, permitting laboratory determination of the quality of the soil solution in situ. Although cups have limitations, at the present time they appear to be the best tool available for sampling unsaturated media, particularly in the field.

Suction cups may be subdivided into three categories: (1) vacuum operated soil-water samplers, (2) vacuum-pressure samplers, and (3) vacuum-pressure samplers with check valves. Soil-water samplers generally consist of a ceramic cup mounted on the end of a small-diameter PVC tube, similar to a tensiometer (see Figure 5-7). The upper end of the PVC tubing projects above the soil surface. A rubber stopper and outlet tubing are inserted into the upper end. Vacuum is applied to the system and soil water moves into the cup. To extract a sample, a small-diameter tube is inserted within the outlet tubing and extended to the base of the cup. The small-diameter tubing is connected to a sample-collection flask. A vacuum is applied via a hand vacuum-pressure pump and the sample is sucked into the collection flask. These units are generally used to sample to depths up to 6 feet from the land surface.

Consequently, they are used primarily to monitor the near-surface movement of pollutants from land disposal facilities or from irrigation return flow.

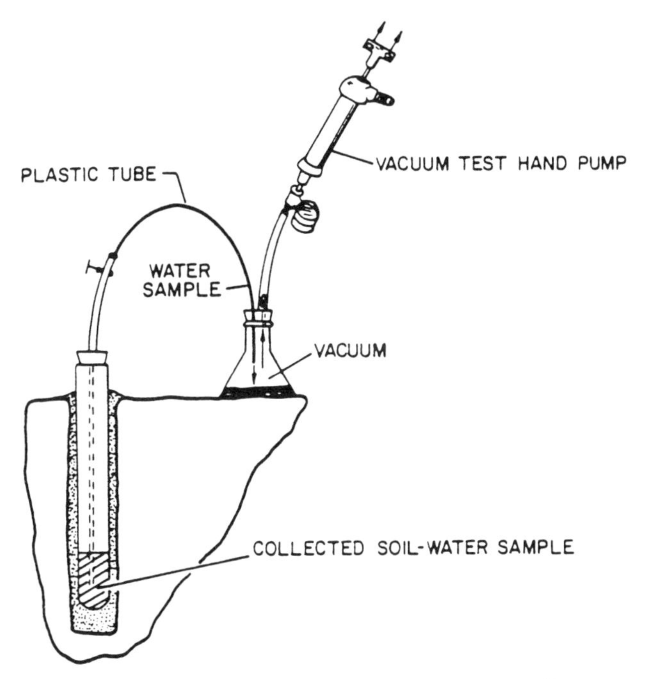

Figure 5-7. Soil-water sampler (courtesy, Soil Moisture Equipment Corporation, 1978).

To extract samples from depths greater than the suction lift of water (about 25 feet), a second type of unit is available, the so-called vacuum-pressure lysimeter. These units were developed by Parizek and Lane (1970) for sampling the deep movement of pollutants from a land disposal project in Pennsylvania. The design of the Parizek and Lane sampler is shown in Figure 5-8. The body tube of the unit is about 2 feet long, holding about 1 liter of sample. Two copper lines are forced through a two-hole rubber stopper sealed into a body tube. One copper line extends to the base of the ceramic cup as shown and the other terminates a short distance below the rubber stopper. The longer line connects to a sample bottle and the shorter line connects to a vacuum-pressure pump. All lines and connections are sealed.

In operation, a vacuum is applied to the system (the longer tube to the sample bottle is clamped shut at this time). When sufficient time has been allowed for the unit to fill with solution, the vacuum is released and the

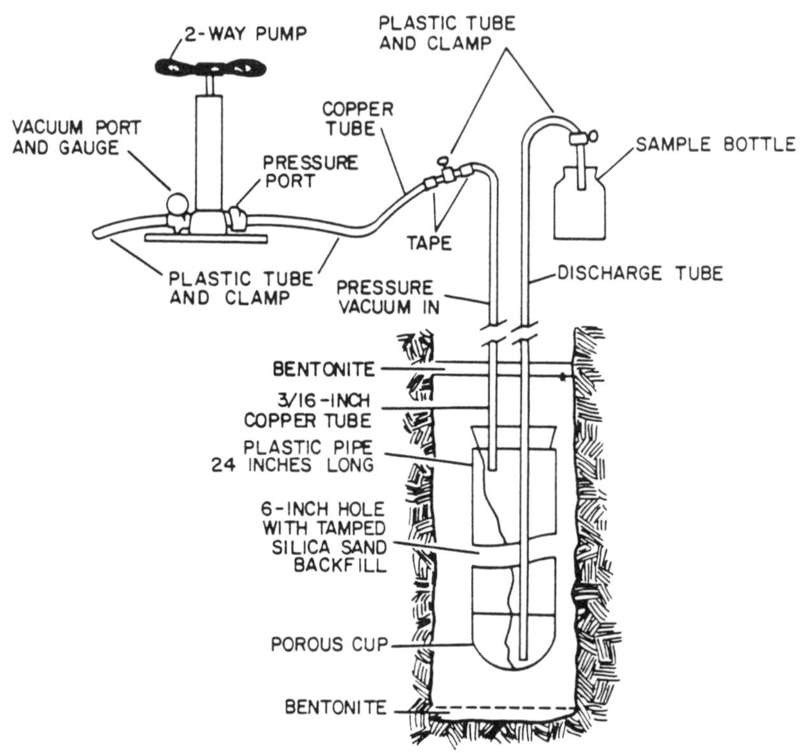

Figure 5-8. Vacuum-pressure sampler (after Parizek and Lane, 1970).

clamp on the outlet line is opened. Air pressure is then applied to the system, forcing the sample into the collection flask. A basic problem with this unit is that when air pressure is applied, some of the solution in the cup may be forced back through the cup into the surrounding pore-water system. Consequently, this type of pressure-vacuum system is recommended for depths only up to about 50 feet below land surface. In addition to the monitoring effort of Parizek and Lane, these units were used by Apgar and Langmuir (1971) to sample leachate movement in the vadose zone underlying a sanitary landfill.

Morrison and Tsai (1981) proposed a modified lysimeter design with the porous material located midway up the sampling chamber instead of at the bottom (see Figure 5-9, after Morrison and Tsai, 1981). This mitigated the basic problem of sample solution being forced back through the cup when air pressure is applied. Polyethylene with 2.5-micron pores has been substituted for

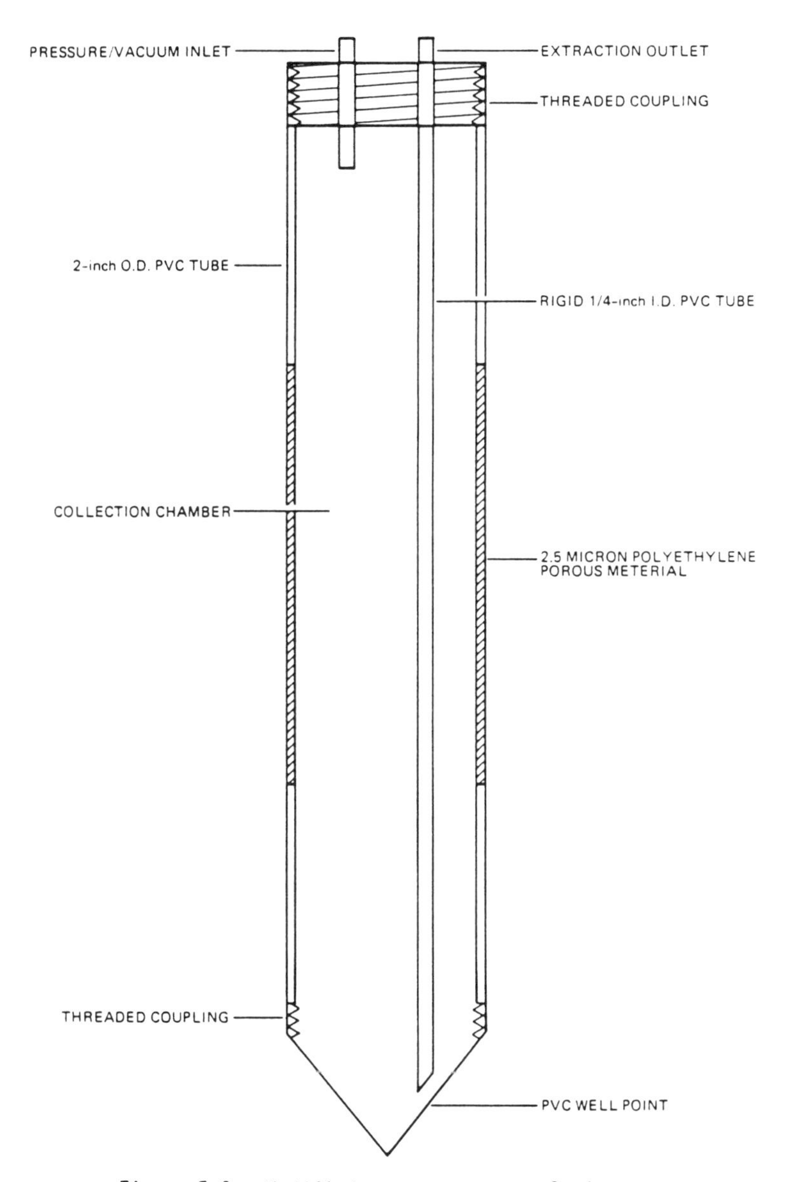

Figure 5-9. Modified pressure-vacuum lysimeter (after Morrison and Tsai, 1981).

ceramic porous material to provide greater sampler durability and comparable or reduced ion attenuation potential.

Wood (1973) reports on a modified version of the design of Parizek and Lane. Wood's design is the third suction sampler discussed in this subsection. Wood's design overcomes the main problem of the simple pressure-vacuum system; namely, that solution is forced out of the cup during application of pressure. A sketch of the sampler is shown in Figure 5-10. The cup ensemble is divided into lower and upper chambers. The two chambers are isolated except for a connecting tube with a check valve. A sample delivery tube extends from the base of the upper chamber to the surface. This tube also contains a check valve. A second shorter tube terminating at the top of the sampler is used to deliver vacuum or pressure. In operation, when a vacuum is applied to the system, it extends to the cup through the open one-way check valve. The second check valve in the delivery tube is shut. The sample is delivered into the upper chamber, which is about 1 liter (0.26 gallon) in capacity. To deliver the sample to the surface, the vacuum is released and pressure (generally of nitrogen gas) is applied to the shorter tube. The one-way valve to the cup is shut and the one-way valve in the delivery tube is opened. Sample is then forced to the surface. High pressures can be applied with this unit without danger of damaging the cup. Consequently, this sampler can be used to depths of about 150 feet below land surface (Soil Moisture Equipment Corporation, 1978). Wood and Signor (1975) used this sampler to examine geochemical changes in water during flow in the vadose zone underlying recharge basins in Texas.

A sampling unit employing a filter candle is described by Duke and Haise (1973). The unit, described as a "vacuum extractor," is installed below plant roots. Figure 5-11 shows an illustrative installation. The unit consists of a galvanized sheet metal trough open at the top. A porous ceramic candle (12 inches long and 1.27 inches in diameter) is placed into the base of the trough. A plastic pipe sealed into one end of the candle is connected to a sample bottle located in a nearby manhole or trench. A small-diameter tube attached to the other end of the candle is used to rewet the candle as necessary. The trough is filled with soil and placed within a horizontal cavity of the same dimensions as the trough. The trough and enclosed filter candle are pressed up against the soil via an air pillow or mechanical jack. In operation, vacuum is applied to the system to induce soil-water flow into the trough and candle at the same rate as in the surrounding soil. The amount of vacuum is determined from tensiometers. Hoffman et al. (1978) used this type of sampler to collect samples of irrigation water leaching beneath the roots of orange trees during return flow studies at Tacna, Arizona.

Cellulose-Acetate Hollow Fiber Samplers--

Jackson, Brinkley, and Bondietti (1976) described a suction sampler constructed of cellulose-acetate hollow fibers. These semipermeable fibers have been used for dialysis of aqueous solutions, functioning as molecular sieves. Soil column studies using a bundle of fibers to extract soil solution showed that the fibers were sufficiently permeable to permit rapid extraction of solution for analysis. Soil solution was extracted at soil-water contents ranging from 50 to 20 percent.

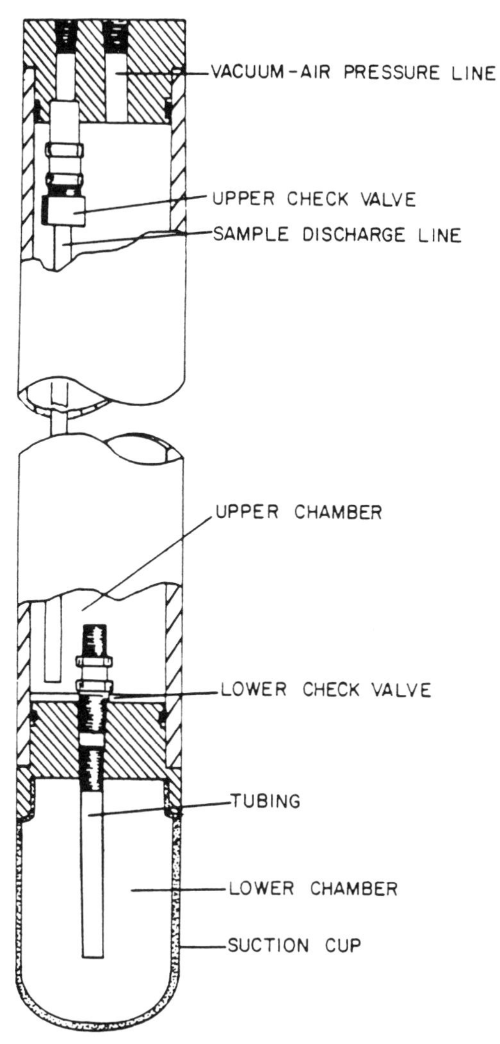

Figure 5-10. "Hi/Pressure-vacuum soil-water sampler" (courtesy, Soil Moisture Equipment Corporation, 1978).

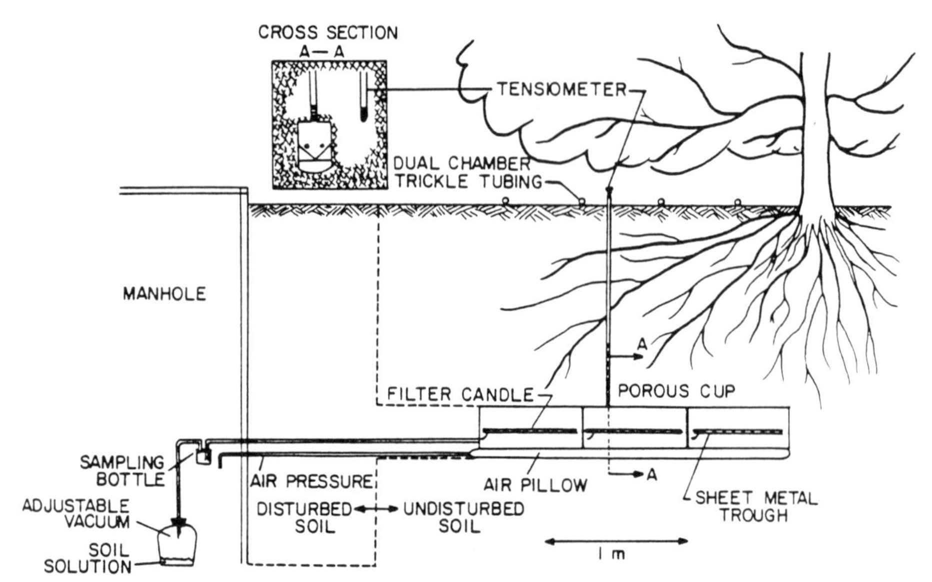

Figure 5-11. Facilities for sampling irrigation return flow via filter candles, for research project at Tacna, Arizona (after Hoffman et al., 1978).

Levin and Jackson (1977) compare ceramic cup samplers and hollow fiber samplers for collecting soil solution samples from intact soil cores. Their conclusion is: "... porous cup lysimeters and hollow fibers are viable extraction devices for obtaining soil solution samples for determining EC, Ca, Mg, and PO_4-P. Their suitability for NO_3-N is questionable." They also conclude that hollow fiber samplers are more suited to laboratory studies, where ceramic samplers are more useful for field sampling.

Membrane Filter Samplers--

Stevenson (1978) presents the design of a suction sampler using a membrane filter and a glass fiber prefilter mounted in a "Swinnex" type filter holder. Figure 5-12 shows the construction of the unit. The membrane filters are composed of polycarbonate or cellulose-acetate. The "Swinnex" filter holders are manufactured by the Millipore Corporation for filtration of fluids delivered by syringe. A flexible tube is attached to the filter holder to permit applying a vacuum to the system and for delivering the sample to a bottle.

The sampler is placed in a hole dug to a selected depth. Sheets of glass fiber "collectors" are placed in the bottom of the hole. Next, two or three smaller glass fiber "wick" discs that fit within the filter holder are placed in the hole. Subsequently, the filter holder is placed in the hole with the glass fiber prefilter in the holder contacting the "wick" discs. The hole is then backfilled.

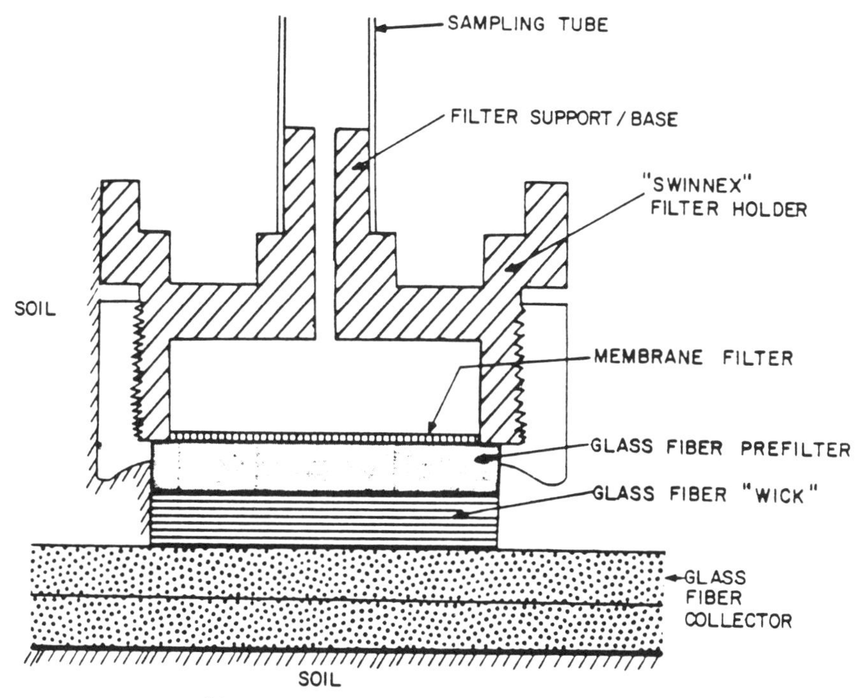

Figure 5-12. Membrane filter sampler (after Stevenson, 1978).

In operation, soil water is drawn into the collector system by capillarity. Subsequently, water flows in the collector sheets toward the glass fiber wicks as a result of the suction applied to the filter holder assembly. The glass fiber prefilter minimizes clogging of the membrane filter by fine material in the soil solution.

During field tests with the sampler, it was observed that sampling rates decreased with decreasing soil-water content. The "wick and collector" system provided contact with a relatively large area of the soil and a favorable sampling rate was maintained even when the "collector" became blocked with fine soil. The basic sampling unit can be used to depths of 4 meters.

Field implementation--Installation of the shallow vacuum-type samplers is similar to the installation of tensiometers. A cavity is first augered to the desired depth. Soil from the base of the hole is wetted to form a slurry (Soil Moisture Equipment Corporation, 1978). The sampler placed in the hole, forcing the cup into the slurry. The entire cup should be covered by the slurry. The hole is then backfilled with soil, tamping to prevent side leakage. An alternative installation method entails placing bentonite clay at the base of the open hole and covering the bentonite with a small amount of 200-mesh silica. The sampler is placed in the hole and covered with a 6-inch layer of sand. Bentonite clay is then placed on top of the sand to seal the cup against side leakage of water. Finally, the hole is filled with native soil.

The pressure-vacuum type lysimeters, either with or without the check valves, can be installed singly in a cavity or clustered with two or more units in a cavity. Figure 5-13 shows an example of a cluster installation. The base of the borehole should be sealed with bentonite. A layer of silica sand is then placed on top of the bentonite plug. The lowermost cup is placed on the silica sand and subsequently surrounded by more sand. Native soil is backfilled in the cavity to an elevation near that of the next sampler. Bentonite is added as a plug. Silica sand is then added, the second unit is installed, and backfilling proceeds as above. In some cases, it may be a good precaution to add a layer of bentonite above the final silica sand layer to minimize the possibility of side leakage.

The aboveground assembly comprises a pressure-vacuum system and collection bottles. For a system with several samplers, a pressure-vacuum manifold assembly is advantageous. Generally, pressure should be added with an inert gas such as nitrogen to preclude oxidation of chemical constituents.

For the shallow soil solution samplers, vacuum can be applied either as a constant vacuum or a falling vacuum. In the falling vacuum type, an initial vacuum is applied to the system and vacuum gradually decreases as the sample is drawn into the cups (Hansen and Harris, 1975). Problems may arise in the operation of the constant vacuum system because of differences in sample collection rate. Some units fill more rapidly than others and solution may overflow into adjoining collection flasks, causing cross-contamination. Chow (1977) presents a design of a mercury-pressure control device to shut off the vacuum to a flask when filled to a preset level.

Vacuum extractor assemblies comprising sheet metal troughs and filter candles can be installed using a technique of the U.S. Salinity Laboratory Staff (1977) during their irrigation return flow studies at Tacna, Arizona. The extractors were installed in a rectangular tunnel formed by first augering a horizontal hole at the desired depth, then forcing a rectangular shaper into the tunnel. The metal troughs containing filter candles were filled with soil augered from the hole. The extractor assembly was then forced against the ceiling of the tunnel by inflating a butyl rubber air pillow. Two tensiometers were installed over each extractor to facilitate determining the vacuum to the filter candles.

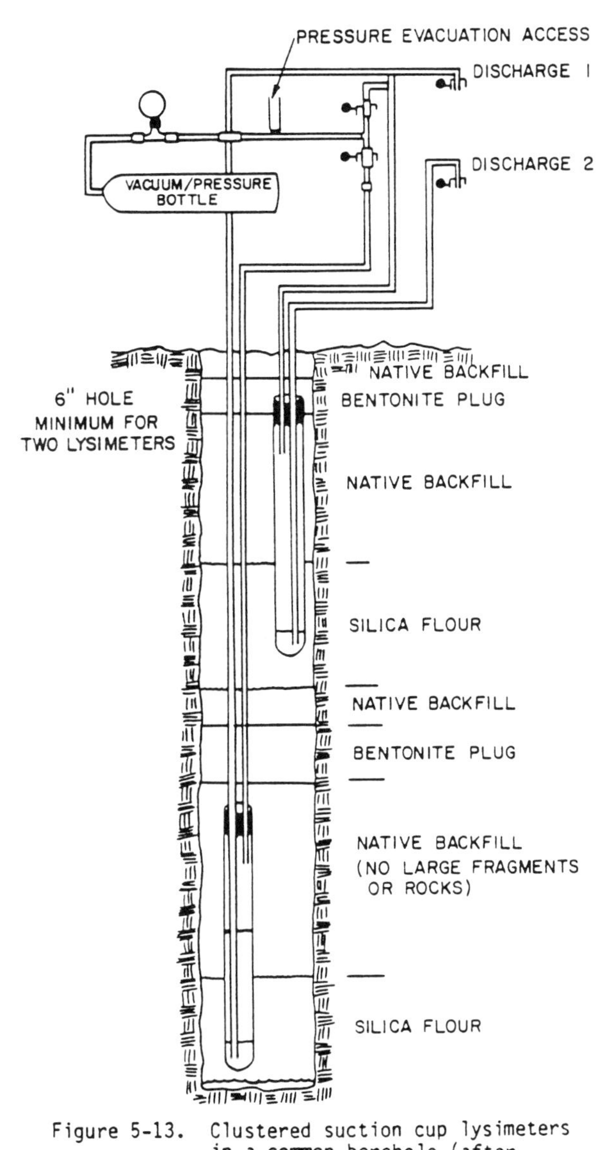

Figure 5-13. Clustered suction cup lysimeters in a common borehole (after Hounslow et al., 1978).

Range of applications and limitations--The range of applications and limitations of the various types of suction samplers used in the vadose zone are similar. Factors that affect the operation of lysimeters may be grouped as follows: (1) soil-physical properties, (2) hydraulic factors, (3) cup-wastewater interactions, and (4) climatic factors. Specific constraints in each of these categories are summarized in Table 5-2.

Physical properties--Soil texture refers to the distribution of constituent soil particles into mixtures of the fundamental grain sizes, namely, gravel, sand, silt, and clay. The successful operation of suction samplers requires a continuity between pore sequences in the cup and those in the surrounding soils. When soils are very coarse-textured, a good contact between a suction cup and the finer pore sequences may be difficult to maintain and the flow continuum may be destroyed.

The problem of contact in coarse-textured soils creates operational problems in using suction cups for soil monitoring, but the problem can be circumvented. For example, the contact problem can be offset by installing cups in a borehole and backfilling around the cups with silica flour (see Figure 5-13).

Soil structure refers to the aggregation of the textural units into blocks. A well-structured soil has two distinct flow regions for liquids applied at the land surface: (1) through the cracks between blocks, i.e., interpedal flow, and (2) through the finer pore sequences inside the blocks, i.e., intrapedal flow. Simpson and Cunningham (1982) have shown that rapid flow of wastewater through interpedal cracks may lessen the renovating capacity of the soil because of reduced surface area and contact face. Similarly, because of the rapid flushing of pollutants through larger interconnected soil pores, the movement of such pollutants into the finer pores of the soil blocks may be limited. Because suction-cup samplers collect water from these finer pore sequences, the resultant samples will not be representative of the bulk flow. Thus, Shaffer, Fritton, and Baker (1979) concluded that "... surface-installed porous cup soil water samplers are unsuited to test leachate water composite when a highly structured soil is kept in a state of high water content."

A primary goal of soil-pore liquid sampling is to detect the presence of fast moving hazardous constituents. Consequently, when soils in the monitoring zone are highly structured, leading to a flow system such as that described in the last paragraph, further testing may be required. For example, at land treatment sites, the structure of a soil profile is best examined by constructing trenches near the proposed monitoring sites, to a depth corresponding to the maximum depth at which lysimeter cups will be installed (i.e., 6 feet below land surface). The extent of large interpedal cracks should be documented at each profile. If such cracks appear to be widespread, the lysimeters may not detect the fast moving hazardous constituents. However, it should be borne in mind that even large cracks frequently diminish in width in deeper reaches of the profile (Beven and Germann, 1982). If it is determined from profile examination that the structural cracks "pinch out" in the vicinity of the proposed monitoring depth, the technique may be applicable.

TABLE 5-2. FACTORS LIMITING THE OPERATION OF SUCTION CUP SAMPLES

Soil Physical Properties				
Texture	Structure	Hydraulic Factors	Cup-Wastewater Interactions	Climatic Factors
1. Contact between cup and soils difficult to maintain in very coarse-textured soils (e.g., gravels).	1. In highly structured soils and fractured material, the composition of fluid in cracks may differ from that in pores. Cups sample only from small pores. Consequently, cup samples may not be representative of "average" fluid.	1. Sampler cannot be obtained when soils dry to the point that soil-water suction is great enough to allow air to enter cups when applying vacuum. 2. For very wet conditions, fluid will move more rapidly in larger pores and cracks. Because of time lag, sample for cups may not be representative.	1. Solids moving with fluid may plug cups. 2. Bacteria may plug cups. 3. Trace metals may be attenuated during flow through cups. 4. Sorption of NH_4-N may occur. 5. Sorption of some organics (e.g., chlorinated hydrocarbons) may occur. 6. Chemical precipitates may clog cup pores.	1. In frozen soils, the tension of unfrozen water is greater than air entry value of cups. 2. In freezing-thawing soils, the unit may shift in the profile and lose contact.

The fact that ceramic-type samplers extract pore water under low suctions also affects the representativeness of the sample. England (1974) describes two of the causative factors:

> First, numerous experimental and theoretical studies have clearly shown that the concentration and the composition of the soil solution are not homogeneous throughout its mass. Cations, in particular, vary widely in degree of dissociation from the surface of electronegative colloidal soil particles. Thus, water drained from large pores at low suctions may have a chemical 'quality' that is very different from that extracted from micropores. A point source of suction, such as the porous cup, samples roughly a sphere, draining different-sized pores as functions of distance from the point, the amount of applied suction, the hydraulic conductivity of the medium, and the soil-water content.
>
> Also, the concentrations of various ions in a soil solution generally do not vary inversely with the soil-water content. Reitemeier ... and many others have presented evidence that the total dissolved quantities of some ions increase on dilution while concurrently, those of other ions may decrease.

In light of these effects, Hansen and Harris (1975) concluded that each group of pore sizes has a unique volume of drainable water, ion concentration, and drainage-rate curve. They concluded that in order to collect a soil-water sample with an ion concentration representative of that draining to the water table, the rate of sample collection (and thus the incremental volumes) should correspond to the pore-water drainage rate.

Hydraulic factors--It is well established (Everett, 1981) that suction-cup samplers fail as the soil-water suction increases to the point that air bubbles enter the cups. The soil-water pressure equivalent to this point is about -1.0 to -2.0 bars. In practice, however, many units fail at a pressure of about -0.5 bar. This appears to be a serious limitation in operating suction samplers. However, as Hillel (1971) pointed out, "Though the suction range of 0 to 0.8 bar is but a small part of the total range of suction variation encountered in the field, it generally encompasses the greater part of the soil wetness range." Thus, the effect of soil-water suction on lysimeter operation is an operational problem to some extent, and if suction is applied to the cup for a sufficiently long period of time, a sample of sufficient volume may be obtained. Nevertheless, because the yield of suction samplers is greatly reduced at the higher suctions, there may be situations in which the time required to obtain a sufficiently large sample exceeds the maximum holding time for analysis. Similarly, there may be cases where the soil is so dry that the units simply will not yield a sample. This may be particularly true in arid regions where rainfall is not great enough to wet up the soil profile.

Field studies by Biggar and Nielsen (1976) illustrate a limitation of point samplers such as ceramic- and fiber-type samplers for estimating the mass flux of solute beneath a given soil depth (mass flux of solute is the mass of solute crossing a unit area per unit time). The soil-water velocity, and thus soil-water flux, may not be normally distributed. In fact, for their test area, Biggar and Nielsen found velocity to be logarithmically normally distributed. Consequently, substantial errors can be made in estimating the mass flux of solute beneath a given soil depth by multiplying average values of the flux of water by average values of the concentration of the soil solution.

An indirect estimate of vertical velocity in the vadose zone beneath a surface source can be obtained by periodically attempting to sample from an array of suction samplers within the vadose zone. Generally, when the surrounding pore-water system is unsaturated, very little, if any, sample will be obtained. When the wetting front reaches a particular cup, samples are more readily obtainable. By observing the response of depthwise units, an apparent vertical velocity can be inferred.

Meyer (personal communication, 1978) used this method to follow the wetting front during deep percolation of irrigation water in the San Joaquin Valley. Signor (personal communication, 1979) used a similar approach during recharge studies in Texas.

Cup-wastewater interactions--For simplicity, the interactions between suction samplers and wastewater can be subdivided into (1) those affecting the operation of the cup (by plugging), and (2) those that attenuate pollutants moving through the porous segment.

An initial concern with lysimeters was that plugging may occur with time, limiting their effective lifetime. Ceramic cups have an effective pore diameter of approximately 1 µm, which would allow colloidal particles to pass through. Levin and Jackson (1977) indicated that clogging of these pores was not a problem. Johnson, Cartwright, and Schuller (1981) conducted laboratory studies to evaluate the plugging of suction lysimeters with particulate matter in landfill leachate. They noted a distinct loss in yield from samplers in direct contact with leachate. In contrast to the discouraging results of these laboratory studies, a concomitant decrease in yield of identical lysimeters placed beneath a landfill was not observed. Apparently, particulate matter moving through the soil profile is filtered out before contacting the cups.

Several studies have been reported involving the use of suction-type samplers for monitoring pollutant movement at land treatment areas. Generally, it appears that the sampling units operated favorably without clogging by particulate matter. Examples of such studies include those by (1) Smith and McWhorter (1977), in which ceramic candles were used to sample pollutant movement in soil during the injection of liquid organic wastes; (2) Grier, Burton, and Tiwari (1977) involving the use of depthwise suction samplers on fields used for disposal of animal wastes; and (3) Smith et al. (1977), in which depthwise suction samples were installed in fields irrigated with wastes from potato processing plants.

Laboratory studies on ceramic cups by Wolff (1967) show that new cups yield several milligrams per liter of Ca, Mg, Na, HCO_3, and SiO_2. Consequently, leaching of cups before installation is recommended. In his studies,

Wood (1973) cleaned new cups as follows: "... the cups were cleaned by letting approximately 1 liter of 8N HCl seep through them. This acid treatment was followed by allowing 15 to 20 liters of distilled water to seep through and rinse the cups thoroughly. The cups are adequately rinsed when there is less than a 2% difference between the specific conductance of the distilled water input and the output from the cup."

Chemical reactions at the surface of a suction sampler may also clog the porous network. One type of chemical reaction is precipitation (e.g., of ferric compounds). However, considering the wide variety of chemical wastes which are disposed of at land treatment sites, other effects are also possible, leading to the inactivation of suction samplers.

Even though suction samplers may fail because of clogging, the problem may still be an operational difficulty which can be overcome. For example, installing silica sand around the cup may filter out particulate matter. Unless this filter becomes clogged, the samplers should continue to operate. This approach may not be sufficient to prevent clogging by chemical interactions.

Dazzo and Rothwell (1974) conducted laboratory studies to determine the extent to which fecal coliforms are strained during suction cup sampling. They observed a 100- to 10,000,000-fold reduction in fecal coliform in manure slurry during sampling, and 65 percent of the cups yielded coliform-free samples. Dazzo and Rothwell concluded that "... porcelain cup water samplers do not yield valid water samples for fecal coliform analyses."

Tsai, Morrison, and Stearns (1980) conducted laboratory studies to determined the attenuation of trace metals during movement of liquid through ceramic cups. They observed that nickel, copper, lead zinc, iron, manganese, and lead were consistently reduced by 5 to 10 percent during percolation through cups. Wagner (1962) found that NO_3-N was not adsorbed by cups, but sorption of NH_4-N did occur. Hansen and Harris (1975) found evidence that phosphorous was strongly adsorbed by cups, but concluded that sorption is only serious when "... The ceramic sorptive capacity is significantly greater than that of the adjacent soil." Ideally, the exact extent of these interactions should be accounted for in order to correct the chemical data obtained from cup samples. However, in light of the variability in the physical composition among cups, it is not possible to develop suitable predictive models at this time.

Because a major concern at disposal areas is the fate of organic pollutants, the effect of organic-cup interactions on sampling results is of paramount importance. According to Morrison and Tsai (1981), "Laboratory studies by Tsai, et al (1980) found that several pesticide species and BOD were substantially reduced when leached through a .32 cm thick ceramic cup with a pore size of 2.5 micron. Concentrations of the chlorinated hydrocarbons pp DDD', pp DDE', and pp DDT' were reduced 90%, 70%, and 94% respectively." Trace metals (e.g., nickel, lead, zinc, manganese, copper, iron, and magnesium) were consistently reduced 5 to 10 percent following percolation through the cups. Because of the attenuation potential of these samplers, polyethylene cups with 2.5- and 5-micron pore openings were tested in leaching experiments by

Morrison and Tsai (1981). Standard solutions of 12 water quality parameters were leached through polyethylene cups. Of the 12 standards leached, An, Ni, Fe, Ca, Ti, Cl, Na, and Ar remained unaffected. Cd and Ag were significantly reduced (5 to 10 ppm) and Se and Cr were slightly reduced (approximately 0.5 ppm). During field studies involving suction sampling versus soil-core sampling, Law Engineering Company (1982) found that organic waste constituents were detected by both techniques, whenever a liquid sample could be obtained in the cups.

Climatic factors--A major factor limiting the operation of suction samplers in frozen soils is that the soil-water suction is about -0.8 bar. This means that the air entry value of the samplers is exceeded and cannot be obtained. Another undocumented problem which conceivably could occur is freezing of samples within the cups and lines, so that the samples cannot be brought to the surface.

Another effect of freezing temperatures is that some soils tend to heave during freezing and thawing. Consequently, suction samplers may be displaced in the soil profile, resulting in a break in contact. In addition, if the cups are full of liquid when frozen, the cups may be fractured as a result of expansion of the frozen liquid.

Water Sampling from Saturated Regions of the Vadose Zone

Saturated regions of the vadose zone that can be monitored for water quality mainly comprise perched groundwater occurring at the interface of regions of varying texture. Perched groundwater bodies generally vary greatly in areal extent and may be either ephemeral or exist for long periods of time. For convenience, perched groundwater is arbitrarily subdivided into two categories: shallow perched groundwater within 30 feet of the surface, and perched groundwater at depths greater than 30 feet.

Since perched groundwater is in essence saturated groundwater, many of the techniques apply to monitoring in perched groundwater (e.g., Todd et al., 1976). A guiding principle for selecting monitoring methods for both the saturated zone and perched groundwater is aptly stated by Vanhof, Weyer, and Whitaker (1979):

> ... for an efficient, long-term operation of an operational monitoring network, the devices to be used must be simple, rugged, and foolproof. The inside diameter of the installed structure should allow for the insertion of down-hole tools for development, pumping, groundwater sampling, hydraulic head measurement, and completion zone maintenance. The installed equipment should be simple enough to be used by trained but not educationally skilled personnel.

Shallow Perched Groundwater Sampling--

Groundwater samples can be obtained from shallow perched groundwater by six methods: (1) tile lines, (2) collection pans and manifolds, (3) wells,

(4) piezometers, (5) multilevel samplers, and (6) groundwater profile samplers. Each is discussed in turn.

Sampling Tile Drain Outflow--

If a tile drainage system has been installed to control the elevation of a perched groundwater table, samples can be collected from the tile outfall. In some cases, it may be desirable to install commercially available composite or discrete sampling devices.

Willardson, Meek, and Huber (1973) discuss a "flow-path groundwater sampler" that enables collection of water in different flow paths around a tile.

Collection Pans and Manifolds--

Collection pans and manifolds can be installed in regions where temporary saturation lenses develop during the percolation of water for overlying surface sources. Generally, pans are installed to permit sampling pore water through a vertical profile, usually 10 to 20 feet in thickness. In construction, pan collectors are similar to the vacuum extractors except that a ceramic filter candle is not included. Thus, the media in the trough must be saturated.

Commonly, the collector pans are connected to a central chamber or trench containing sample bottles and accessories. McMichael and McKee (1966) used a large-diameter culvert as the central chamber in their studies on waste water reclamation at Whittier Narrows. Parizek and Lane (1970) used a trench to collect samples from pans in a forest soil during effluent irrigation studies in Pennsylvania.

The collectors developed by McMichael and McKee (1966) employed pan samplers for the collection of water during periods of saturation in a profile to about 10 feet. These samplers were installed to various depths at a radial distance of about 10 feet from the central well. Individual sampling pans were filled with gravel to prohibit clogging. Each sampler was connected to the central well via tubing. Lateral excavations used for installation of the samplers were carefully backfilled. During sampling, percolate intercepted by the pans flowed by gravity to the central well into collection flasks.

The lysimeter trench described by Parizek and Lane (1970) used pan lysimeters. The largest such lysimeter trench was 47 inches wide by 13 feet long excavated to about 20 feet. The sides of the trench were lined with wood and braced. Sampling pans were installed at 12-inch intervals to a depth of about 20 feet. The pans, constructed from galvanized metal, were 12 inches by 18 inches with copper tubing soldered into one end to permit sample collection. Each unit was placed into the soil a short distance from the side walls of the trench. While applying effluent during land treatment studies, samples of percolate intercepted by the pans were collected in flasks in the lysimeter trench.

A simple manifold collector was installed within a sanitary landfill by Wilson and Small (1973) for the collection of leachate (Figure 5-14). The

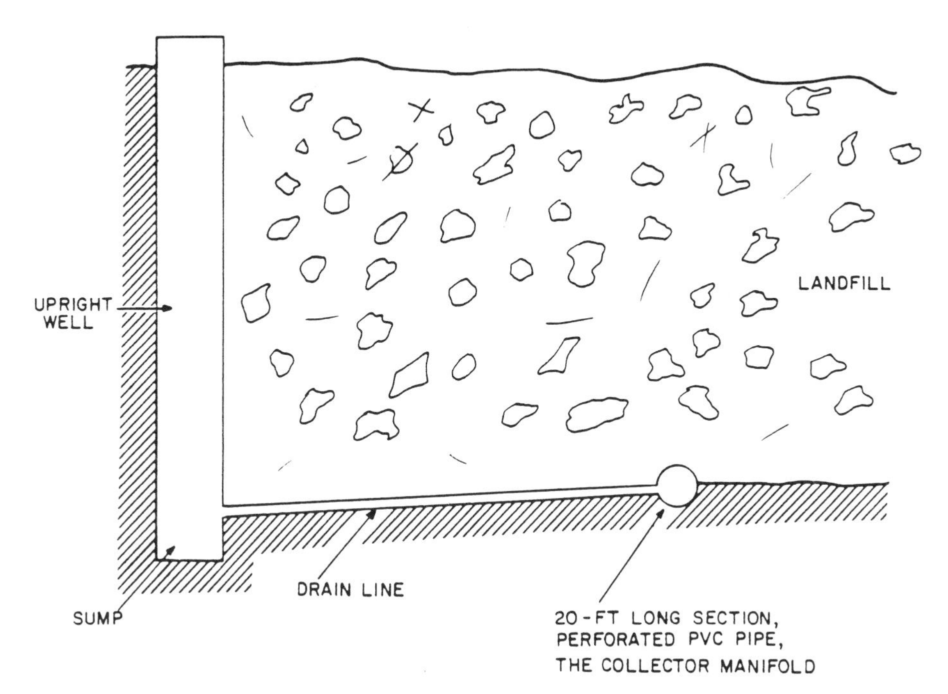

Figure 5-14. Leachate collector installed at base of sanitary landfill.

collector consisted of 20 feet of 2-inch diameter perforated PVC pipe functioning as a horizontal collector manifold connected by a section of PVC pipe to a 4-inch diameter length of vertical PVC pipe. The unit was installed in the base of a new landfill cell before introducing solid waste. The manifold was joined to the upright well about 4 inches above its base. The base of the well was sealed so that the bottom 4 inches functioned as a storage reservoir. The manifold drain line was sloped toward the well to facilitate drainage. Leachate collecting in the reservoir was hand bailed.

A basic problem with collection pans and manifolds is that samples are available only when the soil-water pressure is greater than atmospheric. During drainage, the soil-water system may shift rapidly to the unsaturated state, prohibiting sample collection.

Wells--

Wells installed to monitor water level fluctuations can also be used for sample collection, but they must be thoroughly developed if used for water quality monitoring. Development will ensure removal of drilling mud and foreign water used in the drilling process. Development should be continued until sediment-free water is pumped from the well (Geraghty and Miller, Inc., 1977). Techniques for well development are thoroughly discussed in Manual of Water Well Construction Practices (U.S. EPA, no date). Geraghty and Miller, Inc. also point out that monitoring wells must be correctly oriented with the flow direction of a source of polluted groundwater. A suggested practice is to locate at least one well upgradient of the source, one or more wells on site, and several wells downgradient of the source.

Observation wells should be unperforated throughout the vadose zone overlying the perched groundwater zone. Perforations below the water table should extend throughout the entire saturated thickness for reasons stated by (Geraghty and Miller, Inc., 1977):

> The vertical placement of the monitoring well screen within the aquifer is ... critical. For example, wells screened near the top of the zone of saturation may not detect contamination migrating in natural gradient flow paths along the base of an aquifer. In contrast, lighter-than-water constituents, such as hydrocarbons, may move along the top of the zone of saturation, thereby evading detection by monitoring wells screened only in the bottom portion of an aquifer.

It is generally recognized that wells with fully penetrating perforated sections provide information on the "bulk" or "average" concentration of pollutants in groundwater as a pumped unit (Pickens and Grisak, 1979). In many cases, a profile of pollutant distribution is also important because layering of constituents may occur. Other sampling units (e.g., piezometers, multilevel samplers, and profile samplers) are useful for determining pollutant stratification. Depthwise samples can also be obtained in wells using packer pumps, such as the Casee Sampler. Packer pumps are discussed in detail by Fenn et al. (1977).

Piezometers--

Piezometer wells used for characterizing the vertical head distribution in a perched groundwater mass can also be used for water sampling. Water samples from a piezometer array with individual well points terminating at different depths will indicate the vertical displacement of indigenous groundwater with the recharge source. Similarly, if a number of arrays are distributed in a horizontal transect, samples from individual units terminating at the same elevation will manifest quality changes during the lateral spread of a pollution plume.

Piezometer units in a given array may either be drilled separately or clustered in a common borehole. For the individual-type array, each piezometer should be thoroughly developed and cleaned in accordance with sound well construction practices. For the clustered units, development of the borehole could be difficult, but at any rate the gravel envelope around each well point should effectively prevent the entrance of fines into the well.

The location of individual arrays should be based on a thorough knowledge of the local geohydrology, such as the slope of the water table, hydraulic conductivity, and flow velocity. Generally, wells are installed at three locations: upgradient of the source, onsite, and downgradient of the source.

Multilevel Samplers--

Pickens et al. (1978) present details of a multilevel sampling device, shown in Figure 5-15. The sampler is particularly suited for sampling in cohesionless deposits with shallow groundwater in which flow occurs predominantly in the horizontal direction. The unit is installed in a single borehole and permits groundwater sampling at several levels. Hydraulic head measurements can also be obtained using a mercury manometer.

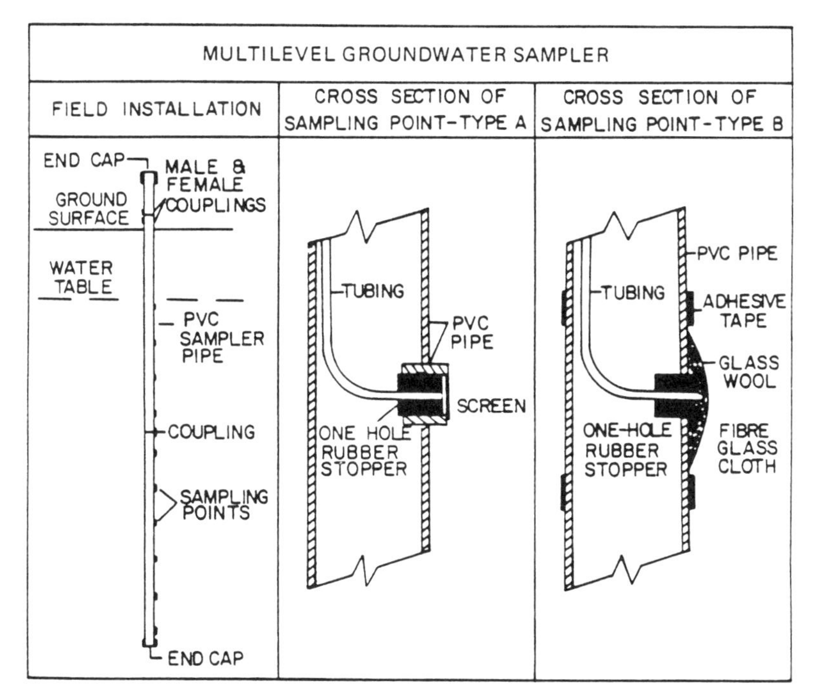

Figure 5-15. Multilevel sampler (after Pickens et al., 1978).

Basically, the sampler consists of PVC pipe, ports or openings at desired incremental depths, screened coverings on the openings, and polypropylene tubing sealed into the openings within the pipe. The polypropylene tubing extends to the surface. A unit may be designed in the field by locating the position of openings using stratigraphic information from drilling.

The pipe can be installed by common drilling techniques such as hollow-stem auger and cable-tool. Using the hollow-stem auger, the casing is installed within the auger when the desired depth is reached and the auger is withdrawn. For cable-tool construction, a steel casing is driven to the desired depth, the PVC pipe installed within the casing, and then the casing pulled out. For both methods, the deposits surrounding the borehole are presumed to collapse around the casing, ostensibly preventing side leakage. Vanhof, Weyer, and Whitaker (1979) question this assumption. However, in their paper, Pickens et al. indicate that sealing should not be a problem with cohesionless deposits. For installation in cohesive deposits, they recommend installation of bentonite or grout between openings to prevent cross-contamination.

A vacuum is applied to the polypropylene tubing to collect samples. The water can be pumped into an air-free cell for measurement of pH and Eh. Quick delivery from the aquifer to the cell probably minimizes degassing of the water.

One of the stated advantages of the multilevel sampler is that concentration profiles can be obtained at a lower cost than by installing piezometer nests. In a subsequent paper, Pickens and Grisak (1979) present comparative cost data to substantiate this contention.

Groundwater Profile Sampler--

Hansen and Harris (1974) designed a depthwise sampler, which they designated a "groundwater profile sampler." The sampler, shown in Figure 5-16, consists of a 1.25-inch diameter well point of optional length, with isolated chambers containing fiberglass probes. The individual chambers are filled with sand and separated by caulking compound. Small-diameter tubing provides surface access to the probes. Positioning of the probes is optional, depending on aquifer materials and desired sampling frequency. In operation, a vacuum is applied to the tubing, pulling groundwater into individual sampling flasks. Hansen and Harris recommend simultaneous extraction of all samples at the same rate to minimize variations in aquifer thickness sampled by the individual probes. Groundwater at depths to 30 feet can be sampled with this unit.

Field implementation--Tile drains, collection pans, and manifolds--Installation of tile drains, collection pans, and manifolds commonly involves substantial excavation depending on the site-specific geology/soil characteristics, goal of the monitoring program, and characteristics of the wastes involved. These monitoring techniques are usually incorporated in the preliminary design of waste disposal sites. Retrofitting a site with tile drains, collection pans, or manifolds is usually cost prohibitive. In addition, a health hazard for the installation crew exists when toxic wastes are present. Specific installation information for tile drains, collection pans, and manifolds is provided or referenced in Water Sampling from Saturated Regions of the Vadose Zone, this section. These sampling methods are commonly utilized for irrigated fields and at land treatment facilities.

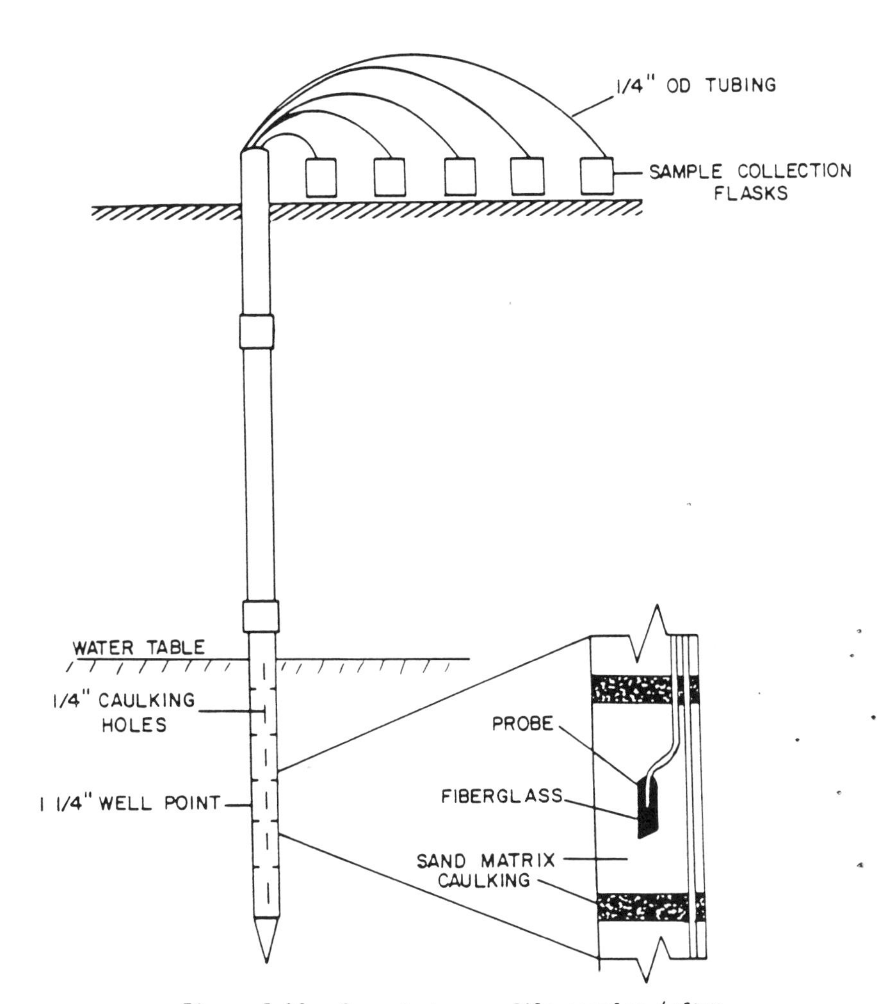

Figure 5-16. Groundwater profile sampler (after Hansen and Harris, 1974).

Wells--A water sample collection well consists of an uncased borehole or perforated pipe extending from ground surface into a perched water table. For shallow water tables, these wells can be installed by hand augers or simple drilling equipment. For deeper wells, standard drilling equipment is necessary. Development of cased and uncased wells by surging or pumping may be necessary to ensure free flow of water into and out of the well bore. Well design and sampling techniques are discussed by Kelly (1982) and Slawson et al. (1982).

Piezometers--Piezometers are cased with small-diameter pipes that are drilled into a saturated zone or a zone in which saturation is expected. Reeve (1965) discusses in detail common techniques for installing and cleaning new piezometers. In general, a tight fit between the outer wall of the piezometer and the surrounding media is essential. Shallow piezometers can be installed by augering and driving with a sledge hammer. Deeper units require jetting or use of standard drilling equipment. It may be necessary to grout the cavity between wells and boreholes to ensure tightness of fit. Piezometers should be developed by pumping or bailing to clean and open up the material at the base of the unit. The new generation of ultra-small-diameter submersible pumps will greatly facilitate cleaning and sample collection from piezometers.

Multilevel samplers--Multilevel samplers are installed using equipment and techniques similar to those for piezometer installation (Reeve, 1965).

Profile samplers--Profile sampler installation is similar to that of the piezometer. Reeve (1965) discusses appropriate installation techniques.

Range of applications and limitations--Each of the six methods for obtaining groundwater samples from shallow perched water tables has geologic, hydrologic, chemical, and physical limitations that restrict its application. Discussions of specific characteristics that limit their utility follow.

Tile drains--In well-structured geologic and soil media, pollutant streams move more rapidly in cracks than in blocks. If drains do not intersect a significant percentage of the fractures, breakthrough of pollutants can be significantly delayed. Pollutants that are heavier than water may move along the bottom of the perched water zone and not appear in the tile drain outflow. Chemical and biological reactions occur at the air-water interface of the tile joints causing precipitation, oxidation, and modification of the chemical nature of the outflow. These reactions at the tile line joints may cause clogging and eventual line failure. Freezing and thawing cause heaving in the soil structure and will affect the flow to and in tile lines. Tile drains cannot be installed in steep terrain and for economic reasons they are limited to shallow depths.

Collection pans and manifolds--In rocky or steep terrain, collection pans or manifolds may be difficult to install. Because of the spatial variability found in rocky and other nonhomogeneous soils, sample representativeness cannot be assured without a large number of data collection points. Trenches used to install these sampling units may affect soil permeability and modify flow patterns, thus further affecting the samples obtained. Collection pans and manifolds, as well as tile drains, will function only when saturated flow conditions exist. These units are all inoperative when flow becomes unsaturated.

Collection pans and manifolds cannot be installed to great depth in the vadose zone and they are quite difficult to install beneath existing waste disposal facilities. Following installation, inlet holes in manifolds may become clogged by chemical precipitates. Liquid wastes can interact with material in collectors, thereby introducing nonrepresentative substances. Field personnel should be aware of these characteristics before selecting collection pans or manifolds as a principal monitoring technique at a given site and during the interpretation of resulting data.

Wells--Wells are a common method of obtaining groundwater samples within the vadose zone. In fractured rock systems, care should be taken to ensure that well installation intercepts as many cracks and water-bearing joints as possible. Generally, several wells are required to determine trends in water levels and direction of flow. Also, in fractured systems, flow trajectories are not necessarily perpendicular to piezometric contours, especially when flow in joints prevails (Sowers, 1979).

In some regions, perched groundwater may be ephemeral. Thus, a well may become dry and unavailable for sampling part of the year. Sampling schedules should be adjusted to meet these conditions.

Proper placement of monitor wells requires information on hydraulic gradient and flow direction of the perched groundwater body. In areas where flow is vertically downward in wells, surface pollutants may be short circuited into deeper zones. In these areas, caution must be used in developing monitoring wells to minimize side leakage through the well annulus. Special well development techniques discussed by Kaufmann et al. (1981) are required to develop wells in existing waste disposal sites or at landfills. In all cases, monitor wells should be perforated throughout the entire saturated region of the vadose zone to ensure that upper and lower pollutant plumes are sampled. Additional packer tests are required to determine layering characteristics if pollutants are detected within a well.

Chemical characteristics of perched groundwater can be altered by monitor well development. Well casings may contribute chemicals. For example, metal well casing could contribute heavy metals. Cadmium, manganese, and possibly organics from joint cement could be introduced by PVC wells. Water quality in shallow perched systems is, in general, highly variable. Heavy rainfall may cause recharge of good quality water, thus diluting a pollutant plume. Exposure of the water table and well water to the atmosphere, are likely to cause oxidation reactions, altering the chemistry of the samples. Adequate well discharge before sampling and appropriate techniques for obtaining representative samples from these types of monitor wells are discussed by Kelly (1982). Conversely, frozen soils during the winter months may prevent the movement of air into the vadose zone, causing reduction conditions and affecting the nature of pollutants. Representative samples may, therefore, be difficult to obtain. Schuller et al. (1981), who have evaluated the oxidation-reduction and dilution characteristics in the vadose zone, recommend frequent sampling throughout the year to obtain truly representative samples.

Piezometers--Many of the problems and limitations associated with well development are shared by piezometers. Locating well points in fractured zones is difficult. Proper placement of the units requires knowledge of the geohydrologic framework, e.g., hydraulic gradients, flow paths, and geologic structure. Piezometers are difficult to install in hard formations without using special drills. Flow trajectories are not necessarily perpendicular to piezometer contours. In addition, sampling may affect the flow regime around the well point. Perched groundwater may be ephemeral with sampling units failing during the dry periods. Conversely, rising water table conditions, may require additional wells for sampling new perched groundwater lenses.

Chemical reactions between the well casing and pollutants may occur, releasing nontypical substances. Oxidation environments may occur near the water table, requiring prolonged pumping before sampling. As with shallow wells, dilution of pollutant plumes with surface infiltration from heavy rains may produce nonrepresentative samples. Without careful construction, freezing soils may breach the borehole-casing seal, allowing surface fluids to penetrate. These fluids could contaminate borehole samples and then move laterally away from the piezometer. In addition, aboveground parts of piezometers completed with PVC should be protected from sunlight. Freezing conditions during winter months may prevent air movement to shallow groundwater zones so that anaerobic reactions occur, affecting the chemical characteristics of the pollutant plumes. Assessments of these conditions are required to characterize the "average" water quality with samples being collected at intervals throughout the year.

Clustered piezometers in a common borehole or separated units installed at varying depths are required to determine layering of pollutants. These wells should be thoroughly developed following installation. Special drilling techniques are required to install piezometers near existing landfills and waste disposal areas.

Multilevel samplers--The assumption that the surrounding medium of multilevel samplers is cohesionless material that collapses around the pipe and prevents side leakage may not be true for all materials. Therefore, installation of these samplers is not recommended. Units are limited to shallow depths, less than the suction lift of water (about 25 feet). Unlike wells or piezometers, there is no effective method for developing multilevel samplers following installation. Thus, because of aquifer disturbance during drilling, a prolonged time period is required to ensure that samples are representative of the undisturbed conditions (Pickens et al., 1981).

Chemical reactions similar to those discussed for wells and piezometers may occur in multilevel samplers. In addition, sample degassing may occur as a result of the sampling technique.

Residual fluids in delivery tubes of multilevel samplers may freeze during winter months. Soil heave from freezing and thawing may loosen soil around the casing, introducing the possibility of cross-contamination of samples and short circuiting of surface fluids to depth. Since sunlight will render PVC brittle, aboveground parts of the casing must be protected.

Multilevel samplers may be unsuitable in fractured material because of side leakage and difficulty in locating sampling ports opposite fractures. They are difficult to install in well-indurated formations. Installation processes may affect permeability near the sampling ports, resulting in long periods of time for return of "natural" conditions.

In regions with fluctuating water table levels, sampling ports can be placed in the vadose zone to sample recharge waves with differing saturation thicknesses without further well installation. Because flow in perched groundwater systems is predominantly horizontal, multilevel samplers can provide valuable information on layering characteristics of pollutant plumes.

Profile sampler--Profile samplers are similar to multilevel samplers but differ in design and construction of the sampling ports. Both units obtain samples by applying a vacuum to the sample port, thus limiting sample depth collection. Degassing of the sample can occur when the vacuum is applied. Originally, profile samplers used fiberglass probes that gradually clogged. New units have alleviated part of this problem by using polyethylene screens.

Installation of these units requires driving a well point to the desired depth. This may not be possible in hard or rocky material. In highly structured geologic settings, alternate probes may sample from either within blocks or fractured zones. Interpretation of the resulting data may be difficult. The units may not be suitable in fractured media because of difficulty in locating probes opposite to joints and fractures.

Chemical reactions between the well casing and the wastewater may occur. These samplers should not be used to sample heavy metals such as iron and zinc (Hansen and Harris, 1980). Cold weather renders units inoperative by freezing residual water in discharge tubing. Freezing and thawing may also loosen soil around wells, causing downward leakage.

In using the units, all ports must be sampled simultaneously to avoid influencing natural flow. However, even with care, fluids may short-circuit between sampling probes as a result of sampling. Installation processes may affect local permeability around probes and, therefore, long time intervals for recovery to "natural" conditions is required. According to Hansen and Harris (1980), this problem is apparently not as severe as for the multilevel sampler. Profile samplers are best suited for regions of horizontal flow.

Well Casing Material--

Two important considerations when designing wells and piezometers for water quality monitoring are casing material and diameter. Materials available for well casing include plain or galvanized steel, PVC, fiberglass, and Teflon. Constitutents being monitored will influence selection of casing materials at specific sites. For example, when monitoring for salinity where only the major constituents are being evaluated, plain or galvanized steel can be used. When monitoring for trace metals, contamination from steel casing may preclude the quantification of specific constituents, such as iron and zinc. Because of cost, PVC is commonly used, but has the disadvantage that it cannot be driven into place when using cable-tool drilling.

The type of casing material to install when monitoring for organic pollutants is controversial. For example, Dunlap et al. (1977) oppose the use of PVC pipe, stating:

> In some earlier work ... PVC casing was utilized for casing of sampling wells. This material is relatively inexpensive and easy to use, but it is less desirable as a casing material than the Teflon tubing-galvanized pipe combination for two reasons. First, organic constituents of groundwater may be adsorbed on the PVC casing. Second, there is evidence that PVC casing may contribute low levels of organic contaminants to the samples, such as phthalic acid esters used as plasticizers in PVC manufacture and solvent from cements used to join lengths of PVC tubing.

For their sampling wells, Dunlap et al. used a combination of galvanized pipe and Teflon tubing. The galvanized pipe extended from ground surface to about 1 foot above the water table. Teflon tubing was used in the portion of the well below the water table.

In contrast to the findings of Dunlap et al. regarding interaction of PVC pipe with organic constituents, Geraghty and Miller, Inc. (1977) indicates the following:

> PVC pipe was used for all well casing. It is light and easy to handle and is more inert toward dissolved organic substances than steel casing. The iron oxide coating that develops on steel casing has an unpredictable and changeable adsorption capacity. However, when the adsorption sites on PVC are saturated, water remains in equilibrium with it. Leakage of organic compounds from PVC is negligible. As a control, samples of pipe and a cemented joint were submitted to the laboratory where they were soaked in water and the water was analyzed. No contaminants were detected.

No general recommendations are available on selecting well casing material when monitoring for microorganisms and virus. Conceivably, metal constituents in steel wells could affect the growth of microorganisms as dissolution of the casing occurs. For this reason, PVC well casings are probably more suitable for sample collection when microbial levels are being evaluated. It is also probably advisable to disinfect the well following well completion. Chlorine in granular or liquid form is commonly used as a disinfectant. Manual of Water Well Construction Practices (U.S. EPA, no date) gives details on disinfection practices. Chlorine must be used carefully when excessive levels of organic constituents are present because of the possible formation of chlorinated hydrocarbons. In addition, the well should be thoroughly pumped after a suitable contact time using sterilized pumping equipment.

Selection of well casing diameter should be based on the water extraction method, quantity of sample required, budget limitations, and other uses of the well. In all cases, the most cost-effective wells are preferred. A number of 1.50-1.88-inch outside diameter submersible pumps are now commercially available. These small-diameter pumps are preferred when: (1) small-diameter wells are in-place, (2) samples of less then 15 gph are required, (3) drilling equipment for small diameter wells is available, and (4) all monitoring requirements can be met. If large sampling quantities are required and a submersible pump is to be installed, the minimum diameter of the casing should be 6 inches, and 8 inches is preferred (Schmidt, personal communication, 1979). If other extraction methods such as hand bailers, gaslift pumps, or peristaltic pumps are suitable, smaller-diameter wells can be used. Peristaltic pumps have the additional requirement that pumping depth be less than 25 feet from the land surface. However, many wells are also used for monitoring water level changes and permanent fixtures such as gas pressure lines may occupy part of the interior space. For practical reasons, therefore, the casing diameter should probably not be less than 4 inches.

Deep Perched Groundwater Sampling--

Several techniques are available for extracting groundwater from deeper parts of the vadose zone. These methods, equally effective for piezometers and wells, have been reviewed by Fenn et al. (1977) and Gibb et al. (1981) in detail. They include (1) hand bailers, (2) airlift and gaslift pumps, (3) suction-lift pumps, (4) piston pumps, (5) submersible pumps, and (6) swabbing.

Hand bailers--The simplest form of hand bailers consists of a weighted tube to overcome buoyancy that is closed at the bottom. To collect a sample, the bailer is lowered into a well below the water table. The filled container is then raised to the surface.

Figure 5-17 shows a more elaborate type of bailer. This unit consists of a tube open at both ends but with a one-way check valve at the base. Basic units, such as the one shown in the figure, use a glass marble as a valve. In operation, the unit is lowered below the water table. Water flows up and into the sampler body. When the unit is raised, the glass marble seats, plugging the end of the bailer. Bailers can be made from various materials such as steel pipe, PVC tubing, or Teflon tubing, depending on the constituents being examined. The unit shown in the figure is made of Teflon to sample highly volatile organic pollutants (Dunlap et al., 1977).

Another type of hand bailer is the "Kemmerer" sampler. This unit is commonly used for sampling in lakes but is also suitable for groundwater sampling. It consists of a cylinder with spring-loaded rubber stoppers at each end; it is lowered to the desired depth below the water table and a weighted "messenger" is sent down a cable to spring shut the stoppers and trap the sample.

Airlift and gaslift pumps--Figure 5-18 illustrates a basic airlift pump. Compressed air introduced into a central pipe bubbles up through the column of water in the well. The specific gravity of the column of water is consequently reduced and an air-water mixture is discharged from an eductor pipe.

In the case shown, the well casing is the eductor pipe. The principles of airlift pumping are discussed in Ground Water and Wells (E.E. Johnson, Inc., 1966) and by Anderson (1977), and Trescott and Pinder (1970). As pointed out in Johnson (1966), the airlift method is particularly effective when the submergence ratio (the percentage of the total length of air line that is below water when pumping) of the air line is about 60 percent. For example, if the air line is 180 feet long and the water level when pumping is at 74 feet, the submergence ratio is (180-74)/180 or 59 percent. Favorable pumping rates are possible with as little as 30-percent submergence.

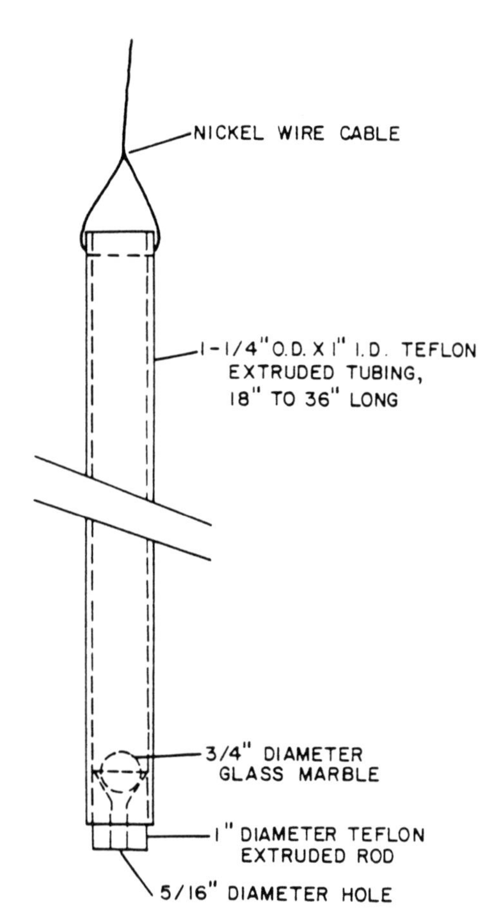

Figure 5-17. Hand bailer made of Teflon (after Dunlap et al., 1977).

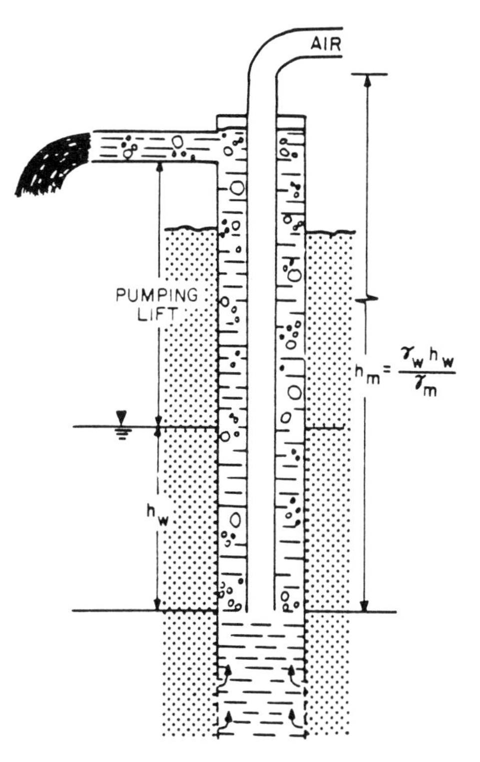

h_m = MAXIMUM HEIGHT TO WHICH THE AIR-WATER MIXTURE WILL RISE

h_w = SUBMERGED LENGTH OF THE AIR LINE

γ_m = SPECIFIC WEIGHT OF THE AIR-WATER MIXTURE

γ_w = SPECIFIC WEIGHT OF WATER

▼ = POTENTIOMETRIC SURFACE

Figure 5-18. Schematic representation of simple airlift pump showing principle of operation (after Trescott and Pinder, 1970).

Trescott and Pinder (1970) developed a simple airlift system for sampling from small-diameter piezometers. An automobile engine was used to provide air pressure for the pump. Components of the system include: a ball check-valve, an air line, a pressure gauge, and a header assembly on the well. The check valve is screwed into a spark plug socket of an engine cylinder. Air from outside the engine is sucked into the cylinder during the inlet stroke. Air is then forced into the air line during the compression stroke. The check valve maintains pressure in the air line. The header assembly at the top of the well casing contains an inlet opening, through which the air line is inserted, and a discharge pipe. This type of device, however, appears inappropriate when monitoring for organic materials because of fuel and lubricant residues from the engine.

The discharge rate for airlift pumps increases with increasing pipe diameter. If sufficient air is available and the submergence rate is favorable, enough water can be extracted by the airlift method for sampling purposes. The system can also be used for the initial development of the well and for flushing prior to sampling.

Gronowski (1979) describes a commercial pump operating on the same principle as the airlift pump except that the pumping energy is obtained from 14-ounce propane cylinders. According to Gronowski, "The unit consists of dual-conductor plastic tubing with a valve assembly on top for attachment to the propane cylinder. A shaped copper tube serves as the intake assembly. Propane injected down one side of the tubing forces water to rise up the other side to ground surface. Several samples of adequate volume can be collected with one 14-ounce propane cylinder."

Bianchi, Johnson, and Haskell (1962) designed an airlift pump, which in effect is an automated hand bailer. As shown in Figure 5-19, the unit consists of a cylinder with a flat check-valve at the base and a two-hole rubber stopper at the upper end. A plastic delivery tube is inserted through one of the holes and extended to the lower end of the cylinder. A second plastic pressure tube is inserted into the other opening. In operation, the cylinder is lowered below the water table and water flows upward through the valve. Air pressure is then applied to the shorter line, the one-way valve shuts, and water is forced up the longer line to the surface. A tire pump can be used to apply pressure. Samples have been lifted with this unit from depths greater than 100 feet.

Suction-lift pumps--The basic principle of suction-lift pumps is that, when a vacuum is created inside a tube or pipe extending below the water table, air pressure on the water outside the tube forces water to rise inside the tube (Anderson, 1977). The practical suction lift with such pumps is about 25 feet. Allison (1971) discusses a simple suction-lift pump for extracting groundwater from auger holes that can also be used for cased wells. The unit consists of plastic tubing of sufficient length to extend below the water table. The upper end of the tube terminates in a collector flask stopper. A second line from the flask stopper is connected to a hand-held vacuum pump. By applying vacuum to the system, water is drawn into the flask.

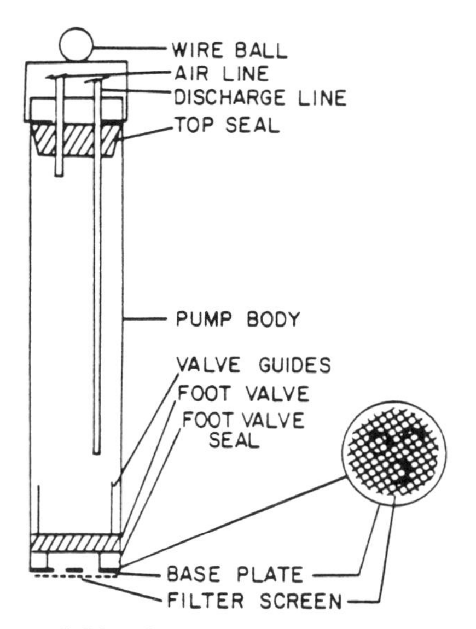

Figure 5-19. Positive action airlift pump (after Bianchi, Johnson, and Haskell, 1962).

A variation of the suction-lift pump is the peristaltic, or squeeze, pump. The pump consists of a rotor equipped with eccentric rotating heads at both ends. The rotor is mounted in a closed housing. Plastic tubing is placed between the rotor heads and the housing such that when the rotor turns, the plastic tubing is squeezed by the rotating heads. The tubing is lowered into a well below the water table and primed with water. When the unit is activated, the squeezing action of the heads produces a negative pressure on the water in the tubing extending into the well. At the same time, a positive pressure is exerted on the water in the tubing leaving the housing.

Piston pumps--Piston pumps consist of a cylinder containing piston that is driven up and down by a rod. The lower end of the cylinder contains an inlet check valve. A one-way check-valve is also mounted in the piston.

The operation of the pump is described in E.E. Johnson, Inc. (1966):

> As the piston moves downward, the intake valve closes at the instant that the pressure above it exceeds the pressure below it; and the discharge valve opens when the pressure below it exceeds that above it. Water thus is trapped in the cylinder during the downstroke of the piston, then is forced upward into the discharge pipe on the next upstroke.

During installation, the cylinder is attached to a length of piston rod inside a length of pipe. The pipe functions as an eductor tube. Successive lengths of pipe and rod are added until the cylinder is located at the desired depth below the water table. When the water table is not in excess of about 10 feet, the pump can be operated by hand. For deeper water tables, a tripod and pulley arrangement may be necessary. Alternatively, a pitcher pump could be installed on the top of the well.

Submersible pumps--Submersible pumps comprise a sealed centrifugal-type electric pumping plant that operates below the water table. Water is delivered to the surface by tubing.

McMillion and Keeley (1968) describe a portable submersible pump suitable for sampling in wells greater than 4.5 inches inside diameter. The components of the unit are a submersible pump, 300 feet of flexible hose, electric winch, and portable generator. The hose is unreeled by the winch to position the pump at the desired depth and the pump is activated. Pumping rates range from 7 to 14 gallons per minute.

Swabbing--Swabbing is commonly used in oil field operations and has been adapted for sampling wells or piezometers. The swabbing technique involves introducing a swabbing cup into the well, which is supported from the surface by a rod. As the leather swabbing cup is lowered to the desired depth, water flows past it. It opens as it is drawn out of the well, lifting the water to the surface.

Field implementation--Sound monitoring practice requires that a monitor well or piezometer be thoroughly flushed before sample collection. Groundwater within a passive monitor well tends to be affected by exposure to the atmosphere and by mixing within the casing. (Passive monitoring is defined as periodically sampling wells in the path of groundwater flow for changes in concentration of the constituents of interest (Hart Associates, 1979)). Consequently, the water in the casing should be completely displaced to obtain a truly representative sample of water from a perched region.

Fenn et al. (1977) thoroughly review procedures for flushing a well. The principal recommendations are cited below:

> To safeguard against collecting nonrepresentative stagnant water in a sample, the following guidelines and techniques should be adhered to during sample withdrawal: As a general rule, all monitoring wells should be pumped or bailed prior to withdrawing a sample. Evacuation of a minimum of one volume of water in the well casing and preferably three to five volumes is recommended for a representative sample. In a high-yielding groundwater formation and where there is no stagnant water in the well above the screened section, evacuation prior to sample withdrawal is not as critical. However, in all cases where the monitoring data are to be used for enforcement actions, evacuation is recommended.

For wells that can be pumped or bailed to dryness with the sampling equipment being used, the well should be evacuated and allowed to recover prior to sample withdrawal. If the recovery rate is fairly rapid and time allows, evacuation of more than one volume of water is preferred.

For high-yielding monitoring wells which cannot be evacuated to dryness, bailing without prepumping the well is not recommended; there is no absolute safeguard against contaminating the sample with stagnant water. The following procedures should be used:

(a) The inlet line of the sampling pump should be placed just below the surface of the well water and three to five volumes of water pumped at a rate equal to the well's recovery rate. This provides reasonable assurance that all stagnant water has been evacuated and that the sample will be representative of the groundwater body at the time. The sample can then be collected directly from the pump discharge line.

(b) The inlet line of the sampling pump (or the submersible pump itself) should be placed near the bottom of the screen section, pumped approximately one well volume of water at the well's recovery rate, and the sample collected directly from the discharge line.

A nonrepresentative sample can also result from excessive prepumping of the monitoring well. Stratification of the leachate concentrations in the groundwater formation may occur, and excessive pumping can dilute or increase the contaminant concentrations from what is representative of the sampling point of interest.

At existing waste disposal sites, standard drilling, coring, augering, and boring for deep vadose zone groundwater sampling wells can cause short circuiting of pollutants to greater depths. Kaufman et al. (1981) discuss a driven pile method for well completion in hazardous waste areas. The method consists of simultaneously driving two 10- and 12-inch diameter steel-cylinder piles, with 0.5-inch wall thickness, one inside the other. The pile assembly is driven to the desired depth or until refusal and the inner pile is withdrawn, allowing space for a 5-inch diameter well completion. The completion is within the pile or, with additional drilling, it can be developed below the base of the pile. Following installation of the sampling well casing, screen, sand pack, and seals, the 12-inch-diameter cylindrical pile is removed. Kaufman et al. (1981) comment on the use of driven piles as a method of penetrating hazardous waste material areas as follows:

1. The method was considered extremely safe insofar as no waste or leachate was removed from the monitoring installation. Routine safety measures for explosive gases were followed when welding or cutting operations were underway.

2. Wastes were readily penetrated for the full thickness of 25 to 60 feet at three locations.

3. Yield of the 5-inch diameter wells, fitted with 6 feet of stainless well screen, was less than 1 gpm and considerably less than expected. Possible causes for this include low yield of the wastes, smearing of the borehole wall as the pile advanced, and/or compaction of the wastes by pile driving or removal.

Following penetration to the desired depth and completion of the sampling well, the preferred sampling method is selected for obtaining the groundwater sample.

In deep vadose zones, such as those of western alluvial basins, encountering one or more regions of perched groundwater in the lower vadose zone is not unusual. Water samples from these deeper saturated regions reflect the quality of water that has moved below the soil zone en route to the water table. In many cases, large quantities of water can be extracted from these deeper perched groundwater zones. Furthermore, the samples may reflect the integrated quality of water draining from an extensive portion of the overlying vadose zone. This will be particularly true when the overlying source is diffuse, such as irrigation return flow. Because of the integrated nature of the water in these perched regions, samples may be obtained using one or more wells at a fraction of the cost of installing batteries of suction cups.

Wilson and Schmidt (1979) present data from a number of case studies on monitoring perched groundwater. The sampling techniques reviewed here are discussed in detail in their paper. Two methods are possible for sampling perched groundwater: cascading water, and special wells. Generally, perched groundwater cascades into wells through poorly welded joints, cracks, or perforations exposed as the water table recedes. Figure 5-20 illustrates a cross section of a well with cascading water. Samples of cascading water could be obtained from abandoned wells in the vicinity of the source being monitored. Alternatively, cascading water could be obtained from operating wells whenever pumps are removed for servicing.

Samples of cascading water can be obtained by lowering a sample bailer to the desired depth. In sampling for chemical constituents, the bailer and sample bottles should be rinsed several times with cascading water during sample collection. To sample for microorganisms, a sterilized container should be used. Because cascading water is exposed to the atmosphere, some problems may be experienced in determining pH, dissolved oxygen, and oxidation-reduction potential.

Range of applications and limitations--The geologic, hydrologic, chemical, and physical applications and limitations for deep perched groundwater sampling are similar to those discussed earlier for shallow wells and piezometers.

Hand bailers are effective for obtaining groundwater samples in small-diameter wells and piezometers. The open-top model shown in Figure 5-17 can

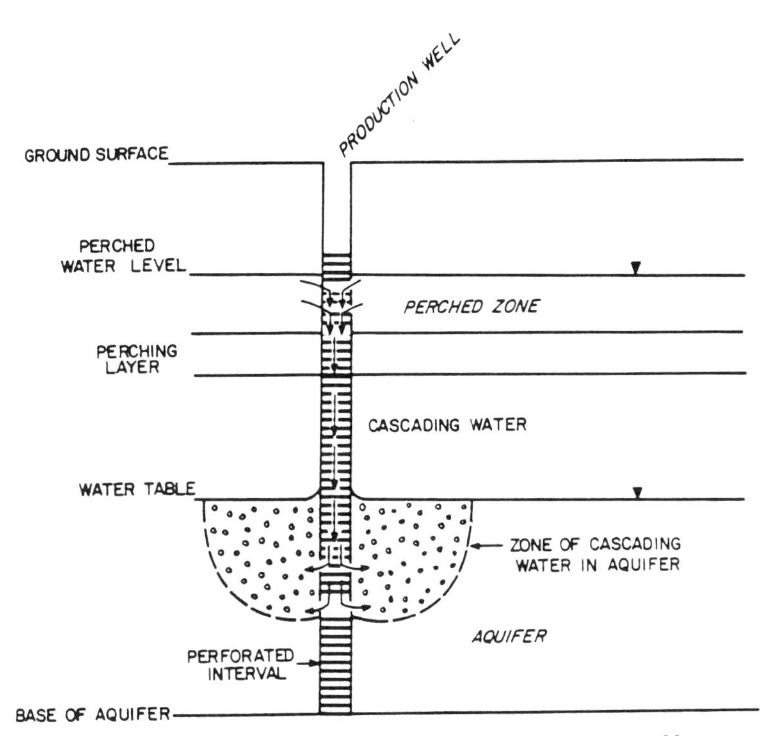

Figure 5-20. Conceptualized cross section of a well showing cascading water from perched zone (after Wilson and Schmidt, 1979).

be used for collecting cascading water as well as depth-specific samples. Although hand bailers can be used at great depth, bailers with power winches greatly reduce field time for multiple deep well sampling.

Field studies by Slawson et al. (1982) have shown extreme variations in water quality data when samples are not obtained at consistent depths. Stratification in water quality is a particular problem for bailers because they are not effective for flushing wells. Further modification of sample chemistry can occur when transferring the sample from the bailer to the sample container.

Before sampling for microorganisms, bailers should be sterilized by autoclaving and properly wrapped for transportation to the field. Field crews should be continuously alert for unnatural constituents introduced through interaction with the ground surface and well casing. Recommended procedures for bailer sampling are given in Slawson et al. (1982).

Air- and gaslift pumps introduce air or other gas into the sample. Consequently, extracted samples are unsuitable for determining dissolved gasses and many unstable constituents in groundwater. For example, Gibb, Schuller, and Griffin (1981) have shown that nitrogen and airlift methods can increase the pH and later the concentrations of several chemical constituents. Zinc and iron are particularly sensitive to this sampling method. The effect on water constituents when using a gaslift propane pump, should be examined before the method is employed in a monitoring program.

When sampling in deep wells using a portable air compressor and the well casing as the conductor tubing, unnatural constituents may be introduced into the sample through interaction with dry and wetted portions of the casing.

Field equipment for a suction-lift pumping is very inexpensive, requiring only a hand-held vacuum pump, sampling flask, and conductor tubing. Although the method is limited to a depth of about 25 feet, it works well in small-diameter wells and piezometers.

An advantage of the suction-lift pump is that water is in contact only with the conductor tubing and sampling flask. Contamination of the water sample with pump accessories is thus minimized. Dunlap et al. (1977) used suction-lift peristaltic pumps to sample groundwater for organic pollutants. Degassing of the sample may occur, however, when the vacuum is applied, thus altering the chemical makeup of unstable constituents. Chemical data from Gibb, Schuller, and Griffin (1981) indicate that degassing is not a problem for the peristaltic pump.

A basic problem with suction-lift pumps, especially peristaltic units, is that discharge rates are very low. An additional problem for peristaltic pumps is adding foreign water when priming the unit unless hand-bailed groundwater is used.

Piston pumps are effective for developing a new well or providing water samples. If the volume of water stored in the well is not excessive, piston pumps can be used for flushing the casing before to sampling. A problem with piston pumps for routine sampling is that installation time is excessive compared to more portable units. An expensive alternative would be to permanently install a separated unit in each monitoring well. Piston pumps may introduce heavy metals from the cylinder, thus contaminating the pumped samples.

Submersible pumps are a common method of sampling perched water in the vadose zone. The units can be used to develop, flush, and provide depth-specific samples. Little maintenance is required and they are easy to install and withdraw from shallow wells. Discharge can be easily controlled and both very low and very high flows can be obtained. They have relatively little effect on the native water quality.

The primary disadvantage of the submersible pump is the minimum size requirements of the well annulus. Standard pumps require a minimum 4-inch-diameter well for shallow sampling efforts and a 6- to 8-inch-diameter well for deeper sampling. This problem has been alleviated by the development of the small-diameter Bennett submersible pumps. Two models are available with outside diameters of 1.8 and 1.4 inches. These units are driven by compressed air and can discharge 0.25 and 0.5 gallons per minute against a 500-foot hydrostatic head at an operation pressure of 150 psi. The pumps were developed for use in standard core holes and small-diameter piezometers and wells.

Swabbing is effective where the depth to water is relatively great and well diameters are relatively small. It is an effective way to remove one well-bore volume from a well, but the volumes of water obtained and discharge rates cannot be regulated. Contamination is common when oil-field equipment is used for deep sampling and potential for cross-contamination exists. The method is difficult to employ and requires a crew of about four men. It has been found to accelerate plugging of piezometer perforations, especially in small-diameter wells. Consistent water quality sample collection is difficult to achieve due to the vertical mixing of well water during extraction. After reviewing its application in the field, Slawson et al. (1981) do not recommend swabbing as a sampling technique.

If wells with cascading water are not present in the vicinity of a perched water source, constructing special wells will be necessary (Figure 5-21). Methods for locating existing or potential perching regions include drillers' logs, grain size of drill cuttings, and borehole geophysical methods.

Preferred casing materials depend on the constituents of interest in the monitoring program. Thus, if metals and possibly organic pollutants are of concern, PVC casing may be advisable. Inasmuch as PVC casing cannot be driven, rotary drilling will be required for installation when the region of interest is deep. If the region is not too deep, however, it may be possible to use cable-tool drilling by driving steel casing into the borehole as drilling progresses and, when the desired depth is reached, lowering PVC casing into the steel casing and removing the steel casing. The PVC casing should be preperforated throughout the region of interest.

The ideal casing size for monitoring wells is 8 inches (Schmidt, personal communication, 1979). This size permits easy installation and removal of a portable submersible pump for sampling and permits installation of a water level recorder.

When several perching layers occur, a cluster of sampling wells can be installed within a common borehole. Methods for clustering wells are discussed in Section 4 (see multiple piezometer installations).

Perched groundwater zones do not occur everywhere and may not be present at a specific area of interest. The detection and location of perched groundwater may be expensive, involving the construction of test wells or requiring the use of geophysical methods. Even when found, some perched groundwater bodies may be ephemeral and certain perching regions may be more responsive to recharge from a given source of pollution than other regions. Thus, the location of successful monitoring wells may, in part, be a matter of chance.

Water in cascading wells may enter the casing from a number of perching regions. Consequently, the water collected below the uppermost region may be a mixture of water from various depths. Such a blend, however, could provide useful information.

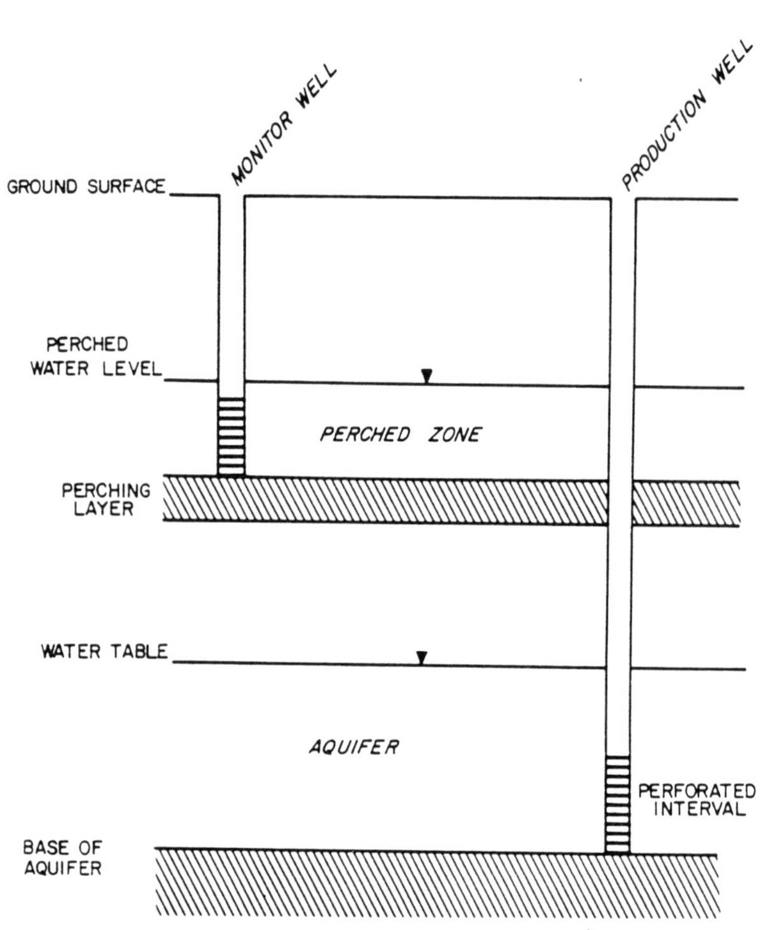

Figure 5-21. Conceptualized monitor well used to sample from a perched zone (after Wilson and Schmidt, 1979).

6. Waste Source Pollutant Characterization

Categories of sources of pollutants were generically reviewed to approximately determine kinds of pollutants that might be expected from point sources. Contamination sources were categorized as follows:

1. Municipal wastes
   - Solid waste disposal sites
   - Wastewater treatment plants
2. Agricultural wastes
   - Herbicide industry
   - Fertilizer industry
   - Feedlots
3. Oil industry wastes
   - Oil production fields
   - Oil refining
4. Industrial wastes
   - Iron and steel manufacturing
   - Inorganic chemicals manufacturing
   - Leather tanning and finishing.

Tables show the typical types of pollutants and their concentrations that might be expected to enter a landfill, waste pile, impoundment, land treatment area, or injection system for each subwaste category. The tables are intended to assist in developing vadose zone monitoring programs in disposal areas.

MUNICIPAL WASTES

Municipal wastes consist of solid waste (e.g., sanitary landfills), and wastewater (e.g., sewage effluent, sludge).

Solid Wastes

Solid waste includes a variety of pollutants. These pollutants and associated leachate vary from landfill to landfill depending on the type of material being accepted at the disposal site and the industrial base in the surrounding area.

Table 6-1 lists substances associated with leachate from solid waste disposal sites. This list is widely used for leachate indicators (Chain and DeWalle, 1975), although all of these substances may not be represented at a single site. A range of concentrations typical for some substances selected from this list is presented in Table 6-2. Three sources of leachate quality data are provided.

Wastewater

Pollutants associated with municipal wastewater collection and treatment facilities result from toxic compounds that interfere with, pass through, or are otherwise incompatible with the treatment plant operations. The chemical compounds or elements that have been identified as toxic are listed in Table 6-3. These chemical compounds or elements are termed priority pollutants and must be investigated during treatment plant operation. Not all of these priority pollutants, however, are encountered in a wastewater treatment facility.

The influent and effluent of several major sewer districts have been sampled and characterized in the literature. This book presents the results of the priority pollutant analyses for eight wastewater treatment plants in: District of Columbia-Department of Environmental Services (Black and Veatch, 1981a); St. Louis, Missouri-Metropolitan Sewer District (Black and Veatch, 1981b); and Laurel, Maryland-Washington Suburban Sanitary Commission (Sverdrup and Parcel, 1981). Table 6-4 presents the results of the sampling programs for the influent to the treatment facilities and Table 6-5 presents the effluent sampling results. Table 6-6 shows the frequency of occurrence for the priority pollutants by analytical groups.

As can be seen from the tables, a wide range of concentrations for priority pollutants is reported. As with the solid waste disposal sites, this variability in concentrations is due primarily to the diversity of materials contributing to the wastewater treatment facility. Consequently, pollutants typical for municipalities as a whole cannot be accurately generalized, although the pollutants listed in Tables 6-4 through 6-6 will probably be present.

AGRICULTURAL WASTES

Agricultural sources of waste include the fertilizer industry, the dairy product industry, and feedlot operations. Discussions of these sources follow.

TABLE 6-1. LEACHATE SUBSTANCES AT SOLID WASTE DISPOSAL SITES

Physical	Chemical		Biological
	Organic	Inorganic	
Appearance	Phenols	Total bicarbonate solids (TSS, TDS)	Biochemical oxygen demand (BOD)
pH	Chemical oxygen demand (COD)	Volatile solids	Coliform bacteria (total, fecal; fecal streptococcus)
Oxidation-reduction potential	Total organic carbon (TOC)	Chloride	Standard plate count
Conductivity	Volatile acids	Sulfate	
Color	Tannins, lignins	Phosphate	
Turbidity	Organic-N	Alkalinity and acidity	
Temperature	Ether-soluble (oil and grease)	Nitrate-N	
Odor	MBAS	Nitrite-N	
	Organic functional groups as required	Ammonia-N	
	Chlorinated hydrocarbons	Sodium	
		Potassium	
		Calcium	
		Magnesium	
		Hardness	
		Heavy metals (Pb, Cu, Ni, Cr, Zn, Cd, Fe, Mn, Si, Hg, As, Se, Ba, Ag)	
		Cyanide	
		Fluoride	

Source: U.S. Environmental Protection Agency, 1977a.

TABLE 6-2. TYPICAL CONCENTRATIONS FOR LEACHATE SUBSTANCES COMMONLY ASSOCIATED WITH SOLID WASTE

Constituent	Range[a] (mg/l)	Range[b] (mg/l)	Range[c] (mg/l)
Chloride	34-2,800	100-2,400	600-800
Iron	0.2-5,500	200-1,700	210-325
Manganese	0.06-1,400	--	75-125
Zinc	0-1,000	1-135	10-30
Magnesium	16.5-15,600	--	160-250
Calcium	5-4,080	--	900-1,700
Potassium	2.8-3,770	--	295-310
Sodium	0-7,700	100-3,800	450-500
Phosphate	0-154	5-130	--
Copper	0-9.9	--	0.5
Lead	0-5.0	--	1.6
Cadmium	--	--	0.4
Sulfate	1-1,826	25-500	400-650
Total N	0-1,416	20-500	--
Conductivity (μmhos/cm)	--	--	6,000-9,000
TDS	0-42,276	--	10,000-14,000
TSS	6-2,685	--	100-700
pH	3.7-8.5	4.0-8.5	5.2-6.4
Alkalinity ($CaCo_3$)	0-20,850	--	800-4,000
Hardness (Total)	0-22,800	200-5,250	3,500-5,000
BOD_5	9-54,610	--	7,500-10,000
COD	0-89,520	100-51,000	16,000-22,000

Sources:

[a] U.S. Environmental Protection Agency, 1973a.

[b] R.C. Steiner et al., 1971.

[c] U.S. Environmental Protection Agency, 1975.

TABLE 6-3. LIST OF PRIORITY POLLUTANTS

Base/Neutral Compounds			
Acenaphthene	Bis(2-chloroisopropyl)ether	Diethylphthalate	Indeno(1,2,3-cd)pyrene
Acenaphthylene	4-Bromophenyl phenyl ether	Dimethylphthalate	Isophorone
Anthracene	Butyl benzyl phthalate	2,4-Dinitrotoluene	Naphthalene
Benzo(a)anthracene	2-Chloronaphthalene	2,6-Dinitrotoluene	Nitrobenzene
Benzo(b)fluoranthene	4-Chlorophenyl phenyl ether	Dioctylphthalate	N-Nitrosodimethylamine
Benzo(k)fluoranthene	Chrysene	1,2-Diphenylhydrazine	N-Nitrosodi-n-propylamine
Benzo(a)pyrene	Dibenzo(a,h)anthracene	Fluoranthane	N-Nitrosodiphenylamine
Benzo(g,h,i)perylene	Di-n-butylphthalate	Fluorene	Phenanthrene
Benzidine	1,3-Dichlorobenzene	Hexachlorobenzene	Pyrene
Bis(2-chloroethyl)ether	1,4-Dichlorobenzene	Hexachlorobutadiene	2,3,7,8-Tetrachlorodibenzo-p-dioxin
Bis(2-chloroethoxy)methane	1,2-Dichlorobenzene	Hexachloroethane	1,2,4-Trichlorobenzene
Bis(2-ethylhexyl)phthalate	3,3-Dichlorobenzidine	Hexachlorocyclopentadiene	

Acid Compounds	
Antimony	Mercury
Arsenic	Nickel
Beryllium	Selenium
Cadmium	Silver
Chromium	Thallium
Copper	Zinc
Lead	

Miscellaneous Inorganics
Asbestos
Cyanide

Pesticides/PCBs		
Aldrin	Dieldrin	Toxaphene
a-BHC	Endosulfan I	PCB-1016
b-BHC	Endosulfan II	PCB-1221
d-BHC	Endosulfan Sulfate	PCB-1232
g-BHC	Endrin	PCB-1242
Chlordane	Endrin Aldehyde	PCB-1248
4,4'-DDD	Heptachlor	PCB-1254
4,4'-DDE	Heptachlor Epoxide	PCB-1260
4,4'-DDT		

Volatile Organics			
Chloromethane	Trans-1,2-Dichloroethene	Trichloroethene	1,1,2,2-Tetrachloroethane
Bromomethane	Chloroform	Dibromochloromethane	Tetrachloroethene
Vinyl chloride	1,2-Dichloroethane	1,1,2-Trichloroethane	1,4-Dichlorobutane (SS)
Chloroethane	1,1,1-Trichloroethane	Dis-1,3-Dichloropropene benzene	Toluene
Methylere chloride	Carbon tetrachloride	Benzene	Chlorobenzene
Trichlorofluoromethane	Bromodichloromethane	2-Chloroethylvinyl ether	Ethylbenzene
1,1-Dichloroethene	1,2-Dichloropropane	2-Bromo-1-chloropropane (SS)	Acrolein
Bromochloromethane (SS)	Trans-1,3-Dichloropropene	Bromoform	Acrylonitrile
1,1-Dichlorosthane			

TABLE 6-4. PRIORITY POLLUTANTS OBSERVED IN INFLUENT SAMPLES ($\mu g/l$ except asbestos, = 10^6 fibers/l)

Pollutant	Range[a]	Detection Limit
Base/neutral compounds		
Bis(2-ethylhexyl)phthalate	BDL-440	10
Diethyl phthalate	BDL-38	10
Di-n-butyl phthalate	BDL-41	10
Butyl benzyl phthalate	BDL-11	10
Naphthalene	BDL-48	10
Acid compounds		
4-Nitrophenol	BDL-500	25
Phenol	BDL-80	25
Volatiles		
Benzene	BDL-1,600	10
Chlorobenzene	BDL->300	10
Chloroform	BDL-88	10
1,2-Dichloroethane	BDL-540	10
Ethylbenzene	BDL-120	10
Methylene chloride	BDL->10,000	10
Tetrachloroethylene	BDL-650	10
1,2-Trans-dichloroethylene	BDL-220	10
Toluene	BDL-280	10
1,1,1-Trichloroethane	BDL->10,000	10
Trichloroethyelene	BDL-240	10
Metals		
Antimony	BDL-450	100
Arsenic	BDL-200	30
Cadmium	BDL-20	10
Chromium	BDL-570	10
Copper	BDL-510	10
Lead	BDL-520	10
Mercury	BDL-5.2	5
Nickel	BDL-650	10
Selenium	BDL-200	5
Silver	BDL-70	10
Thallium	BDL-590	20
Zinc	BDL-1,900	10
Miscellaneous inorganics		
Asbestos	BDL-0.7	0.1
Cyanide	BDL-210	5

Source: Miller and Barch, 1981.

Note:

[a]BDL = below detection limit.

TABLE 6-5. PRIORITY POLLUTANTS OBSERVED IN EFFLUENT SAMPLES (μg/l except asbestos, = 10^6 fibers/l)

Pollutant	Range	Detection Limit
Base/neutral compounds		
Bis(2-ethylhexyl)phthalate	BDL-150	10
Diethyl phthalate	BDL-34	10
Di-n-butyl phthalate	BDL-22	10
Butyl benzyl phthalate	BDL-11	10
Naphthalene	BDL-24	10
Acid compounds		
4-Nitrophenol	BDL-280	25
Pentachlorophenol	BDL-34	25
Phenol	BDL-120	25
Volatiles		
Benzene	BDL-2,500	10
Chlorobenzene	BDL-190	10
Chloroform	BDL-74	10
1,2-Dichloroethane	BDL-550	10
Ethylbenzene	BDL-26	10
Methylene chloride	BDL-1,100	10
Tetrachloroethylene	BDL-90	10
Toluene	BDL-110	10
1,1,1-Trichloroethane	BDL-410	10
Trichloroethyelene	BDL-16	10
Metals		
Antimony	BDL-300	100
Arsenic	BDL-76	30
Cadmium	BDL-19	10
Chromium	BDL-280	10
Copper	BDL-300	10
Lead	BDL-180	10
Nickel	BDL-240	10
Selenium	BDL-270	5
Silver	BDL-20	10
Thallium	BDL-260	20
Zinc	BDL-4,600	10
Miscellaneous inorganics		
Asbestos	BDL-0.01	0.1
Cyanide	BDL-130	5

Source: Miller and Barch, 1981.

BDL = below detection limit.

TABLE 6-6. NUMBER OF PRIORITY POLLUTANTS DETECTED IN INFLUENT SAMPLES

Base/neutral compounds	5 out of 47
Acid compounds	2 out of 11
Pesticides/PCBs	None out of 25
Volatile organic compounds	11 out of 28
Metals	12 out of 13
Miscellaneous inorganics	2 out of 2

Source: Miller and Barch, 1981.

Fertilizer Industry

The fertilizer industry primarily involves the production of nitrogen, phosphate, and potassium oxide fertilizers. Of these, nitrogen and phosphate predominate.

Nitrogen fertilizers comprise three products: ammonia, urea, or ammonium nitrate. The four processes of operation to manufacture these products result in the following sources of waste (U.S. EPA, 1973b):

- Cooling tower blowdown
- Boiler blowdown
- Process condensate.

Typical concentrations of contaminants for these sources are shown in Tables 6-7, 6-8, and 6-9.

Phosphate-based fertilizers comprise four products: phosphatic acid, normal superphosphate, triple superphosphate, and ammonium phosphates. The process operations associated with phosphate fertilizer production have the following sources of waste: closed-loop cooling tower blowdown and boiler blowdown. Typical concentrations for the effluent of each of these sources are shown in Tables 6-10 and 6-11.

Dairy Product Industry

The dairy product industry is defined as a plant producing milk or milk byproducts, including whey and buttermilk. Most material in dairy plant wastewaters is organic in nature, consisting of milk solids or organic components of cleaners, sanitizers, and lubricants. The contaminants associated with dairy product raw wastewater are shown in Table 6-12.

TABLE 6-7. TYPICAL RANGE OF CONCENTRATIONS OF CONTAMINANTS IN NITROGEN FERTILIZER PLANT COOLING TOWER BLOWDOWN EFFLUENT

Contaminant	Range (mg/l)
Chromate	0-250
Ammonia	5-100
Sulfate	500-3,000
Chloride	0-40
Phosphate	10-50
Zinc	0-30
Oil	10-1,000
Total dissolved solids	500-10,000

Source: U.S. Environmental Protection Agency, 1973b.

TABLE 6-8. TYPICAL RANGE OF CONCENTRATIONS OF CONTAMINANTS IN NITROGEN FERTILIZER PLANT BOILER BLOWDOWN EFFLUENT

Contaminant	Range (mg/l)
Phosphate	0-250
Sulfite	0-100
Total dissolved solids	500-3,500
Zinc	0-10
Total suspended solids	10-50
Alkalinity	50-700
Hardness	50-500
Silica (as SiO_2)	10-50

Source: U.S. Environmental Protection Agency, 1973b.

TABLE 6-9. AMMONIA DISCHARGE FOR EACH NITROGEN PROCESS CONDENSATE

Process	Range of Ammonia
Ammonia condensate	2,400-3,500 lb/1,000 ton
Urea condensate[a]	100-200 gal/ton
Ammonium nitrate condensate	50-110 gal/ton
Nitrate acid process	None

Source: U.S. Environmental Protection Agency, 1973b.

Note:

[a]Condensate can include minor amounts of carbon dioxide.

TABLE 6-10. TYPICAL RANGE OF CONCENTRATIONS OF CONTAMINANTS IN PHOSPHATE FERTILIZER PLANT COOLING TOWER BLOWDOWN

Contaminant	Range (mg/l)
Chromate	0-250
Sulfate	500-3,000
Chloride	35-160
Phosphate	10-50
Zinc	0-30
Total dissolved solids	500-10,000
Total suspended solids	0-50
Biocides	0-100

Source: U.S. Environmental Protection Agency, 1973b.

TABLE 6-11. TYPICAL RANGE OF CONCENTRATIONS OF CONTAMINANTS IN PHOSPHATE FERTILIZER PLANT BOILER BLOWDOWN EFFLUENT

Contaminant	Range (mg/l)
Phosphate	5-50
Sulfite	0-100
Zinc	0-10
Total dissolved solids	500-3,500
Alkalinity	50-700
Hardness	50-500
Silica (as SiO_2)	25-80

Source: U.S. Environmental Protection Agency, 1973b.

TABLE 6-12. TYPICAL RANGE OF CONCENTRATIONS OF EFFLUENT PARAMETERS IN DAIRY PRODUCT RAW WASTEWATER

Parameter	Range (mg/l)
Biochemical oxygen demand	40-1,000
Total suspended solids	400-2,000
pH (units)	4.0-10.8
Temperature (Celsius)	8-38
Phosphorus	9-210
Ammonia nitrogen	1.0-13.4
Nitrogen (total)	1.0-115

Source: U.S. Environmental Protection Agency, 1974a.

Feedlot Operations

Feedlots consist of high concentrations of animals held in a small area for the production of meat, milks, eggs, or breeding stock; and stabling of horses. Beef cattle feedlots are the largest producers of waste by volume (as compared to poultry, swine, etc.); the feedlot discussion will be limited to this operation.

Beef feedlot wastes generally include the following components (U.S. EPA, 1973c):

- Bedding or litter (if used) and animal hair
- Water and milking center wastes
- Spilled feed
- Undigested or partially digested food or feed additives
- Digestive juices
- Biological products of metabolism
- Microorganisms from the digestive tract
- Cell and cell debris from the digestive tract wall
- Residual soil and sand.

The greatest influences on the generation and concentration of these wastes are the type of facility utilized and the diet of the cattle. The effluent parameters given in Tables 6-13 and 6-14 are for open dirt lots with flat to moderate slope and open dirt lots with steep slope, respectively. The slope characteristics tend to have different degrees of influence on the concentrations of the parameters shown in the tables.

OIL PRODUCTION AND REFINING INDUSTRY

The oil industry can potentially have two sources of contamination: oil field operations, and petroleum refinery operations. These are discussed separately.

Oil Field Operations

The primary contamination associated with oil field operations is brine, although minor amounts of hydrocarbons and oily water are common. Brine evaporation pits are the most common methods of disposal for these wastes. Although these pits are fairly effective disposal methods, waste excursions are in the form of salts, primarily consisting of sodium and calcium as the cations, with chloride, sulfate, and nitrate as the anions. Typically, the total dissolved solids content of these wastes is over 100,000 parts per million (ppm).

TABLE 6-13. TYPICAL RANGE OF CONCENTRATIONS FOR EFFLUENT PARAMETERS IN OPEN DIRT FEEDLOTS WITH FLAT TO MODERATE SLOPE

Parameter	Range (mg/l)
Total suspended solids	1,000-5,000
pH (units)	5.1-9.4
Biochemical oxygen demand	1,000-5,000
Chemical oxygen demand	3,000-20,000
Total nitrogen	20-1,100
Ammonia nitrogen	0-500
Nitrate nitrogen	0-120
Total phosphorus	14-200
Total potassium	20-900
Magnesium	70-120
Sodium	65-700

Source: U.S. Environmental Protection Agency, 1973c.

TABLE 6-14. TYPICAL RANGE OF CONCENTRATIONS FOR EFFLUENT PARAMETERS IN OPEN DIRT FEEDLOTS WITH STEEP SLOPE

Parameter	Range (mg/l)
Total suspended solids	1,150-5,750
pH (units)	5.1-9.4
Biochemical oxygen demand	1,150-5,750
Chemical oxygen demand	3,450-23,000
Total nitrogen	23-1,265
Ammonia nitrogen	0-575
Nitrate nitrogen	0-138
Total phosphorus	16-230
Total potassium	23-1,035
Magnesium	81-138
Sodium	75-805

Source: U.S. Environmental Protection Agency, 1973c.

Petroleum Refinery Operations

The petroleum refining industry involves several operations engaged in the separation of crude molecular constituents, molecular cracking, molecular rebuilding, and solvent finishing. The groups of refining operations include storage and transportation, hydrocarbon processing, petrochemical operations, lube manufacturing processes, treating and finishing, asphalt production, and auxiliary activities (U.S. EPA, 1974b).

The number of values, ranges, and concentrations of toxic pollutants found during a screening study of 17 petroleum refineries is shown in Table 6-15. Intake water, raw wastewater, and final DAF (Dissolved Air Flotation) effluent samples for these refineries were sampled for three consecutive 24-hour periods. The raw wastewater was defined as the API separator, which is an integral part of refinery process operations for product and raw material recovery before final wastewater treatment (U.S. EPA, 1977b).

INDUSTRIAL WASTES

Industrial wastewaters are quite variable and dependent upon the industry. This subsection characterizes the wastewaters, both before and after treatment, from various industrial facilities. The industries discussed include:

TABLE 6-15. CONCENTRATIONS OF TOXIC POLLUTANTS FOUND DURING A SCREENING STUDY IN PETROLEUM REFINING WASTEWATER (μ/l)

	Intake			API Separator Effluent			DAF Effluent		
Toxic Pollutants	Number of Values[b]	Range	Median	Number of Values[b]	Range	Median	Number of Values[b]	Range	Median
Metals and inorganics									
Antimony[c]	17	<1-<25	<25	10	<1-360	<13	7	1-<25	<25
Arsenic[c]	18	3-35	<20	14	3-480	<20	7	<4-<20	20
Asbestos[d]	4		ND	3	ND-3.4	ND			
Beryllium[c]	85	<1-<20	<2	50	<1-<20	<2	35	<1-<20	<2
Cadmium[c]	85	<1-<200	<20	50	<1-<200	<20	35	<1-<200	<2
Chromium[c]	85	1-3,000	<24	58	1-2,000	<240	38	<5-2,000	270
Copper[c]	86	1-300	10	54	2-1,400	32	35	3-400	9
Cyanide	52	10-60	20	36	<5-1,500	<60	20	10-3,000	45
Lead[c]	88	<1-700	<60	54	2-960	<60	38	<15-<600	<60
Mercury[c]	70	<0.1-7	<0.5	53	<0.1-78	0.8	20	<0.1-1.1	<0.5
Nickel[c]	88	<1-790	<50	53	<1-770	<50	35	1-<500	<18
Selenium[c]	23	2-<20	<10	25	<4-<20	9	10	5-<20	<12
Silver[c]	85	<1-<250	<25	50	<1-<250	<25	30	<1-<250	<2.5
Thallium[c]	34	<1-<25	<2	25	<1-<15	<1	11	<1-<15	<15
Zinc[c]	90	<1-2,800	62	66	24-3,400	280	38	30-3,000	91
Pthalates									
Bis(2-ethylhexyl)phthalate	6	ND-1,100	550	6	ND-700	300	1		1,100
Di-n-butyl prthalate	5	ND-30	0.4	3	ND-1.3	ND	2		ND
Diethyl phthalate	2		ND	2	ND-12	6			
Dimethyl phthalate	3	ND-20	ND	2		ND	1		ND
Phenols									
2-Chlorophenol	1		ND	1		315			
2,4-Dichlorophenol	1		ND	1		ND			
2,4-Dinitrophenol	3		ND	2	110-11,000	5,600	1		2,700
2,4-Dimethylphenol	9		ND	5	ND-1,200	>100	4	>100-18,000	6,000
2-Nitrophenol	1		<10	1		1,400			
4-Nitrophenol	4	ND-<10	ND	2	20-5,800	2,900	2	ND-1,400	700
Pentachlorophenol	1		ND	1		ND			
Phenol	14	ND-10	ND	10	13-4,900	250	5	ND-34,000	1,900
4,6-Dinitro-o-cresol	1		ND	1		60			
Parachlorometa cresol	3		ND	1		ND	2		ND

(continued)

TABLE 6-15. (continued)

Toxic Pollutants	Intake			API Separator Effluent			DAF Effluent		
	Number of Values[b]	Range	Median	Number of Values[b]	Range	Median	Number of Values[b]	Range	Median
Aromatics									
Benzene	16	ND-14	<1	12	ND-2,400	>100	4[e]	ND-2,000	>100
1,2-Dichlorobenzene	1		<0.5				1		ND
1,4-Dichlorobenzene	1		<0.5				1		ND
Ethylbenzene	10		ND	9	ND-810	>100	1		>100
Toluene	14	ND-<10	ND	10	ND-12,000	>100	4	<10-76,000	<100
Polycrylic aromatic hydrocarbons									
Acenaphthene	7	ND-29	ND	5	ND-520	37	2	150-390	270
Acenaphthylene	5	ND-0.2	ND	4	ND-660	41	1		530
Anthracene	2		ND	1		660	1		1,800
Benzo(a)pyrene	2	ND-33	17	1		190			
Chrysene	8	ND-49	ND	5	0.1-40	20	3	ND-0.3	ND
Fluoranthene	8	ND-29	ND	5	ND-40	8	3		ND
Fluorene	4	ND-1	ND	2	ND-270	140	2	110-495	300
Naphthalane	11	ND-2	ND	9	ND-3,200	302	3	106-3,700	700
Phenanthrene	11	ND-160	ND	8	5-1,100	165	3	50-1,800	600
Pyrene	6	ND-140	<0.1	3	ND-16	11	1		5
Polychlorinated biphenyls and related compounds									
Aroclor 1016	7		ND	5	ND-40	1.9	3	7.9-<10	<10
Aroclor 1221	7		ND	4	ND-<10	<5	2		<10
Aroclor 1232	8		ND	6	ND-<10	0.7	3	3.5-<10	<10
Aroclor 1242	8	ND-0.2	ND	6	ND-<10	3.2	4	0.2-<10	<5
Aroclor 1248	4		ND	2		<10	2		<10
Aroclor 1254	4		ND	2		<10	2		<10
Aroclor 1260	4		ND	2		<10	2		<10

(continued)

TABLE 6-15. (continued)

Toxic Pollutants	Intake			API Separator Effluent			DAF Effluent		
	Number of Values[b]	Range	Median	Number of Values[b]	Range	Median	Number of Values[b]	Range	Median
Halogenated aliphatics									
Carbon tetrachloride	4	ND->50	<5	1		ND	2	ND-<10	<5
Chloroform	8[e]	ND-70	<8	9	ND-100	10	1[e]		13
Dichlorobromomethane	1		ND	1		24			
1,2-Dichloroethane	2		ND	2	ND-16	8			
1,2-Trans-dichloroethylene	3	ND-11	ND	3	ND-20	ND			
Methylene chloride	10[e]	ND-130	<85	7[e]	ND-1,600	>100	3[e]	ND-560	30
1,1,2,2-Tetrachloroethane	2	ND-<10	<5	2		ND			
Tetrachloroethylene	4	ND-50	<10	3	ND->50	ND	1		ND
1,1,1-Trichlcroethane	1		<50						
Trichloroethylene	2	<10-20	<15	1		ND	1		ND
Pesticides and metabolites									
Aldrin	2		ND	2	<5-12	<8			
α-BHC	1		ND				1		<10
β-BHC	3		ND	1		<5	1		<5
δ-BHC	2		ND	2	<5-12	<8			
γ-BHC	1		ND				1		<5
Chlordane	1		2.8						
4,4'-DDE	1		ND	1		7			
4,4'-DDD	1		ND				1		<5
α-Endosulfan	1		ND	1		ND	1		0.1
β-Endosulfan	1		ND	1		13			
Endosulfan sulfate	1		ND				1		<5
Heptachlor	2		ND	1		<5	1		<5
Heptachlor epoxide	2		ND	2	ND-<5	<2.5			
Isophorone	2		ND	1		3,600	1		2,500

Source: U.S. Environmental Protection Agency, 1978b and 1979f.

Notes:

[a]Values are corrected for blanks when blank values are reported.

[b]Values include samples that contained nondetectable quantities.

[c]Values include 3-day composite samples.

[d]Units of million fibers per liter.

[e]Not all values counted because values in blanks greater than values in sample(s).

Blanks mean compound data not available or, if in "range" column, range is not defined. ND = Not detected in sample.

- Iron and steel manufacturing
- Inorganic chemicals manufacturing
- Leather tanning and finishing
- Textile milling
- Nonferrous metals manufacturing.

Although several additional industries are sources of wastewater, the above industries are probably the largest wastewater producers as a group. This selection provides a representative sampling of industrial wastewaters.

Iron and Steel Manufacturing

The iron and steel manufacturing industry is involved in the production of iron, steel, and ferrous products that do not require machining. This industry also includes ancillary processes such as coke production, scale removal, pickling, and alkaline cleaning. These categories include the following manufacturing processes (U.S. EPA, 1980a):

- Coke making: production of metallurgical coke as a basic raw material for operating iron-making blast furnaces.
- Sintering: an agglomeration process in which iron-bearing material is mixed with finely divided fuel.
- Blast furnaces: pig iron and ferromanganese iron production via combustion of a mixture of iron ore, coke, and limestone.
- Basic oxygen furnaces: refining hot metal (iron) and metallics into steel by oxidizing and removing the elements such as silicon, phosphorus, etc.
- Electric arc furnaces: production of high-quality and alloy steels in refractory-lined cylindrical furnaces utilizing an electrical current (arcing) for melting cold steel scrap and fluxes.
- Vacuum degassing: subjecting molten steel to low pressures to eliminate gases absorbed by the steel that may reduce its quality.
- Hot forming primary mill: rolling steel to a desired size. This also includes mills that process steel into finished products (e.g., section rolling mills and hot forming flat mills).
- Sulfuric acid pickling: chemical removal of surface oxides from metal by immersion in a heated sulfuric acid solution. Pickling can also include the use of a hydrochloric acid solution.

- Hot coating: immersion of clean steel into baths of molten metal (e.g., galvanizing) to plate a thin layer of metal onto the steel surfaces.

- Continuous alkaline cleaning: removal of mineral and animal fats and oils via immersion into an alkaline solution.

Although these are not all of the processes involved in iron and steel manufacturing, they are the major preparation procedures. Tables 6-16 through 6-26 provide the median, maximum, and average toxic pollutant concentrations for the raw and treated wastewater for the above manufacturing processes.

Inorganic Chemical Manufacturing

Inorganic chemical manufacturing can considerably vary in wastewater composition. In general, toxic pollutants consist of metals except in the cases where organic products are also produced onsite. Tables 6-27 through 6-36 characterize raw wastewater information for 11 inorganic chemical industries that can have discharges after treatment. These industries include:

- Aluminum fluoride

- Chrome pigments

- Hydrogen cyanide

- Hydrofluoric acid

- Nickel sulfate

- Sodium bisulfate

- Sodium dichromate

- Titanium dioxide

- Copper sulfate.

Leather Tanning and Finishing Industry

Leather tanning encompasses numerous processing steps in converting animal skins or hides into leather. Three primary hide or skin types are used to manufacture leather: cattle hides, sheepskins, and pigskins. In addition, goatskin, horsehide, deerskin, and elkskin can be used.

The U.S. EPA (1979c) developed the following criteria for categorizing the industry: (1) type or condition of animal hide processed, (2) the method of hair removal, (3) the type of tanning agent used, and (4) the extent of finishing performed. The seven categories derived from this criteria are:

TABLE 6-16. WASTEWATER CHARACTERIZATION OF TOXIC POLLUTANTS FOR THE COKEMAKING PROCESS

	Raw Wastewater				Treated Effluent				
		Concentration (μg/l)				Concentration (μg/l)			
Toxic Pollutant	Number Detected	Median	Maximum	Average	Number Detected	Median	Maximum	Average	Average Percent Removal
Metals and inorganics									
Antimony	3	33	335	120	3	41	130	60	50
Arsenic	3	660	170,000	57,000	2	210	400	210	99
Cyanide	7	26,000	190,000	47,000	5	2,500	22,000	8,800	81
Selenium	4	410	2,600	860	3	640	650	430	50
Silver	6	25	670	130	4	17	25	17	87
Zinc	5	130	470	200	3	130	220	110	45
Nitrogen compounds									
Acrylonitrile	5	2,700	4,700	2,700	2	1,600	3,000	1,600	40
Phenols									
2,4-Dimethylphenol	3	5,000	84,000	23,000	0				
2-Nitrophenol	2	770	1,500	770	1	<5			99
Pentachlorophenol	1	395			1	49			88
Phenol	7	120,000	670,000	240,000	5	48	53,000	11,000	95
2,4,6-Trichlorophenol	1	400			0				
p-Chloro-m-cresol	2	2,200	4,300	2,200	2	33	64	33	94
4,6-Dinitro-o-cresol	2	530	970	530	1	<5			99
Aromatics									
Benzene	7	27,000	86,000	29,000	5	260	140,000	30,000	--[a]
2,4-Dinitrotoluene	1	1,900			1	510			78
2,6-Dinitrotoluene	1	240			1	140			42
Ethylbenzene	5	300	640	340	4	27	6,600	1,700	--[a]
Toluene	5	5,700	17,000	6,700	5	73	11,000	2,600	61
Polycyclic aromatics									
Acenaphthylene	7	3,200	6,400	3,000	5	7	1,600	330	89
Benzo(a)anthracene	3	150	1,200	490	3	5	260	88	82
Benzo(a)pyrene	4	360	1,100	480	3	13	13	9	98
Chrysene	5	320	1,500	550	0				
Fluoranthene	7	950	3,100	1,200	5	8	500	110	91
Fluorene	7	370	2,500	700	5	10	190	47	87
Naphthalene	7	27,500	39,000	25,000	4	700	5,900	1,800	95
Pyrene	6	760	2,600	910	5	8	280	64	
Halogenated aliphatics									
Chloroform	5	120	1,400	400	3	200	280	180	55
1,1-Dichloroethylene	2		3	3				0	
Pesticides and metabolites									
Isophorone	2	2,000	4,000	2,000	1	170			92

Source: U.S. Environmental Protection Agency, 1980a.

Note:

[a]Treated effluent concentration exceeds raw wastewater concentration.

Blanks indicate no data available; dashes indicate negligible removal.

TABLE 6-17. WASTEWATER CHARACTERIZATION OF TOXIC POLLUTANTS FOR THE SINTERING PROCESS

Toxic Pollutant	Raw Wastewater				Treated Effluent				Average Percent Removal
	Number Detected	Concentration (μg/l)			Number Detected	Concentration (μg/l)			
		Median	Maximum	Average		Median	Maximum	Average	
Metals and inorganics									
Cadmium	2	690	1,300	690	2	420	770	420	39
Chromium	3	98	620	250	2	50	90	50	80
Copper	3	520	600	400	3	270	550	410	--[a]
Cyanide	4	260	15,000	3,900	3	160	1,100	430	89
Lead	2	5,600	5,900	5,600	3	800	5,500	3,200	43
Nickel	2	110	200	110	2	74	130	70	36
Silver	2	12	13	12	2	10	10	10	17
Zinc	3	940	8,700	3,400	3	940	5,000	1,900	44
Phthalates									
Butyl benzyl phthalate	3	85	290	130	3	580	990	520	--[a]
Di-n-butyl phthalate	3	120	250	130	3	170	420	200	--[a]
Di-n-octyl phthalate	3	20	370	130	3	350	490	280	--[a]
Phenols									
2,4-Dinitrophenol	1	14			1	140			--[a]
Phenol	3	56	1,000	380	3	630	990	370	2
Polycyclic aromatics									
Benzo(a)anthracene	2	260	516	260	3	150	260	140	46
Benzo(a)pyrene	2	220	430	220	3	190	240	140	36
Chrysene	2	160	320	160	3	53	410	160	--[a]
Fluoranthene	2	130	254	130	3	310	860	390	--[a]
Pyrene	3	7	320	110	3	300	1,100	470	--[a]

Source: U.S. Environmental Protection Agency, 1980a.

Note:

[a]Treated effluent concentration exceeds raw wastewater concentrations.

Blanks indicate no data available; dashes indicate negligible removal.

TABLE 6-18. WASTEWATER CHARACTERIZATION OF TOXIC POLLUTANTS FOUND IN BLAST FURNACE-IRON PROCESS

Toxic Pollutant	Raw Wastewater				Treated Effluent				Average Percent Removal
	Number Detected	Concentration (μg/l)			Number Detected	Concentration (μg/l)			
		Median	Maximum	Average		Median	Maximum	Average	
Metals and inorganics									
Antimony	1	37			1	15			59
Arsenic	1	46			1	6			87
Cadmium	4	100	200	100	3	10	11	10	90
Chromium	4	300	630	330	3	23	54	29	91
Copper	4	240	1,200	420	4	28	170	60	85
Lead	2	21,000	43,000	18,000	4	81	3,100	830	95
Nickel	3	230	1,200	480	3	60	94	54	89
Selenium	1	63			1	4			94
Silver	3	57	73	47	3	10	26	14	--[a]
Zinc	4	25,000	90,000	36,000	3	1,200	32,000	8,500	76
Phthalates									
Bis(2-ethylhexyl)phthalate	4	100	3,200	860	4	320	11,000	2,900	--[a]
Butyl benzyl phthalate	4	95	340	130	4	8	350	94	28
Di-n-butyl phthalate	4	320	9,800	2,600	4	94	190	73	98
Diethyl phthalate	2	10	16	10	2	86	170	86	--[a]
Dimethyl phthalate	1	47			3	3	120	30	36
Di-n-octyl phthalate	4	82	12,000	3,000	3	36	86	32	99
Phenols									
2,4-Dichlorophenol	1	240			2	30	44	30	88
2,4-Dimethylphenol	3	3	53	18	2	83	163	83	--[a]
Phenol	3	640	2,800	1,200	4	590	1,800	770	38
Aromatics									
Hexachlorobenzene	1	103			0				>99
Polycyclic aromatics									
Benzo(a)pyrene	3	7	9,500	3,200	2	5	8	5	99
Chrysene	3	15	310	110	3	7	74	28	75
Fluoranthene	3	82	11,000	3,600	4	15	230	65	98
Fluorene	2	15	21	15	4	8	29	12	20
Naphthalene	3	14	19	14	3	3	15	5	64
Pyrene	3	53	10,000	3,400	3	12	41	20	99
Halogenated aliphatics									
Chloroform	4	12	48	20	4	31	54	34	--[a]

Source: U.S. Environmental Protection Agency, 1980a.

Note:

[a]Treated effluent concentration exceeds raw wastewater concentration.

Blanks indicate no data available; dashes indicate negligible removal.

TABLE 6-19. WASTEWATER CHARACTERIZATION OF TOXIC POLLUTANTS FOR THE BASIC OXYGEN FURNACE PROCESS

	Wet-Open Combustion								
	Raw Wastewater				Treated Effluent				
		Concentration (μg/l)				Concentration (μg/l)			
Toxic Pollutant	Number Detected	Median	Maximum	Average	Number Detected	Median	Maximum	Average	Average Percent Removal
Metals and inorganics									
Antimony	1	17		0.01					
Arsenic	2	60	70	60	2	12	17	12	80
Cadmium	3	174	260	150	3	10	488	170	--[a]
Chromium	3	360	17,000	5,800	4	540	30,100	7,800	--[a]
Copper	4	825	1,200	600	4	217	476	230	62
Lead	4	370	12,000	3,300	4	455	942	517	84
Mercury	3	33	34	16.8	4	0.05	0.30	0.15	99
Nickel	3	47	675	244	4	530	2,020	773	--[a]
Selenium	2	16	28	16	2	20	31	20	--[a]
Silver	2	27	43	27	2	175	339	175	--[a]
Thallium	2	11	15	11	2	70	80	70	--[a]
Zinc	4	3,400	48,500	14,300	4	706	2,140	970	93
Phthalates									
Bis(2-ethylhexyl)phthalate	2		120	60	4	72	317	118	--[a]
Aromatics									
Benzene	2	1.5	3	1.5					
Polycyclic aromatics									
Chrysene	2	13	23	13					
Fluoranthene	1	34							
Pyrene	1	32							
Halogenated aliphatics									
Chloroform					4	56	122	62	--[a]

(continued)

TABLE 6-19. (continued)

Toxic Pollutant	Wet-Suppressed Combustion								Average Percent Removal
	Raw Wastewater				Treated Effluent				
	Number Detected	Concentration (μg/l)			Number Detected	Concentration (μg/l)			
		Median	Maximum	Average		Median	Maximum	Average	
Metals and inorganics									
Antimony	1	4							
Arsenic	1			0.01					
Cadmium	2	62	91	62	2	9	10	9	85
Chromium		603	1,050	603	3	12	13	11.6	98
Copper	3	63	310	3	2	10	100	60	--
Cyanide	1	1							
Lead	3	700	27,000	13,850	2	645	822	645	95
Mercury	1	0.2							
Nickel	2	174	327	174	3	10	691	237	--[a]
Selenium	1	3							
Silver	2	12	19	12	2	12.5	15	12.5	4.2
Zinc	1	8.3			3	227	281	203	--[a]
Phthalates									
Bis(2-ethylhexyl)phthalate	2	447	868	447	3	29	298	112	75
Butyl benzyl phthalate	1	9							
Di-n-butyl phthalate	2		11	7	2	10	10	10	--[a]
Phenols									
Phenol	1	8							
Aromatics									
Toluene	2	2	3	2					
Polycyclic aromatics									
Pyrene	1	5							

Source: U.S. Environmental Protection Agency, 1980a.

Note:

[a]Treated effluent concentration exceeds raw wastewater concentration.

Blanks indicate no data available; dashes indicate negligible removal.

TABLE 6-20. WASTEWATER CHARACTERIZATION OF TOXIC POLLUTANTS FOR ELECTRIC ARC FURNACE PROCESS

Toxic Pollutant	Raw Wastewater				Treated Effluent				Average Percent Removal
	Number Detected	Concentration (μg/l)			Number Detected	Concentration (μg/l)			
		Median	Maximum	Average		Median	Maximum	Average	
Metals and inorganics									
Antimony	1	670			1	10			99
Arsenic	1	120			1	11			91
Cadmium	1	3,300			1	1,500			55
Chromium	1	4,300			1	550			87
Copper	1	1,300			1	80			94
Lead	1	9			1	1,500			--[a]
Nickel	1	43			1	10			77
Silver	1	63			1	10			84
Zinc	3	100,000	190,000	97,000	2	29,000	38,000	29,000	70
Phthalates									
Bis(2-ethylhexyl)phthalate	3	160	170	110	3	110	330	150	--[a]
Butyl benzyl phthalate	3	57	150	70	2	51	95	51	27
Di-n-butyl phthalate	3	17	65	30	3	11	21	12	60
Phenols									
4-Nitrophenol	2	19	31	19	0				>99
Pentachlorophenol	2	22	40	22	1	14			36
Aromatics									
Benzene	3	10	25	14	3	12	28	15	--[a]
Polycyclic aromatics									
Fluoranthene	2	30	58	30	2	7	10	7	77
Pyrene	2	28	53	28	2	72	150	72	--[a]

Source: U.S. Environmental Protection Agency, 1980a.

Note:

[a]Treated effluent concentration exceeds raw wastewater concentration.

Blanks indicate no data available; dashes indicate negligible removal.

TABLE 6-21. WASTEWATER CHARACTERIZATION OF TOXIC POLLUTANTS FOR THE VACUUM DEGASSING PROCESS

	Raw Wastewater				Treated Effluent				
		Concentration (μg/l)				Concentration (μg/l)			
Toxic Pollutant	Number Detected	Median	Maximum	Average	Number Detected	Median	Maximum	Average	Average Percent Removal
Metals and inorganics									
Chromium	3	130	3,000	1,100	3	26	3,000	1,000	10
Copper	3	90	440	190	3	210	440	230	--[a]
Lead	3	300	2,000	830	3	90	2,000	720	13
Nickel	2	32	40	32	2	22	30	22	31
Zinc	3	2,000	30,000	10,800	3	330	30,000	10,000	8
Phthalates									
Butyl benzyl phthalate	2	34	57	34	2	28	53	28	18
Di-n-butyl phthalate	2	31	43	31	2	260	500	260	--[a]

Source: U.S. Environmental Protection Agency, 1980a.

Note:

[a]Treated effluent concentration exceeds raw wastewater concentration.

Blanks indicate no data available.

TABLE 6-22. WASTEWATER CHARACTERIZATION OF TOXIC POLLUTANTS FOR HOT FORMING-PRIMARY MILL PROCESS

	Raw Wastewater				Treated Effluent				
		Concentration (μg/l)				Concentration (μg/l)			Average
Toxic Pollutant	Number Detected	Median	Maximum	Average	Number Detected	Median	Maximum	Average	Percent Removal
Metals and inorganics									
Cadmium	5	<10	<10	<10					
Chromium	5	50	130	80	5	40	130	62	22
Copper	5	300	970	440	5	40	760	180	59
Cyanide	2	2	2	2					
Lead	5	300	810	330	5	50	320	10	97
Nickel	5	220	570	310	5	20	480	120	61
Silver	5	20	20	20					
Zinc	5	100	140	90	5	30	100	48	47
Phthalates									
Bis(2-ethylhexyl)phthalate	3	18	149	71					
Di-n-butyl phthalate	2	6	10	6					
Di-n-octyl pnthalate	3	7	7	6					
Halogenated aliphatics									
Chloroform	6	<10	13	<10					
Methylene chloride	2	2	2	2					
Trichloroethylene	3	100	270	63					

Source: U.S. Environmental Protection Agency, 1980a.

Blanks indicate no data available.

TABLE 6-23. WASTEWATER CHARACTERIZATION OF TOXIC POLLUTANTS FOR THE SULFURIC ACID PICKLING PROCESS

Toxic Pollutant	Spent Concentrate				Rinsewater				Treated Wastewater			
	Number Detected	Concentration (μg/l)			Number Detected	Concentration (μg/l)			Number Detected	Concentration (μg/l)		
		Median	Maximum	Average		Median	Maximum	Average		Median	Maximum	Average
Metals and inorganics												
Arsenic	1	170			3	<10	173	64				
Cadmium	2	270	280	270	3	<10	302	107				
Hexavalent chromium	2	4										
Chromium, total	2	232,000	260,000	232,000	3	50	2,000	680				
Copper	2	3,600	4,700	3,600	3	140	2,400	860				
Cyanide	2	13	17	13	2	11	11	11				
Lead	2	800	1,600	800	3	40	1,000	360				
Nickel	2	25,000	27,000	25,000	3	60	13,800	4,600				
Silver	2	51	600	51								
Zinc	2	74,000	133,000	74,000	3	90	1,800	640				
Phenols												
2,4,6-Trichlorophenol	1	41										
Aromatics												
Benzene					1	<10			2	<10	<10	<10
Toluene	1	<10										
Polycyclic aromatics												
Acenaphthylene				5				1				
Fluoranthene									1	<10		
Naphthalene					1	<10			2	<10	<10	<10
Pyrene					2	<10	<10	<10				
Halogenated aliphatics												
Chloroform	1	20			2	<10	<10	<10	5	20	25	22
Methylene chloride	2	33	52	33	3	43	165	73	5	154	230	140
Trichloroethylene	1	<10										

Source: U.S. Environmental Protection Agency, 1980a.

Blanks indicate no data available.

TABLE 6-24. WASTEWATER CHARACTERIZATION OF TOXIC POLLUTANTS FOR THE HYDROCHLORIC ACID PICKLING PROCESS

Toxic Pollutant	Absorber Vent Scrubber Raw Wastewater			Fume Hood Scrubber Raw Wastewater				Rinse Wastewater			
		Concentration (µg/l)			Concentration (µg/l)				Concentration (µg/l)		
	Number Detected	Maximum	Average	Number Detected	Median	Maximum	Average	Number Detected	Median	Maximum	Average
Metals and inorganics											
Antimony	1	<210		2	175	200	170	3	<100	190	110
Arsenic	1	<23		3	66	75	50	3	233	290	190[a]
Beryllium	1	<20		1	<20			1	<20		--
Cadmium	1	<20		4	<12	<200	59	6	<15	<200	45
Chromium	2	<200	<140	3	150	<330	190	6	390	840	300
Copper	2	1,600	850	4	120	390	180	5	690	1,500	770
Cyanide	2	18	10	4	6	12	6	6	8	75	20
Lead	1	<600		3	<100	<600	260	6	260	6,200	1,250
Mercury	1	32		1	2						
Nickel	2	790	420	3	150	<500	230	6	690	1,300	700
Selenium	1	<35		3	<10	<10	7	3	<10	200	70
Silver	1	<250		4	<20	<250	75	6	21	<250	58
Thallium	1	<70		1	<50						
Zinc	2	1,300	670	4	87	270	120	6	380	1,500	520
Phenols											
Pentachlorophenol					26	43	26				
Aromatics											
Benzene									12	14	12
Chlorobenzene				1	26						
Polycyclic aromatics											
Fluoranthene								6	<10	65	19
Pyrene								6	<10	75	22
Halogenated aliphatics											
Chlorodibromomethane				1	13						
Chloroform	1	26		4	<10	16	12	5	<10	37	16
1,1-Dichloroethylene				1	12						
1,2-Trans-dichloroethylene	1	23									
Methylene chloride	2	1,100	550	3	<10	82	34	6	11	3,600	690
Tetrachloroethylene	1	14						3	22	40	24
Trichloroethylene								2	37	65	37

(continued)

TABLE 6-24. (continued)

Toxic Pollutant	Spent Pickle Liquor				Discharge Wastewater			
	Number Detected	Concentration (µg/l)			Number Detected	Concentration (µg/l)		
		Median	Maximum	Average		Median	Maximum	Average
Metals and inorganics								
Antimony	2	2,100	4,100	2,100	2	100	190	100
Arsenic	2	35	45	35	3	230	260	170
Beryllium					1	<20		
Cadmium	4	140	280	150	5	<20	240	96
Chromium	5	8,700	37,000	13,000	5	440	2,300	770
Copper	5	11,000	22,000	11,000	5	680	900	620
Cyanide	5	8	11	8	5	14	74	23
Lead	4	1,700	1,500,000	390,000	5	420	33,000	7,000
Mercury					1	<2		
Nickel	5	13,000	22,000	10,000	5	640	860	540
Selenium	4	<10	170	50	4	6	20	9
Silver	4	290	390	250	5	23	250	70
Thallium	1	180			1	<50		
Zinc	5	4,200	61,000	15,000	5	600	290,000	58,000
Aromatics								
Benzene					2	12	14	12
Polycyclic aromatics								
Fluoranthene	5	<10	65	21	5	<10	56	20
Pyrene	4	<10	75	26	5	<10	65	22
Halogenated aliphatics								
Chloroform	5	<10	100	28	5	<10	36	15
Methylene chloride	5	14	3,500	720	5	15	3,600	740
Tetrachloroethylene	3	31	40	27	2	29	37	29
Trichloroethylene					2	50	90	50

Source: U.S. Environmental Protection Agency, 1980a.

Note:

[a]Detected but not quantified.

Blanks indicate no data available.

TABLE 6-25. WASTEWATER CHARACTERIZATION OF TOXIC POLLUTANTS FOR THE HOT COATING-GALVANIZING PROCESS

Toxic Pollutant	Raw Wastewater				Treated Effluent				Average Percent Removal
	Number Detected	Concentration (μg/l)			Number Detected	Concentration (μg/l)			
		Median	Maximum	Average		Median	Maximum	Average	
Metals and inorganics									
Antimony	1	3			1	3			--
Arsenic	3	14	40	21	3	10	10	9	51
Beryllium	1	<20				<20		0.02	--
Cadmium	6	20	200	45	6	20	20	17	62
Chromium	6	15	10,200	2,113	6	35	200	77	96
Copper	6	90	2,500	487	6	30	170	52	90
Cyanide	6	12	19	12	6	10	21	9.8	18
Lead	6	310	25,000	4,390	6	145	600	270	94
Selenium	3	10	10	8.4	3	10	12	11	--[b]
Silver	6	20	2,500	65	6	20	250	70	--[b]
Thallium	1	50			1	50			--
Zinc	6	9,100	88,900	26,760	6	130	770	1,390	95
Phenols									
2-Chlorophenol	2	7	10	7	2	4	5	4	43
2,4-Dichlorophenol	3	5	5	5	1	10			--[a]
2,4-Dinitrophenol	1	5			1	5			--
2-Nitrophenol				ND	2	5	5	5	--[b]
Pentachlorophenol	3	5	22	10	3	5	5	4	60
Phenol	1	5			2	8	10	8	--[b]
2,4,6-Trichlorophenol	2	8	10	8	1	5			38
4,6-Dinitro-o-cresol				ND	2	10	20	10	--[b]
Aromatics									
Benzene	3	5	11	11	4	6	10	7	36
1,3-Dichlorobenzene	1	144						ND	>99
Toluene	2	8	10	8	3	10	10	8	--[b]

(continued)

TABLE 6-25. (continued)

Toxic Pollutant	Raw Wastewater				Treated Effluent				Average Percent Removal
	Number Detected	Concentration (μg/l)			Number Detected	Concentration (μg/l)			
		Median	Maximum	Average		Median	Maximum	Average	
Polycyclic aromatics									
Acenaphthene	1	5			5	5	5	5	--
Acenaphthylene	1	10			5	5	10	7	30
Anthracene	1	5			1	7			--[b]
Benzo(a)anthracene				ND	1	5			--[b]
Benzo(a)pyrene				ND	4	5	10	6	--[b]
Chrysene					1	5			--[b]
Fluoranthene	4	7	24	11	6	9	10	7	36
Fluorene	3	5	10	10	4	5	10	6	14
Naphthalene				ND	2	10	10	10	--[b]
Phenanthrene	1	5			1	7			--[b]
Pyrene	5	5	21	12	5	7	10	7	42
2-Chloronaphthalene					2	4	5	4	--[b]
Halogenated aliphatics									
Chloroform	6	13	106	37	6	12	48	18	51
Dichlorobromomethane	1	10						ND	>99
1,2-Trans-dichloroethylene	1	5						ND	>99
Methylene chloride	5	16	12	134	5	13	230	60	--[b]
Tetrachloroethylene	5	10	17	10	2	8	8	8	20
1,1,1-Trichloroethane	2	39	67	39	2	19	32	19	51
Trichloroethylene	1	46			2	7	10	7	85

Source: U.S. Environmental Protection Agency, 1980a.

Notes:

[a]Indicates water quality of central treatment effluent.

[b]Treated effluent concentration exceeds raw wastewater concentration.

ND = Not detected in sample.

Blanks indicate no data available; dashes indicate negligible removal.

TABLE 6-26. WASTEWATER CHARACTERIZATION OF TOXIC POLLUTANTS FOUND IN THE CONTINUOUS ALKALINE CLEANING PROCESS

Toxic Pollutant	Raw Wastewater				Treated Effluent				Average Percent Removal
	Number Detected	Concentration (μg/l)			Number Detected	Concentration (μg/l)			
		Median	Maximum	Average		Median	Maximum	Average	
Metals and inorganics									
Nickel	2	20	35	20	2	4,175	7,000	4,175	--[a]
Phenols									
Phenol	1	24							
Aromatics									
2,6-Dinitrotoluene	1	47							
Fluoranthene	1	24							
Pyrene	1	32							
Halogenated aliphatics									
Chloroform	2	48.5	52	48.5	3	64.5	65	64.5	--[a]
Tetrachloroethylene	2	37	49	37	1	52			--[a]

Source: U.S. Environmental Protection Agency, 1980a.

Note:

[a]Treated effluent concentration exceeds raw wastewater concentration.

Blanks indicate no data available; dashes indicate negligible removal.

TABLE 6-27. SUMMARY OF RAW WASTEWATER CHARACTERISTICS--ALUMINUM FLUORIDE INDUSTRY

Pollutant	Raw Waste Loadings						Number of Plants Averaged
	Minimum (kg/d)	Average (kg/d)	Maximum (kg/d)	Minimum (kg/Mg)	Average (kg/Mg)	Maximum (kg/Mg)	
Toxic pollutants							
Arsenic	0.071	0.078	0.086	0.0007	0.0016	0.002	3
Cadmium		0.010			0.0002		1
Chromium	0.072	0.16	0.25	0.0016	0.0035	0.0054	2
Copper	0.02	0.16	0.33	0.0002	0.0033	0.0071	3
Nickel	0.025	0.13	0.26	0.00025	0.003	0.0056	3
Mercury	0.0013	0.0041	0.0095	0.000027	0.00005	0.00009	3
Selenium	0.051	0.11	0.17	0.001	0.0015	0.002	2
Conventional pollutants							
TSS	751	2,920	5,510	16.3	53.7		
Fluorine	493	727	986	9.71	11.9		
Aluminum	98.4	220	352	0.97	4.40		

Source: U.S. Environmental Protection Agency, 1979b.

Blanks indicate data not available.

TABLE 6-28. SUMMARY OF RAW WASTEWATER CHARACTERISTICS--CHROME PIGMENTS INDUSTRY

Pollutant	Raw Waste Loadings						Number of Plants Averaged
	Minimum (kg/d)	Average (kg/d)	Maximum (kg/d)	Minimum (kg/Mg)	Average (kg/Mg)	Maximum (kg/Mg)	
Toxic pollutants							
Antimony	5.90	51.7	98.0	0.14	0.87	1.61	2
Cadmium	0.87	5.44	10.0	0.02	0.16	0.09	2
Chromium	698	1,020	1,330	11.5	21.5	30.8	2
Copper	6.08	50.8	95.2	0.14	0.86	1.58	2
Lead	237	347	458	5.46	6.49	7.62	2
Nickel	1.38	1.71	2.03	0.032	0.0325	0.033	2
Zinc	52.2	381	712	0.86	8.63	16.4	2
Cyanide	3.11	24.4	45.8	0.072	0.41	0.75	2
Organics							
Pherols		0.93			0.015		1
Pherolics		8.80			0.14		1
Convertional pollutants							
TSS		3,050			70.4		1

Source: U.S. Environmental Protection Agency, 1979b.

Blanks indicate data not available.

TABLE 6-29. SUMMARY OF RAW WASTEWATER CHARACTERISTICS--HYDROGEN CYANIDE INDUSTRY

	Raw Waste Loadings						Number of Plants Averaged
Pollutant	Minimum (kg/d)	Average (kg/d)	Maximum (kg/d)	Minimum (kg/Mg)	Average (kg/Mg)	Maximum (kg/Mg)	
Toxic pollutants							
Total cyanide	173	205	237	0.81	1.20	1.60	2
Free cyanide	106	113	120	0.49	0.65	0.81	2
Conventional pollutants							
TSS	152	383	614	1.02	1.94	2.87	
NH_3^-N	3,880	5,790	7,700	26.2	31.1	36.0	
BOD_5	24.5	4,320	8,620	0.16	20.2	40.3	

Source: U.S. Environmental Protection Agency, 1979b.

Blanks indicate data not available.

TABLE 6-30. SUMMARY OF RAW WASTEWATER CHARACTERISTICS--HYDROFLUORIC ACID INDUSTRY

Pollutant	Raw Waste Loadings						Number of Plants Averaged
	Minimum (kg/d)	Average (kg/d)	Maximum (kg/d)	Minimum (kg/Mg)	Average (kg/Mg)	Maximum (kg/Mg)	
Toxic pollutants							
Antimony	0.015	1.63	6.44	0.0003	0.03	0.12	4
Arsenic	0.012	0.46	1.12	0.003	0.0056	0.012	3
Cadmium	0.0036	0.011	0.017	0.0001	0.00027	0.00031	3
Chromium	0.14	1.73	5.49	0.0043	0.024	0.06	4
Copper	0.60	1.42	2.80	0.015	0.028	0.051	4
Lead	0.10	1.74	5.62	0.003	0.046	0.165	4
Mercury	0.0027	0.056	0.20	0.00008	0.00065	0.002	4
Nickel	0.14	3.90	13.0	0.0004	0.051	0.14	4
Selenium	0.016	0.066	0.12	0.0005	0.001	0.002	3
Thallium	0.0054	0.084	0.16	0.00016	0.0021	0.003	2
Zinc	0.49	21.1	72.1	0.014	0.41	1.33	4
Conventional pollutants							
TSS	13,600	133,000	247,000	170	2,710	5,700	
Fluorine	497	2,970	7,890	14.6	45.4	86.9	

Source: U.S. Environmental Protection Agency, 1979b.

Blanks indicate data not available.

TABLE 6-31. SUMMARY OF RAW WASTEWATER CHARACTERISTICS--NICKEL SULFATE INDUSTRY

Pollutant	Raw Waste Loadings						Number of Plants Averaged
	Minimum (kg/d)	Average (kg/d)	Maximum (kg/d)	Minimum (kg/Mg)	Average (kg/Mg)	Maximum (kg/Mg)	
Toxic pollutants							
Cadmium	0.000014	0.0015	0.0045	0.000002	0.00017	0.0005	3
Chromium	0.00023	0.00091	0.0018	0.00001	0.00025	0.0005	2
Copper	0.0011	0.039	0.11	0.0001	0.01	0.03	3
Lead	0.000082	0.0014	0.0028	0.00002	0.0001	0.0003	3
Mercury		0.000027			0.00003		1
Nickel	0.27	10.8	31.5	0.035	1.20	3.45	3
Selenium	0.00027	0.00059	0.00091	0.00003	0.000035	0.00004	2
Thallium		0.000032			0.000009		1
Conventional pollutants							
TSS	0.34	31.2	92.5	0.031		10.1	

Source: U.S. Environmental Protection Agency, 1979b.

Blanks indicate data not available.

TABLE 6-32. SUMMARY OF RAW WASTEWATER CHARACTERISTICS--SODIUM BISULFITE INDUSTRY

Pollutant	Raw Waste Loadings						Number of Plants Averaged
	Minimum (kg/d)	Average (kg/d)	Maximum (kg/d)	Minimum (kg/Mg)	Average (kg/Mg)	Maximum (kg/Mg)	
Toxic pollutants							
Antimony	0.00045	0.0018	0.0041	0.000007	0.000052	0.00008	2
Cadmium	0.00023	0.0003	0.00041	0.000004	0.00001	0.000017	3
Chromium	0.018	0.54	1.05	0.0003	0.011	0.022	2
Copper	0.005	0.011	0.015	0.00007	0.00046	0.001	2
Lead	0.000091	0.0045	0.0095	0.000007	0.000092	0.0002	3
Mercury	0.000091	0.00021	0.00045	0.000001	0.000006	0.00001	2
Nickel	0.0032	0.0068	0.0091	0.00005	0.00031	0.0007	3
Zinc	0.016	0.18	0.42	0.0002	0.0053	0.0088	3
Conventional pollutants							
TSS	3.20	12.9	25.4	0.21	0.27	0.38	
COD	54.4	117	234	1.33	2.94	4.04	

Source: U.S. Environmental Protection Agency, 1979b.

Blanks indicate data not available.

TABLE 6-33. SUMMARY OF RAW WASTEWATER CHARACTERISTICS--SODIUM DICHROMATE INDUSTRY

	Raw Waste Loadings						Number of
Pollutant	Minimum (kg/d)	Average (kg/d)	Maximum (kg/d)	Minimum (kg/Mg)	Average (kg/Mg)	Maximum (kg/Mg)	Plants Averaged
Toxic pollutants							
Chromium	82.1	132	181	0.95	1.17	1.39	2
Hex. chromium	27.5	1,210	3,105	0.466	15.7	43.9	3
Copper	0.0091	0.32	0.92	0.00005	0.0046	0.013	3
Nickel	0.27	4.26	8.98	0.006	0.034	0.049	3
Silver		0.058			0.0009		1
Zinc	0.067	0.22	3.91	0.0009	0.002	0.003	3
Selenium		0.23			0.003		1
Arsenic		0.005			0.00008		1
Conventional pollutants							
TSS	26,600	131,000	236,000	140	2,070	4,000	

Source: U.S. Environmental Protection Agency, 1979b.

Blanks indicate data not available.

TABLE 6-34. SUMMARY OF RAW WASTEWATER CHARACTERISTICS--TITANIUM DIOXIDE INDUSTRY (CHLORIDE PROCESS)

Pollutant	Raw Waste Loadings Minimum (kg/d)	Average (kg/d)	Maximum (kg/d)	Minimum (kg/Mg)	Average (kg/Mg)	Maximum (kg/Mg)	Number of Plants Averaged
Toxic pollutants							
Chromium	1.76	64.4	127	0.024	0.79	1.55	2
Lead	0.0032	2.0	4.0	0.0004	0.024	0.049	2
Nickel	0.14	2.04	3.93	0.002	0.025	0.048	2
Zinc	0.75	1.47	2.19	0.01	0.019	0.027	2
Conventional pollutants							
TSS	442	4,140	7,830	6.06	51.0	95.9	
Iron	7.57	768	1,530	0.10	9.40	18.7	

Source: U.S. Environmental Protection Agency, 1979b.

Blanks indicate data not available.

TABLE 6-35. SUMMARY OF RAW WASTEWATER CHARACTERISTICS--TITANIUM DIOXIDE INDUSTRY (SULFATE PROCESS)

Pollutant	Raw Waste Loadings						Number of Plants Averaged
	Minimum (kg/d)	Average (kg/d)	Maximum (kg/d)	Minimum (kg/Mg)	Average (kg/Mg)	Maximum (kg/Mg)	
Toxic pollutants							
Antimony	7.66	18.0	28.3	0.08	0.21	0.32	2
Arsenic		1.31			0.104		1
Cadmium	0.091	2.40	6.85	0.0009	0.027	0.078	3
Chromium	132	200	327	1.36	2.11	3.37	3
Copper	8.30	11.6	15.1	0.094	0.12	0.16	3
Lead	3.28	8.56	12.4	0.037	0.089	0.13	3
Nickel	8.30	11.5	14.7	0.086	0.12	0.15	2
Thallium		0.76			0.0078		1
Zinc	53.4	55.3	57.1	0.55	0.57	0.59	2
Organics							
Phenol	0.20				0.002		
Conventional pollutants							
TSS		20,500			211		
Iron		58,500			602		

Source: U.S. Environmental Protection Agency, 1979b.

Blanks indicate data not available.

TABLE 6-36. SUMMARY OF RAW WASTEWATER CHARACTERISTICS FOUND AT A COPPER SULFATE PLANT

Pollutant	Raw Waste Loadings	
	Average (kg/d)	Average (kg/Mg)
Toxic pollutants		
Antimony	0.014	0.00069
Arsenic	0.16	0.0078
Cadmium	0.039	0.0019
Copper	83.9	4.11
Lead	0.0079	0.00039
Nickel	5.08	0.25
Zinc	0.50	0.024
Conventional pollutants		
TSS	1.78	0.087

Source: U.S. Environmental Protection Agency, 1979b.

1. Hair pulp/chrome tan/retan-wet finish--facilities that primarily process raw or cured cattle or cattle-like hides into finished leather by chemically dissolving the hair (hair pulp), tanning with chrome, and retanning and wet finishing.

2. Hair save/chrome tan/retan-wet finish--facilities that primarily process raw or cured cattle or cattle-like hides into finished leather by chemically loosening and mechanically removing the hair, tanning with chrome, and retanning and wet finishing.

3. Hair save/nonchrome tan/retan-wet finish--facilities that process raw or cured cattle or cattle-like hides into finished leather by chemically loosening and mechanically removing the hair; tanning, primarily with vegetable tannins, alum, syntans, oils, or other chemicals; and retanning and wet finishing.

4. Retan-wet finish--facilities that process previously unhaired and tanned hides or splits into finished leather through retanning and wet finishing processes including coloring, fatliquoring, and mechanical conditioning.

5. No beamhouse--facilities that process previously unhaired and pickled cattle hides, sheepskins, or pigskins into finished leather by tanning with chrome or other agents, followed by re-tanning and wet finishing.

6. Through-the-blue--facilities that process raw or cured cattle or cattle-like hides into the blue-tanned state only, by chemically dissolving or loosening the hair and tanning with chrome, with no retanning or wet finishing.

7. Shearling--facilities that process raw or cured sheep or sheep-like skins into finished leather by retanning the hair on the skin, tanning with chrome or other agents, and retanning and wet finishing.

The concentrations of toxic pollutants in the pretreated wastewater for these categories are listed in Tables 6-37 through 6-43.

Textile Milling

Textile manufacturing and milling facilities are principally engaged in receiving and preparing fibers; transforming fibers into yarn, thread, or webbing; converting the yarn and web into fabric or related products; and finishing these materials at various stages of the production. Although a majority of the mills involve "dry" processes and do not discharge, there are approximately 2,000 "wet" processing mills discharge wastewater (U.S. EPA, 1978c). The toxic pollutants found in detectable concentrations for wet mill plant water supply, raw wastewater, and secondary effluents are presented in Table 6-44.

Nonferrous Metal Manufacturing Industry

The nonferrous metals industry entails primary and secondary smelting and refining of nonferrous metals. Primary processes involve final recovery of pure or usable metal from metal ore. Secondary processes refer to nonferrous metal recovery from scrap.

The nonferrous metals industry can be divided into numerous categories depending on the type of metal to be smelted and/or refined. However, only 10 of these categories can be accurately characterized with respect to raw wastewater because of insufficient data for the remaining industries. The industries that are discussed are:

- Primary aluminum
- Secondary aluminum
- Primary columbium and tantalum
- Primary copper
- Secondary copper

TABLE 6-37. WASTEWATER CHARACTERIZATION OF TOXIC POLLUTANTS FOR THE HAIR PULP/CHROME TAN/RETAN-WET FINISH CATEGORY

Toxic Pollutants	Number of Samples	Number Detected	Concentration (µg/l) Range	Mean
Metals and inorganics				
Chromium	3	3	43,000-180,000	80,000
Copper	3	3	50-380	173
Cyanide	2	2	20-60	40
Lead	3	3	1,100-2,400	1,700
Nickel	3	3	20-60	40
Zinc	3	3	200-580	430
Ethers				
Bis(2-chloroisopropyl)ether	3			ND
Phthalates				
Bis(2-ethylhexyl)phthalate	3	1		51
Butyl benzyl phthalate	3			ND
Di-n-butyl phthalate	3			ND
Diethyl phthalate	3			ND
Dimethyl phthalate	3	1		120
Nitrogen compounds				
Benzidine	3	1		27
3,3'-Dichlorobenzidine	3			ND
1,2-Diphenylhydrazine	3			ND
N-nitrosodiphenylamine	3			ND
Phenols				
2,4-Dichlorophenol	3			ND
2,4-Dimethylphenol	3	1		Present
Pentachlorophenol	3			ND
Phenol	3	3	3,000-4,000	3,700
2,4,6-Trichlorophenol	3	2	880-5,900	3,400
Aromatics				
Benzene	3	3	10-20	15
Chlorobenzene	3			ND
1,2-Dichlorobenzene	3	1		260
1,3-Dichlorobenzene	3			ND
1,4-Dichlorobenzene	3	1		54
Ethylbenzene	3	2	88	88
Hexachlorobenzene	3			ND
Nitrobenzene	3	1		430
Toluene	3	3	150-400	280
1,2,4-Trichlorobenzene	3			ND
Polycyclic aromatic hydrocarbons				
Acenaphthene	3	1		32
Acenaphthylene	3	1		16
Chrysene	3			ND
Fluoranthene	3			ND
Fluorene	3			ND
Naphthalene	3	2	24-67	46
Phenanthrene/anthracene	3	1		94
Pyrene	3			ND
Halogenated aliphatics				
Chlorodibromomethane	3			ND
Chloroform	3	1		20
Dichlorobromomethane	3	3	10	10
1,1-Dichlproethane	3	1		20
1,2-Dichloroethane	3			ND
1,2-Trans-dichloroethylene	3	1		30
1,1,2,2-Tetrachloroethane	3	1		10
1,1,1-Trichloroethane	3	1		Present
1,1,2-Trichloroethane	3	1		10
Trichlorofluoromethane	3			ND
Pesticides and metabolites				
Chlordane	3			ND
Isophorone	3			ND

Source: U.S. Environmental Protection Agency, 1979c.

ND = Not detected.

TABLE 6-38. WASTEWATER CHARACTERIZATION OF TOXIC POLLUTANTS FOR THE HAIR SAVE/CHROME TAN/RETAN-WET FINISH CATEGORY

Toxic Pollutants	Number of Samples	Number Detected	Concentration (μg/l)	
			Range	Mean
Metals and inorganics				
Chromium	2	2	31,000-150,000	90,500
Cyanide	2	2	20-50	35
Lead	2	2	100-1,300	700
Nickel	2	2	5-40	22
Zinc	2	2	240-400	315
Ethers				
Bis(2-chloroisopropyl)ether	2			ND
Phthalates				
Bis(2-ethylhexyl)phthalate	2	1		32
Butyl benzyl phthalate	2			ND
Di-n-butyl phthalate	2			ND
Diethyl phthalate	2			ND
Dimethyl phthalate	2			ND
Nitrogen compounds				
Benzidine	2			ND
3,3'-Dichlorobenzidine	2			ND
1,2-Diphenylhydrazine	2			ND
N-nitrosodiphenylamine	2			ND
Phenols				
2,4-Dichlorophenol	2	1		114
2,4-Dimethylphenol	2			ND
Pentachlorophenol	2	1		6,200
Phenol	2	2	252-5,500	2,900
2,4,6-Trichlorophenol	2	1		4,800
Aromatics				
Benzene	2	1		10
Chlorobenzene	2	1		10
1,2-Dichlorobenzene	2			ND
1,3-Dichlorobenzene	2			ND
1,4-Dichlorobenzene	2			ND
Ethylbenzene	2	1		150
Hexachlorobenzene	2			ND
Nitrobenzene	2			ND
Toluene	2	2	10-150	80
1,2,4-Trichlorobenzene	2			ND
Polycyclic aromatic hydrocarbons				
Acenaphthene	2			ND
Acenaphthylene	2			ND
Chrysene	2			ND
Fluoranthene	2	1		2
Fluorene	2			ND
Naphthalene	2	1		49
Phenanthrene/anthracene	2	1		56
Pyrene	2	1		1
Halogenated aliphatics				
Chloroform	2	2	10-41	26
Dichlorobromomethane	2			ND
1,1-Dichloroethane	2			ND
1,2-Dichloroethane	2			ND
1,1,2,2-Tetrachloroethane	2			ND
1,1,1-Trichloroethane	2	1		10
1,1,2-Trichloroethane	2			ND
Trichlorofluoromethane	2			ND
Pesticides and metabolites				
Chlordane	2			ND
Isophorone	2			ND

Source: U.S. Environmental Protection Agency, 1979c.

ND = Not detected.

TABLE 6-39. WASTEWATER CHARACTERIZATION OF TOXIC POLLUTANTS FOR THE HAIR SAVE NONCHROME TAN/RETAN-WET FINISH CATEGORY

Toxic Pollutants	Number of Samples	Number Detected	Concentration (μg/l)	
			Range	Mean
Metals and inorganics				
Chromium	4	4	430-10,000	5,100
Copper	4	4	100-740	380
Cyanide	3	2	60-100	80
Lead	4	4	100-200	140
Nickel	4	4	40-95	61
Zinc	4	4	300-700	490
Ethers				
Bis(2-chloroisopropyl)ether	4			ND
Phthalates				
Bis(2-ethylhexyl)phthalate	4			ND
Butyl benzyl phthalate	4			ND
Di-n-butyl phthalate	4	1		Present
Diethyl phthalate	4	1		Present
Dimethyl phthalate	4			ND
Nitrogen compounds				
Benzidine	4			ND
3,3'-Dichlorobenzidine	4			ND
1,2-Diphenylhydrazine	4			ND
N-nitrosodiphenylamine	4			ND
Phenols				
2,4-Dichlorophenol	4			ND
2,4-Dimethylphenol	4			ND
Pentachlorophenol	4	2	10-2,900	1,500
Phenol	4	4	51-25,000	9,000
Aromatics				
Benzene	4	3	10-10	10
Chlorobenzene	4			ND
1,2-Dichlorobenzene	4	3	49-200	126
1,3-Dichlorobenzene	4			ND
1,4-Dichlorobenzene	4	3	19-20	20
Ethylbenzene	4	3	10-120	58
Hexachlorobenzene	4			ND
Nitrobenzene	4			ND
Toluene	4	4	10-15	12
1,2,4-Trichlorobenzene	4			ND
Polycyclic aromatic hydrocarbons				
Acenaphthene	4			ND
Acenaphthylene	4			ND
Chrysene	4			ND
Fluoranthene	4			ND
Fluorene	4			ND
Naphthalene	4	3	6-59	32
Phenanthrene/anthracene	4	1		8
Pyrene	4			ND
Halogenated aliphatics				
Chloroform	4	1		24
Dichlorobromomethane	4	1		10
1,1-Dichloroethane	4			ND
1,2-Dichloroethane	4			ND
1,2-Trans-dichloroethylene	4			ND
1,1,2,2-Tetrachloroethane	4	1		10
Tetrachloroethylene	4	1		23
1,1,1-Trichloroethane	4			ND
1,1,2-Trichloroethane	4			ND
Trichloroethylene	4			ND
Trichlorofluoromethane	4			ND
Pesticides and metabolites				
α-BHC	4			ND
β-BHC	4			ND
Chlordane	4			ND
Isophorone	4			ND

Source: U.S. Environmental Protection Agency, 1979c.

ND = Not detected.

TABLE 6-40. WASTEWATER CHARACTERIZATION OF TOXIC POLLUTANTS FOR THE RETAN-WET FINISH CATEGORY

Toxic Pollutants	Number of Samples	Number Detected	Concentration (μg/l)	
			Range	Mean
Metals and inorganics				
Chromium	3	3	16,000-130,000	89,000
Copper	3	3	160-330	250
Cyanide	2	1		30
Lead	3	3	100-3,500	1,300
Nickel	3	3	6-100	45
Zinc	3	3	150-280	198
Ethers				
Bis(2-chloroisopropyl)ether	3			ND
Phthalates				
Bis(2-ethylhexyl)phthalate	3			ND
Butyl benzyl phthalate	3			ND
Di-n-butyl phthalate	3	1		Present
Diethyl phthalate	3	1		Present
Dimethyl phthalate	3			ND
Nitrogen compounds				
Benzidine	3			ND
3,3'-Dichlorobenzidine	3			ND
1,2-Diphenylhydrazine	3			ND
N-nitrosodiphenylamine	3	1		250
Phenols				
2,4-Dichlorophenol	3			ND
2,4-Dimethylphenol	3			ND
Pentachlorophenol	3			ND
Phenol	3	2	3,200	3,200
2,4,6-Trichlorophenol	3	2	570	570
Aromatics				
Benzene	3	2		10
Chlorobenzene	3			ND
1,2-Dichlorobenzene	3			ND
1,3-Dichlorobenzene	3			ND
1,4-Dichlorobenzene	3			ND
Ethylbenzene	3	3	10-150	90
Hexachlorobenzene	3			ND
Nitrobenzene	3			ND
Toluene	3	3	10-11	10
1,2,4-Trichlorobenzene	3			ND
Polycyclic aromatic hydrocarbons				
Acenaphthene	3	1		Present
Acenaphthylene	3			ND
Chrysene	3			ND
Fluoranthene	3			ND
Fluorene	3			ND
Naphthalene	3	1		Present
Phenanthrene/anthracene	3	2	110-140	120
Pyrene	3			ND
Halogenated aliphatics				
Chloroform	3	2	10-10	10
Dichlorobromomethane	3			ND
1,1-Dichloroethane	3			ND
1,2-Dichloroethane	3			ND
1,2-Trans-dichloroethylene	3			ND
1,1,2,2-Tetrachloroethane	3			ND
Tetrachloroethylene	3			ND
1,1,1-Trichloroethane	3			ND
1,1,2-Trichloroethane	3			ND
Trichloroethylene	3			ND
Trichlorofluoromethane	3			ND
Pesticides and metabolites				
α-BHC	3			ND
β-BHC	3			ND
Chlordane	3			ND
Isophorone	3			ND
TCDD	3	1		ND

Source: U.S. Environmental Protection Agency, 1979c.

ND = Not detected.

TABLE 6-41. WASTEWATER CHARACTERIZATION OF TOXIC POLLUTANTS FOR THE NO BEAMHOUSE CATEGORY

Toxic Pollutants	Number of Samples	Number Detected	Concentration (μg/l)	
			Range	Mean
Metals and inorganics				
Chromium	3	3	16,000-170,000	74,000
Copper	3	3	140-260	190
Cyanide	3			ND
Lead	3	3	60-1,600	790
Nickel	3	3	6-30	15
Zinc	3	3	96-2,600	1,000
Ethers				
Bis(2-chloroisopropyl)ether	3			ND
Phthalates				
Bis(2-ethylhexyl)phthalate	3			ND
Butyl benzyl phthalate	3			ND
Di-n-butyl phthalate	3			ND
Diethyl phthalate	3			ND
Dimethyl phthalate	3			ND
Nitrogen compounds				
Benzidine	3			ND
3,3'-Dichlorobenzidine	3			ND
1,2-Diphenylhydrazine	3			ND
N-nitrosodiphenylamine	3			ND
Phenols				
2,4-Dichlorophenol	3			ND
2,4-Dimethylphenol	3			ND
Pentachlorophenol	3	2	3,400-3,700	3,600
Phenol	3	1		6,200
2,4,6-Trichlorophenol	3	3	2,400-4,200	3,300
Aromatics				
Benzene	3	2	10-150	80
Chlorobenzene	3			ND
1,2-Dichlorobenzene	3	1		36
1,3-Dichlorobenzene	3			ND
1,4-Dichlorobenzene	3	1		13
Ethylbenzene	3	2	10-150	80
Hexachlorobenzene	3			ND
Nitrobenzene	3			ND
Toluene	3	2	10-150	80
1,2,4-Trichlorobenzene	3			ND
Polycyclic aromatic hydrocarbons				
Acenaphthene	3			ND
Acenaphthylene	3			ND
Chrysene	3			ND
Fluoranthene	3			ND
Fluorene	3			ND
Naphthalene	3	2	5-49	27
Phenanthrene/anthracene	3	2	111-130	120
Pyrene	3			ND
Halogenated aliphatics				
Chloroform	3	3	2-18	10
Dichlorobromomethane	3	1		10
1,1-Dichloroethane	3			ND
1,2-Dichloroethane	3			ND
1,2-Trans-dichloroethylene	3			ND
1,1,2,2-Tetrachloroethane	3			ND
Tetrachloroethylene	3	1		40
1,1,1-Trichloroethane	3			ND
1,1,2-Trichloroethane	3			ND
Trichloroethylene	3	1		10
Trichlorofluoromethane	3			ND
Pesticides and metabolites				
α-BHC	3			ND
β-BHC	3			ND
Chlordane	3			ND
Isophorone	3			ND

Source: U.S. Environmental Protection Agency, 1979c.

ND = Not detected.

TABLE 6-42. WASTEWATER CHARACTERIZATION OF TOXIC POLLUTANTS FOR THE THROUGH-THE-BLUE CATEGORY

Toxic Pollutants	Number of Samples	Number Detected	Concentration (μg/l)	
			Range	Mean
Metals and inorganics				
Chromium	1	1		550,000
Copper	1	1		100
Lead	1	1		28
Nickel	1	1		160
Zinc	1	1		980
Ethers				
Bis(2-chloroisopropyl)ether	1			ND
Phthalates				
Bis(2-ethylhexyl)phthalate	1			ND
Butyl benzyl phthalate	1			ND
Di-n-butyl phthalate	1			ND
Diethyl phthalate	1			ND
Dimethyl phthalate	1			ND
Nitrogen compounds				
Benzidine	1			ND
3,3'-Dichlorobenzidine	1			ND
1,2-Diphenylhydrazine	1			ND
N-nitrosodiphenylamine	1			ND
Phenols				
2,4-Dichlorophenol	1			ND
2,4-Dimethylphenol	1	1		Present
Pentachlorophenol	1			ND
Phenol	1	1		Present
2,4,6-Trichlorophenol	1	1		Present
Aromatics				
Benzene	1			ND
Chlorobenzene	1			ND
1,2-Dichlorobenzene	1	1		ND
1,3-Dichlorobenzene	1			Present
1,4-Dichlorobenzene	1	1		Present
Ethylbenzene	1	1		Present
Hexachlorobenzene	1			ND
Nitrobenzene	1			ND
Toluene	1	1		Present
1,2,4-Trichlorobenzene	1			ND
Polycyclic aromatic hydrocarbons				
Acenaphthene	1			ND
Acenaphthylene	1	1		Present
Chrysene	1			ND
Fluoranthene	1			ND
Fluorene	1	1		Present
Naphthalene	1	1		Present
Phenanthrene/anthracene	1	1		Present
Pyrene	1			ND
Halogenated aliphatics				
Chloroform	1	1		Present
Dichlorobromomethane	1			ND
1,1-Dichloroethane	1			ND
1,2-Dichloroethane	1			ND
1,2-Trans-dichloroethylene	1			ND
1,1,2,2-Tetrachloroethane	1			ND
Tetrachloroethylene	1			ND
1,1,1-Trichloroethane	1	1		Present
1,1,2-Trichloroethane	1			ND
Trichloroethylene	1			ND
Trichlorofluoromethane	1			ND
Pesticides and metabolites				
α-BHC	1			ND
β-BHC	1			ND
Chlordane	1			Present
Isophorone	1			ND

Source: U.S. Environmental Protection Agency, 1979c.

ND = Not detected.

TABLE 6-43. WASTEWATER CHARACTERIZATION OF TOXIC POLLUTANTS FOR THE SHEARLING CATEGORY

Toxic Pollutants	Number of Samples	Number Detected	Concentration (µg/l)	
			Range	Mean
Metals and inorganics				
Chromium	2	2	2,000-53,000	36,500
Copper	2	2	35-120	78
Cyanide	2	2	10	10
Lead	2	2		75
Nickel	2	2	20-27	24
Zinc	2	2	190-500	340
Ethers				
Bis(2-chloroisopropyl)ether	2			ND
Phthalates				
Bis(2-ethylhexyl)phthalate	2	1		93
Butyl benzyl phthalate	2			ND
Di-n-butyl phthalate	2			ND
Diethyl phthalate	2			ND
Dimethyl phthalate	2			ND
Nitrogen compounds				
Benzidine	2			ND
3,3'-Dichlorobenzidine	2			ND
1,2-Diphenylhydrazine	2			ND
N-nitrosodiphenylamine	2			ND
Phenols				
2,4-Dichlorophenol	2			ND
2,4-Dimethylphenol	2			ND
Pentachlorophenol	2	1		400
Phenol	2	1		91
2,4,6-Trichlorophenol	2			ND
Aromatics				
Benzene	2	2	5-10	8
Chlorobenzene	2			ND
1,2-Dichlorobenzene	2	1		61
1,3-Dichlorobenzene	2			ND
1,4-Dichlorobenzene	2	2	19-20	20
Ethylbenzene	2			ND
Hexachlorobenzene	2			ND
Nitrobenzene	2			ND
Toluene	2	2	9-10	10
1,2,4-Trichlorobenzene	2			ND
Polycyclic aromatic hydrocarbons				
Acenaphthene	2			ND
Acenaphthylene	2			ND
Chrysene	2			ND
Fluoranthene	2			ND
Fluorene	2			ND
Naphthalene	2	2		26
Phenanthrene/anthracene	2	1		36
Pyrene	2			ND
Halogenated aliphatics				
Chloroform	2	2	12-20	16
Dichlorobromomethane	2			ND
1,1-Dichloroethane	2			ND
1,2-Dichloroethane	2			ND
1,2-Trans-dichloroethylene	2			ND
1,1,2,2-Tetrachloroethane	2	1		18
Tetrachloroethylene	2			ND
1,1,1-Trichloroethane	2			ND
1,1,2-Trichloroethane	2			ND
Trichloroethylene	2			ND
Trichlorofluoromethane	2			ND
Pesticides and metabolites				
α-BHC	2			ND
β-BHC	2			ND
Chlordane	2			ND
Isophorone	2			ND

Source: U.S. Environmental Protection Agency, 1979c.

ND = Not detected.

TABLE 6-44. CONCENTRATIONS OF TOXIC POLLUTANTS FOUND IN TEXTILE MILL WASTEWATER

	Concentrations Observed (μg/l)								
	Water Supply			Raw Wastewater			Secondary Effluent		
Toxic Pollutant	Number of Plants	Median	Maximum	Number of Plants	Median	Maximum	Number of Plants	Median	Maximum
Metals and inorganics									
Antimony	6	<5	48	23	7	170	16	4.5	680
Arsenic	4	<5	<5	14	10	200	8	39	160
Asbestos	--[a]	--	--	--	--	--	--	--	--
Beryllium	4	<5	<5	5	<5	40	5	<5	<5
Cadmium	5	<10	<10	22	<5	46	15	6	13
Chromium	5	<5	<5	37	14	880	27	20	1,800
Copper	6	10	47	40	40	2,400	28	32	290
Cyanide	4	11	22	10	8.0	39	5	12	980
Lead	6	<5	45	26	35	750	16	46	120
Mercury	4	0.2	0.8	10	0.6	4	7	0.4	0.7
Nickel	6	<5	47	32	54	300	18	70	150
Selenium	6	<5	23	10	35	740	4	47	97
Silver	6	<5	17	26	32	130	15	25	140
Thallium	4	3	3	5	3	9	4	3	18
Zinc	12	60	4,540	45	190	7,900	30	200	38,000
Phthalates									
Bis(2-ethylhexyl)phthalate	6	8.2	39	27	26	860	23	18	231
Butyl benzyl phthalate				2	42	73			
Di-n-butyl phthalate	1		1.6	7	16	67	1		3.6
Diethyl phthalate	3	2.1	5.5	10	6.0	86	4	1.5	9.4
Dimethyl phthalate				4	12	14	1		1.0
Nitrogen compounds									
Acrylonitrile				1		1,600	1		400
1,2-Diphenylhydrazine				1		22			
N-nitrosodiphenylamine				3	15	72			
N-nitroso-di-n-propylamine							2	10	19

(continued)

TABLE 6-44. (continued)

Toxic Pollutant	Concentrations Observed (μg/l)								
	Water Supply			Raw Wastewater			Secondary Effluent		
	Number of Plants	Median	Maximum	Number of Plants	Median	Maximum	Number of Plants	Median	Maximum
Phenols									
2-Chlorophenol				1		78	1		5.9
2,4-Dichlorophenol				2	26	41			
2,4-Dimethylphenol							1		8.0
2-Nitrophenol							1		4.1
4-Nitrophenol							1		<10
Pentachlorophenol				11	52	940	2	12	15
Phenol	5	10	36	25	55	4,900	7	14	50
2,4,6-Trichlorophenol				4	20	27	1		19
Cresols									
-Chloro- -cresol				1		170	1		32
Monocyclic aromatics									
Benzene	2	<4	<5	10	<5	200	4	<5	64
Chlorobenzene				5	25	300	1		3.5
1,2-Dichlorobenzene				7	2.0	290	4	10	20
1,4-Dichlorobenzene				2	110	215	2	0.8	1.5
2,6-Dinitrotoluene				1		54			
Ethylbenzene				20	54	2,840	8	63	3,000
Hexachlorobenzene				2	1.3	2			
Toluene	4	0.8	2.4	25	26	620	16	14	1.400
1,2,4-Trichlorobenzene				8	410	2,700	4	610	1,580
Polycyclic aromatics									
Acenaphthene				3	8.7	12	1		0.5
Anthracene	3	0.2	0.4	1		0.1	1		4.4
Benzo(b)fluoranthene				1		<10			
Benzo(k)fluoranthene				1		<10			
Fluorene	2	0.2	0.4	1		15			
Naphthalene				19	44	110	5	22	255
Pyrene				1		0.9	4	0.2	0.3

(continued)

TABLE 6-44. (continued)

Toxic Pollutant	Concentrations Observed (μg/l)								
	Water Supply			Raw Wastewater			Secondary Effluent		
	Number of Plants	Median	Maximum	Number of Plants	Median	Maximum	Number of Plants	Median	Maximum
Polychlorinated biphenyls and related compounds									
2-Chloronaphthalene				1		<10			
Halogenated aliphatics									
Chloroform	6	39	1,360	11	48	640	6	8.5	58
Dichlorobromomethane	2	<5	<5	1		6.6			
1,1-Dichloroethane				1		14			
1,2-Dichloroethane				1		<5			
1,1-Dichloroethylene				1		<5			
1,2-Dichloropropane				1		100			
1,3-Dichloropropane	1		0.8						
Methyl chloride				1		<5			
Methylene chloride	2	<5	<5	3	47	110	3	<5	<5
Tetrachloroethylene				7	<5	2,100	2	11	17
1,1,1-Trichloroethane	1		<5	4	7.8	17	1		<5
Trichloroethylene				10	47	840	4	4.9	87
Trichlorofluoromethane				3	90	2,140			
Vinyl chloride				1		11			
Pesticides and metabolites									
4,4'-DDT							1		0.5
Dieldrin							1		0.2
TCDD	--[a]	--	--	--	--	--	--	--	--

Source: U.S. Environmental Protection Agency, 1978c.

Note:

[a]Dashes indicate no analysis for pollutant.

- Primary lead
- Secondary lead
- Secondary silver
- Primary tungsten
- Primary zinc.

The concentrations of toxic pollutants found in raw wastewaters of these categories are provided in Tables 6-45 through 6-54.

TABLE 6-45. CONCENTRATIONS OF TOXIC POLLUTANTS FOUND IN PRIMARY ALUMINUM RAW WASTEWATER

Toxic Pollutant	Number of		Concentration (μg/l)[a,b]		
	Analyses	Times Detected	Range	Median	Mean
Metals and inorganics					
Antimony	3	2	ND-770	99	290
Arsenic	3	2	ND-260	130	130
Asbestos	1	1		2.2×10^{10}[c]	
Beryllium	3	2	ND-76	33	36
Cadmium	3	3	2.3-200	24	75
Chromium	3	2	ND-2,200	86	760
Copper	3	3	13-140	77	77
Cyanide	3	2	<0.004-29	0.022	9.7
Lead	3	3	0.58-780	650	480
Mercury	3	2	<0.1-1.3	0.40	0.60
Nickel	3	3	500-770	660	640
Selenium	3	2	ND-450	0.20	150
Silver	3	2	ND-<250	0.40	83
Thallium	3	1	ND-<50	ND	<17
Zinc	3	2	ND-540	25	188
Phthalates					
Bis (2-ethylhexyl) phthalate	7	5	ND-40		82
Butyl benzyl phthalate	7	2	ND-86		22
Di-n-butyl phthalate	7	1	ND-120		19
Diethyl phthalate	7	1	ND-2.5		0.4
Dimethyl phthalate	7	0			
Di-n-octyl phthalate	7	0			
Phenols					
Phenol	6	1	ND-70	12	12

(continued)

TABLE 6-45. (continued)

Toxic Pollutant	Number of Analyses	Number of Times Detected	Concentration (μg/l)[a,b] Range	Median	Mean
Aromatics					
Benzene	8	1	ND-6.0	0.8	0.8
Ethylbenzene	8	0			
Toluene	8	1	ND-1.0	0.2	0.2
Polycyclic aromatic hydrocarbons					
Acenaphthene	7	1	ND-50	8.4	8.4
Acenaphthylene	7	1	ND-30	7.6	5.6
Anthracene	7	4	ND-150	8.6	40
Benz(a)anthracene	7	3	ND-180		38
Benzo(a)pyrene	7	3	ND-570		95
Benzo(b)fluoranthene	7	1	ND-260		37
Benzo(k)fluoranthene	7	2	ND-210		39
Chrysene	7	2	ND-230		40
Dibenze(ah)anthracene	7	1	ND-110		16
Fluoranthene	7	4	ND-320	49	95
Fluorene	7	1	ND-50		7.4
Indeno(1,2,3-cd)pyrene	7	2	ND-350		53
Naphthalene	7	1	ND-20		3.0
Pyrene	7	4	ND-219		70
Polychlorinated biphenyls and related compounds					
Aroclor 1248	7	0			
Aroclor 1254	7	0			

(continued)

TABLE 6-45. (continued)

Toxic Pollutant	Number of		Concentration (μg/l)[a,b]		
	Analyses	Times Detected	Range	Median	Mean
Halogenated aliphatics					
Chloroform	8	1	ND-6.0		0.8
1,2-Dichloroethane	8	0			
1,1-Dichloroethylene	8	0			
Methylene chloride	9	1	ND-15		3.0
1,1,2,2-Tetrachloroethane	8	0			
Tetrachloroethylene	8	0			
Trichloroethylene	8	0			
Pesticides and metabolites					
Aldrin	7	0			
δ-BHC	7	0			
γ-BHC	7	1	ND-0.01		
Chlordane	7	0			
4,4'-DDT	7	0			
Dieldrin	7	0			
Endrin aldehyde	7	0			
Heptachlor	7	0			
Heptachlor epoxide	7	0			
Isophorone	7	1	ND-1.5		0.2

Source: U.S. Environmental Protection Agency, 1979a.

Notes:

[a]Except asbestos, which is given in fibers/l.

[b]All concentrations except for those for asbestos were calculated by multiplying the concentrations of the various wastestreams by the normalized percentage of the total flow and then subtracting the concentration present in the intake.

[c]Maximum value.

ND = Not Detected in sample.

Blanks indicate insufficient data.

TABLE 6-46. CONCENTRATIONS OF TOXIC POLLUTANTS FOUND IN RAW WASTEWATER OF THE SECONDARY ALUMINUM CATEGORY

Toxic Pollutant	Number of Analyses	Number of Times Detected	Concentration (μg/l)[a,b] Range	Median	Mean
Metals and inorganics					
Antimony	4	2	ND-950	150	31
Arsenic	4	3	ND-4,000	32	1,000
Asbestos	1	1		7.5×10^8	
Beryllium	4	4	<7.0-310	97	130
Cadmium	4	4	<35-2,000	240	630
Chromium	4	4	<5-1,200	380	490
Copper	4	4	<70-6,100	575	1,800
Cyanide	4	4	<0.001-0.008	0.004	0.004
Lead	4	3	ND-5,600	1,000	1,900
Mercury	4	3	ND-6.4	0.38	1.8
Nickel	4	3	ND-620	28	170
Selenium	4	1	ND-200	0	50
Silver	4	2	ND-30	<12	14
Thallium	3	1	ND-540	ND	180
Zinc	4	4	<2,000-5,900	2,200	3,000
Phthalates					
Bis (2-ethylhexyl) phthalate	6	4	ND-2,000	46	380
Butyl benzyl phthalate	6	2	ND-98		19
Di-n-butyl phthalate	6	3	ND-44		16
Dimethyl phthalate	6	1	ND-56	9.5	9.5
Di-n-octyl phthalate	6	1	ND-25	4.2	4.2
Nitrogen compounds					
3,3'-Dichlorobenzidine	6	1	ND-2.0	0.3	0.3
Aromatics					
Benzene	10	1	ND-94		9.4
Chlorobenzene	10	0			
1,4-Dichlorobenzene	6	1	ND-26		4.3

(continued)

TABLE 6-46. (continued)

Toxic Pollutant	Number of		Concentration (μg/l)[a,b]		
	Analyses	Times Detected	Range	Median	Mean
Aromatics (continued)					
Ethylbenzene	10	0			
1,2,4-Trichlorobenzene	6	0			
Polycyclic aromatic hydrocarbons					
Acenaphthylene	6	1	ND-17		2.8
Anthracene	6	1	ND-4.0		0.7
Benzo(a)pyrene	6	1	ND-12		2.0
Benzo(b)fluoranthene	6	0			
Benzo(ghi)perylene	6	0			
Benzo(k)fluoranthene	6	0			
Chrysene	6	1	ND-190		32
Fluoranthene	6	2	ND-12		
Naphthalene	6	1	ND-1.0		0.2
Phenenthrene	6	1	ND-10		1.7
Pyrene	6	1	ND-24		4.5
Polychlorinated biphenyls and related compounds					
Aroclor 1248	6	1	ND-0.3		0.1
Aroclor 1254	6	1	ND-0.9		0.4
Halogenated aliphatics					
Bromoform	10	0			
Carbon tetrachloride	10	1	ND-10		1.0
Chlorodibromomethane	10	0			
Chloroform	10	6	ND-34		3.4
Dichlorobromomethane	10	1	ND-19		1.9
1,1-Dichloroethane	10	0			
1,2-Dichloroethane	10	0			
1,1-Trans-dichloroethylene	10	5	ND-57		19

(continued)

TABLE 6-46. (continued)

Toxic Pollutant	Number of		Concentration (µg/l)[a,b]		
	Analyses	Times Detected	Range	Median	Mean
Halogenated aliphatics (continued)					
Methylene chloride	10	1	ND-93		9.3
1,1,2,2-Tetrachloroethane	10	0			
Tetrachloroethylene	10	1	ND-310		32
1,1,1-Trichloroethane	10	0			
1,1,2-Trichloroethane	10	0			
Trichloroethylene	10	5	ND-530		61
Pesticides and metabolites					
Aldrin	6	0			
α-BHC	6	1	ND-0.1		
β-BHC	6	1	ND-0.4		
γ-BHC	6	1	ND-0.1		
Chlordane	6	1	ND-0.3		0.1
4,4'-DDE	6	1	ND-0.01		
4,4'-DDD	6	0			
4,4'-DDT	6	1	ND-0.02		
Dieldrin	6	1	ND-0.2		
α-Endosulfan	6	0			
Endrin	6	1	ND-0.01		
Endrin aldehyde	6	1	ND-0.4	0.1	0.1
Heptachlor	6	1	ND-0.4		
Heptachlor epoxide	6	1	ND-0.2		

Source: U.S. Environmental Protection Agency, 1979a.

Notes:

[a]Except asbestos, which is given in fibers/l.

[b]All concentrations except for those for asbestos were calculated by multiplying the concentrations of the various wastestreams by the normalized percentage of the total flow and then subtracting the concentration present in the intake.

ND = Not Detected in sample.

Blanks indicate insufficient data.

TABLE 6-47. CONCENTRATIONS OF TOXIC POLLUTANTS FOUND IN RAW WASTEWATER OF THE PRIMARY COLUMBIUM AND PRIMARY TANTALUM CATEGORIES

Toxic Pollutant	Number of		Concentration (μg/l)[a,b]		
	Analyses	Times Detected	Range	Median	Mean
Metals and inorganics					
Antimony	3	2	ND-11,000	10	3,700
Arsenic	3	3	180-14,000	380	4,900
Asbestos	3	1		1.4×10^{10}	
Beryllium	3	3	20-190	89	100
Cadmium	3	3	8.0-19,000	48	6,400
Chromium	3	3	3,000-520,000	3,000	180,000
Copper	3	3	400-260,000	500	87,000
Cyanide	3	3	0.002-0.012	0.004	0.006
Lead	3	3	$3{,}000-2.7 \times 10^{7}$	3,000	9.0×10^{6}
Mercury	3	3	<0.1-36	6.0	14
Nickel	3	3	600-2,600	2,000	1,700
Selenium	3	2	ND-24,000	<10	8,000
Silver	3	3	<20-610	60	230
Thallium	3	3	ND-<100	24	41
Zinc	3	3	540-700,000	6,000	240,000
Phthalates					
Bis(2-ethylhexyl)phthalate	15	12	ND-1,100	22	150
Butyl benzyl phthalate	15	2	ND-47		6.3
Di-n-butyl phthalate	15	5	ND-60		12
Diethyl phthalate	15	1	ND-17		1.7
Dimethyl phthalate	15	2	ND-39		4.1
Di-n-octyl phthalate	15	1	ND-95		6.6
Phenols					
Pentachlorophenol	8	1	ND-17		2.1

(continued)

TABLE 6-47. (continued)

Toxic Pollutant	Number of Analyses	Number of Times Detected	Concentration (μg/l)[a,b] Range	Median	Mean
Aromatics					
Benzene	22	2	ND-44		4.4
Chlorobenzene	22	0			
2,4-Dinitrotoluene	15	1	ND-16		1.7
2,6-Dinitrotoluene	15	1	ND-16		1.7
Ethylbenzene	22	0			
Nitrobenzene	15	2	ND-163		18
Toluene	22	0			
1,2,4-Trichlorobenzene	15	2	ND-260		22
Polycyclic aromatic hydrocarbons					
Acenaphthene	15	1	ND-17		1.1
Acenaphthylene	15	1	ND-2.0		0.2
Anthracene	15	1	ND-2.0		0.3
Benz(a)anthracene	15	1	ND-1.0		0.1
Benzo(a)pyrene	15	1	ND-1.0		0.1
Benzo(b)fluoranthene	15	0			
Benzo(ghi)perylene	15	1	ND-2.0		0.2
Benzo(k)fluoranthene	15	0			
2-Chloronaphthalene	15	1	ND-3.0		0.3
Chyrsene	15	1	ND-45		3.1
Dibenz(ah)anthracene	15	1	ND-4.0		0.3
Fluoranthene	15	1	ND-7.2		1.1
Fluorene	15	2	ND-2.0		1.3
Indeno(1,2,3-cd)pyrene	15	1	ND-4.0		0.3
Naphthalene	15	1	ND-84		6.1
Phenanthrene	15	1	ND-2.0		0.3
Pyrene	15	1	ND-3.0		0.5

(continued)

TABLE 6-47. (continued)

Toxic Pollutant	Number of		Concentration (μg/l)[a,b]		
	Analyses	Times Detected	Range	Median	Mean
Polychlorinated biphenyls and related compounds					
Aroclor	15	1	ND-32		2.6
Aroclor	15	1	ND-52		4.1
Halogenated aliphatics					
Bromoform	22	1	ND-21		1.2
Carbon tetrachloride	22	2	ND-74		5.1
Chlorodibromomethane	22	3	ND-81		5.2
Chloroform	22	7	ND-140		7.8
Dichlorobromomethane	22	1	ND-13		0.6
1,2-Dichloroethane	22	6	ND-150		13
1,1-Dichloroethylene	22	1	ND-22		1.4
1,2-Trans-dichloroethylene	22	6	ND-480		49
Hexachloroethane	15	1	ND-23		1.5
Methylene chloride	22	1	ND-88,000		4,000
1,1,2,2-Tetrachloroethane	15	1	ND-6.0		0.5
Tetrachloroethylene	22	1	ND-65		3.6
1,1,1-Trichloroethane	22	2	ND-40		2.5
1,1,2-Trichloroethane	22	2	ND-29		2.1
Trichloroethylene	22	3	ND-230		21

(continued)

TABLE 6-47. (continued)

Toxic Pollutant	Number of Analyses	Times Detected	Concentration (μg/l)[a,b] Range	Median	Mean
Pesticides and metabolites					
Aldrin	15	1	ND-4.0		0.3
α-BHC	15	1	ND-0.04		
β-BHC	15	1	ND-4.5		0.4
δ-BHC	15	1	ND-4.0		0.3
γ-BHC	15	1	ND-0.03		
Chlordane	15	1	ND-0.8		0.1
4,4'-DDE	15	1	ND-0.4		
4,4'-DDT	15	1	ND-1.0		0.1
Dieldrin	15	1	ND-0.1		
α-Endosulfan	15	1	ND-0.01		
Endosulfan sulfate	15	1	ND-0.03		
Endrin	15	1	ND-5.4		0.4
Endrin aldehyde	15	1	ND-0.2		
Heptachlor	15	1	ND-0.5		
Heptachlor epoxide	15	1	ND-0.1		
Isophorone	15	1	ND-29		2.1
Toxaphene	15	1	ND-0.1		

Source: U.S. Environmental Protection Agency, 1979a.

Notes:

[a]Except asbestos, which is given in fibers/l.

[b]Concentrations were calculated by multiplying the concentrations of the various wastestreams by the normalized percentage of the total flow and then subtracting the concentration present in the intake.

ND = Not Detected in sample.

Blanks indicate insufficient data.

TABLE 6-48. CONCENTRATIONS OF TOXIC POLLUTANTS FOUND IN RAW WASTEWATER FROM THE PRIMARY COPPER CATEGORY

Toxic Pollutant	Number of		Concentration (μg/l)[a]		
	Analyses	Times Detected	Range	Median	Mean
Metals and inorganics					
Antimony	3	3	<50-3,300	100	1,200
Arsenic	3	3	<2.0-310,000	9,300	110,000
Beryllium	3	3	<2-7.7	6.0	5.2
Cadmium	3	3	<5-9,600	7.0	3,200
Chromium	3	3	<10-73	51	45
Copper	3	3	1,600-450,000	2,300	150,000
Cyanide	2	2	<0.001-<0.02		0.01
Lead	3	3	<20-170,000	470	56,000
Mercury	3	3	<0.5-48	4.6	18
Nickel	3	3	<20-1,100	340	490
Selenium	3	3	6.0-310	15	110
Silver	3	3	20-480	54	185
Thallium	3	3	21-<100	<100	74
Zinc	3	3	30-150,000	400	50,000
Phthalates					
Bis(2-ethylhexyl)phthalate	11	5	ND-78		17
Butyl benzyl phthalate	11	1	ND-1.0		0.1
Di-n-butyl phthalate	11	2	ND-75	0.7	7.6
Di-n-octyl phthalate	11	1	ND-3.0		0.3
Phenols					
2,4-Dimethylphenol	2	1	ND-14		7.0
Aromatics					
Benzene	11	1	ND-3.0		0.7
Chlorobenzene	11	1	ND-40		8.4
Toluene	11	1	ND-1.0		0.2

(continued)

TABLE 6-48. (continued)

Toxic Pollutant	Number of Analyses	Number of Times Detected	Concentration (μg/l)[a] Range	Median	Mean
Polycyclic aromatic hydrocarbons					
Acenaphthylene	11	1	ND-3.0		0.3
Anthracene	11	4	ND-21		6.1
Benz(a)anthracene	11	1	ND-1.0		0.1
Chrysene	11	0			
Fluoranthene	11	1	ND-1.0		0.3
Fluorene	11	1	ND-1.0		0.1
Phenanthrene	11	4	ND-21	7.0	7.1
Pyrene	11	1	ND-1.0		0.4
Polychlorinated biphenyls and related compounds					
Aroclor	9	1	ND-0.6		0.1
Aroclor	11	1	ND-0.7		0.1
Halogenated aliphatics					
Carbon tetrachloride	11	4	ND-40		8.4
Chlorodibromomethane	11	2	ND-13		1.2
Chloroform	11	4	ND-93	5.0	16
Dichlorobromomethane	11	2	ND-14		1.3
1,2-Dichloroethane	11	1	ND-7.0		0.6
1,1-Dichloroethylene	11	0			
Methylene chloride	11	1	ND-6.8		0.6
1,1,2,2-Tetrachloroethane	11	4	ND-12		1.9
Tetrachloroethylene	11	5	ND-15	4.0	5.4
1,1,1-Trichloroethane	11	0			
1,1,2-Trichloroethane	11	1	ND-2.0		0.2
Trichloroethylene	11	1	ND-9.0		1.5

(continued)

TABLE 6-48. (continued)

Toxic Pollutant	Number of		Concentration (μg/l)[a]		
	Analyses	Times Detected	Range	Median	Mean
Pesticides and metabolites					
β-BHC	11	1	ND-0.01		
γ-BHC	11	1	ND-0.04		
Chlordane	11	1	ND-0.2		
4,4'-DDT	11	1	ND-0.01		
4,4'-DDT	11	1	ND-0.02		
Dieldrin	11	1	ND-0.02		
α-Endosulfan	11	0			
β-Endosulfan	11	1	ND-0.01		
Endosulfan sulfate	11	0			
Endrin	11	1	ND-0.1		
Endrin aldehyde	11	1	ND-0.4		0.1
Heptachlor	11	1	ND-0.01		
Heptachlor epoxide	11	1	ND-0.01		0.01
Isophorone	11	1	ND-3.0		0.3

Source: U.S. Environmental Protection Agency, 1979a.

Note:

[a]Concentrations were calculated by multiplying the concentrations of the various wastestreams by the normalized percentage of the total flow and then subtracting the concentration present in the intake.

ND = Not Detected in sample.

Blanks indicate insufficient data.

TABLE 6-49. CONCENTRATIONS OF TOXIC POLLUTANTS FOUND IN RAW WASTEWATER FROM THE SECONDARY COPPER CATEGORY[a]

Toxic Pollutant	Number of		Concentration (μg/l)[b]		
	Analyses	Times Detected	Range	Median	Mean
Metals and inorganics					
Antimony	5	2	ND-11,000	ND	2,200
Arsenic	5	3	ND-4,200	100	940
Asbestos	2	2	3.3×10^7-1.0×10^{11}		5.0×10^{10}
Beryllium	5	4	ND-160	30	58
Cadmium	5	5	5.0-1,200	50	390
Chromium	5	5	5.0-2,100	<240	640
Copper	5	5	620-2.1×10^6	40,000	450,000
Cyanide	4	4	<0.001-0.026	0.006	0.010
Lead	5	5	450-53,000	10,000	17,000
Mercury	5	5	ND-0.6	0.53	0.35
Nickel	5	5	7.0-3.1×10^6	3,000	620,000
Selenium	5	2	ND-270	ND	98
Silver	5	3	ND-1,600	<10	370
Thallium	5	2	ND-53	ND	21
Zinc	5	5	1,400-1.5×10^6	40,000	330,000
Phthalates					
Bis(2-ethylhexyl)phthalate	12	10	ND-7,000	53	1,100
Butyl benzyl phthalate	12	2	ND-56		5.3
Di-n-butyl phthalate	12	6	ND-390	9.5	56
Diethyl phthalate	12	3	ND-83		11
Dimethyl phthalate	12	0			
Di-n-octyl phthalate	12	2	ND-67	5.8	
Aromatics					
Benzene	10	2	ND-13		1.3
Ethylbenzene	10	1	ND-4.0		0.4
Hexachlorobenzene	12	1	ND-5,000		420
Nitrobenzene	12	0			
Toluene	10	1	ND-10		1.7

(continued)

TABLE 6-49. (continued)

Toxic Pollutant	Number of		Concentration (μg/l)[b]		
	Analyses	Times Detected	Range	Median	Mean
Polycyclic aromatic hydrocarbons					
Acenaphthene	12	3	ND-36		4.6
Acenaphthylene	12	4	ND-120		23
Anthracene	12	3	ND-3,000		260
Benzo(a)pyrene	12	1	ND-1.0		0.1
Benzo(b)fluoranthene	12	0			
Benzo(k)fluoranthene	12	0			
Chrysene	12	3	ND-10,000		840
Dibenz(ah)anthracene	12	0			
Fluoranthene	12	4	ND-3,000	1.0	280
Fluorene	12	4	ND-94		14
Indeno(1,2,3-cd)pyrene	12	0			
Naphthalene	12	4	ND-5,000		550
Phenanthrene	12	4	ND-3,000		260
Pyrene	12	4	ND-7,000		610
Polychlorinated biphenyls and related compounds					
Aroclor 1248	14	1	ND-2.0		0.5
Aroclor 1254	14	1	ND-3.0		0.5
Halogenated aliphatics					
Carbon tetrachloride	10	2	ND-120		12
Chloroform	10	6	ND-1,000	7.0	130
Dichlorobromomethane	10	0			
1,2-Dichloroethane	10	1	ND-32		3.2
1,1-Dichloroethylene	10	3	ND-530		57
1,2-Trans-dichloroethylene	10	1	ND-5.0		0.5
Methylene chloride	10	3	ND-510		80
1,1,2,2-Tetrachloroethane	10	1	ND-4.0		0.4
Tetrachloroethylene	10	3	ND-72		8.8

(continued)

TABLE 6-49. (continued)

Toxic Pollutant	Number of		Concentration (μg/l)[b]		
	Analyses	Times Detected	Range	Median	Mean
Halogenated aliphatics (continued)					
Trichloroethylene	10	2	ND-70		7.1
Pesticides and metabolites					
Aldrin	14	1	ND-0.2		
α-BHC	14	1	ND-0.2		
β-BHC	14	1	ND-0.02		
δ-BHC	14	1	ND-0.2		
γ-BHC	14	1	ND-0.04		
Chlordane	14	1	ND-0.7		0.1
4,4'-DDE	14	1	ND-0.02		
4,4'-DDD	14	1	ND-0.1		
4,4'-DDT	14	1	ND-0.03		
Dieldrin	14	1	ND-0.03		
α-Endosulfan	14	1	ND-0.3		
β-Endosulfan	14	1	ND-0.3		
Endrin	14	1	ND-0.4		
Endrin aldehyde	14	1	ND-0.3		
Heptachlor	14	1	ND-0.02		
Heptachlor epoxide	14	0			
Toxaphene	14	1	ND-0.4		

Source: U.S. Environmental Protection Agency, 1979a.

Notes:

[a]Some numbers in this table do not represent the concentration of the combined total wastewater from the plant but instead represent only one or several wastestreams. This is due to one or more of the streams not having concentration values reported.

[b]Except asbestos, which is given in fibers/L.

ND = Not Detected in sample.

Blanks indicate insufficient data.

TABLE 6-50. CONCENTRATIONS OF TOXIC POLLUTANTS FOUND IN RAW WASTEWATER OF THE PRIMARY LEAD CATEGORY[a]

	Number of		Concentration (μg/l)[b]		
Toxic Pollutant	Analyses	Times Detected	Range	Median	Mean
Metals and inorganics					
Antimony	3	2	ND-<330	3.1	110
Arsenic	3	3	58-96	93	82
Beryllium	3	1	ND-6.7	ND	2.2
Cadmium	3	3	690-2,700	1,300	1,600
Chromium	3	3	9.1-30	14	18
Copper	3	3	100-5,300	610	2,000
Cyanide	2	2	<0.02-0.13		0.08
Lead	3	3	7,900-24,000	10,000	14,000
Mercury	3	3	0.29-7.5	0.68	2.8
Nickel	3	3	50-150	130	110
Selenium	3	3	3.1-<13	5.4	7.2
Silver	3	2	ND-<20	7.0	9.0
Thallium	3	2	ND-<100	15	38
Zinc	3	3	2,700-20,000	5,300	9,300
Polycyclic aromatic hydrocarbons					
Pyrene	3	1	ND-7.0		2.3
Halogenated aliphatics					
Methylene chloride	4	2	ND-25	3.0	7.8

Source: U.S. Environmental Protection Agency, 1979a.

Notes:

[a]Some numbers in this table do not represent the concentration of the combined total wastewater from the plant but instead represent only one or several wastestreams. This is due to one or more of the streams not having concentration values reported.

[b]Concentrations were calculated by multiplying the concentrations of the various wastestreams by the normalized percentage of the total flow and then subtracting the concentration in the intake.

ND = Not Detected in sample.

Blanks indicate insufficient data.

TABLE 6-51. CONCENTRATIONS OF TOXIC POLLUTANTS FOUND IN RAW WASTEWATER OF THE SECONDARY LEAD CATEGORY

Toxic Pollutant	Number of		Concentration (μg/l)[a,b]		
	Analyses	Times Detected	Range	Median	Mean
Metals and inorganics					
Antimony	4	4	1,700-80,000	38,000	39,000
Arsenic	3	3	3,000-13,000	7,000	7,700
Asbestos	1	1		1.3×10^{11}	
Beryllium	3	3	1.0-30	3.2	11
Cadmium	4	4	220-1,900	800	930
Chromium	4	4	110-1,000	480	520
Copper	4	4	220-8,200	3,200	3,700
Cyanide	4	4	0.002-<0.01	0.006	0.006
Lead	4	4	$7,000-1.8 \times 10^6$	22,000	460,000
Mercury	4	4	0.6-12	0.78	3.5
Nickel	4	4	210-2,000	960	1,000
Selenium	3	1	ND-<2.0	ND	0.67
Silver	3	3	90-250	100	150
Thallium	3	3	50-620	350	340
Zinc	4	4	790-15,000	3,600	5,700
Phthalates					
Bis(2-ethylhexyl)phthalate	5	5	ND-580	30	180
Butyl benzyl phthalate	5	2	ND-85		17
Di-n-butyl phthalate	5	4	ND-27	13	12
Dimethyl phthalate	5	3	ND-13		2.6
Di-n-octyl phthalate	5	3	ND-27	2.0	9.0
Nitrogen compounds					
Benzidine	5	1	ND-6.0		1.2
Aromatics					
Benzene	10	1	ND-2.0		0.2
Chlorobenzene	10	1	ND-5.0		0.5
Ethylbenzene	10	1	ND-1.2		0.3

(continued)

TABLE 6-51. (continued)

Toxic Pollutant	Number of		Concentration (µg/l)[a,b]		
	Analyses	Times Detected	Range	Median	Mean
Aromatics (continued)					
Nitrobenzene	5	2	ND-16		3.2
Toluene	10	0			
Polycyclic aromatic hydrocarbons					
Acenaphthylene	5	2	ND-35	3.0	8.6
Anthracene	5	2	ND-20		4.0
Benzo(a)pyrene	5	2	ND-10		2.0
Benzo(b)fluoranthene	5	1	ND-5.3		1.6
Benzo(ghi)perylene	5	0			
Benzo(k)fluoranthene	5	1	ND-5.3		1.6
Chrysene	5	3	ND-540	40	140
Fluoranthene	5	3	ND-27	1.0	7.6
Fluorene	5	1	ND-2.0		0.4
Indeno(1,2,3-cd)pyrene	5	1	ND-1.0		0.2
Naphthalene	5	1	ND-4.0		0.8
Phenanthrene	5	2	ND-20		4.6
Pyrene	5	3	ND-38	1.0	10
Polychlorinated biphenyls and related compounds					
Aroclor 1248	5	1	ND-3.1	1.3	1.4
Aroclor 1254	5	1	ND-2.6	1.8	1.3
Halogenated aliphatics					
Bromoform	10	2	ND-49		5.7
Chloroform	10	4	ND-31	3.0	6.9
1,2-Dichloroethane	10	2	ND-10	4.0	4.0
1,1-Dichloroethylene	10	2	ND-10	2.0	3.7
1,2-Trans-dichloroethylene	10	0			
1,1,2,2-Tetrachloroethane	10	1	ND-4.0		1.0

(continued)

TABLE 6-51. (continued)

Toxic Pollutant	Number of Analyses	Number of Times Detected	Concentration (μg/l)[a,b] Range	Median	Mean
Halogenated aliphatics (continued)					
Tetrachloroethylene	10	1	ND-5.0		1.1
1,1,2-Trichloroethane	10	0			
Trichloroethylene	10	1	ND-6.0		0.8
Pesticides and metabolites					
Aldrin	5	1	ND-0.1		
α-BHC	5	1	ND-0.2		
β-BHC	5	1	ND-0.3	0.1	0.1
γ-BHC	5	1	ND-0.1		
Chlordane	5	1	ND-0.2	0.2	0.2
4,4'-DDE	5	1	ND-0.2		
4,4'-DDT	5	1	ND-0.1		
Dieldrin	5	1	ND-0.2		
α-Endosulfan	5	1	ND-0.2		
β-Endosulfan	5	0			
Endrin	5	1	ND-4.0		1.0
Endrin aldehyde	5	1	0.6		0.1
Heptachlor	5	1	ND-0.3	0.1	0.1
Heptachlor epoxide	5	1	ND-0.2	0.1	0.1
Isophorone	5	1	ND-2.7		1.8

Source: U.S. Environmental Protection Agency, 1979a.

Notes:

[a]Except asbestos, which is given in fibers/l.

[b]All concentrations except for asbestos were calculated by multiplying the concentrations of the various wastestreams by the normalized percentage of the total flow and then subtracting the concentration present in the intake.

ND = Not Detected in sample.

Blanks indicate insufficient data.

TABLE 6-52. CONCENTRATIONS OF TOXIC POLLUTANTS FOUND IN RAW WASTEWATER OF THE SECONDARY SILVER CATEGORY

Toxic Pollutant	Number of Analyses	Number of Times Detected	Concentration (μg/l)[a,b] Range	Median	Mean
Metals and inorganics					
Antimony	3	1	ND-25,000	ND	8,300
Arsenic	3	3	40-900	40	330
Asbestos	1	1	5.8×10^8		
Beryllium	3	2	ND-<20	19	13
Cadmium	3	3	1,000-80,000	3,200	28,000
Chromium	3	3	2,000-27,000	20,000	16,000
Copper	3	3	7,400-70,000	60,000	46,000
Cyanide	3	3	0.001-2.1	0.05	0.72
Lead	3	3	4,000-50,000	4,200	19,000
Mercury	3	1	ND-5.5	ND	1.8
Nickel	3	3	1,100-800,000	30,000	280,000
Selenium	3	1	ND-590	ND	200
Silver	3	3	<250-4,700	410	1,800
Thallium	3	1	ND-510	ND	170
Zinc	3	3	8,400-2.0×10^6	20,000	680,000
Phthalates					
Bis(2-ethylhexyl)phthalate	5	5	7.0-34	11	18
Butyl benzyl phthalate	5	2	ND-53		11
Di-n-butyl phthalate	5	5	ND-300	15	75
Diethyl phthalate	5	2	ND-38		7.6
Di-n-octyl phthalate	5	4	ND-58	33	30

(continued)

TABLE 6-52. (continued)

Toxic Pollutant	Number of Analyses	Number of Times Detected	Concentration (μg/l)[a,b] Range	Median	Mean
Aromatics					
Benzene	6	6	3.0-160	66	75
Chlorobenzene	6	1	ND-9.0	0.5	2.8
Ethylbenzene	6	4	ND-21		9.2
Toluene	6	6	3.0-55	18	21
Polycyclic aromatic hydrocarbons					
Acenaphthene	5	2	ND-10		2.0
Anthracene	5	2	ND-4.0		0.8
Fluoranthene	5	0			
Naphthalene	5	1	ND-1.0		0.2
Phenanthrene	5	2	ND-4.0		0.8
Pyrene	5	1	ND-2,100		430
Polychlorinated biphenyls and related compounds					
Aroclor 1248	3	1	ND-0.5		0.2
Aroclor 1254	3	1	ND-0.7		0.2
Halogenated aliphatics					
Bromoform	6	1	ND-65		11
Carbon tetrachloride	6	2	ND-2,300		380
Chlorodibromomethane	6	2	ND-64		11
Chloroform	6	1	ND-890	8.5	160
1,2-Dichloroethane	6	4	ND-560	21	120
1,1-Dichloroethylene	6	3	ND-6,100		1,100
1,2-Trans-dichloroethylene	6	0			
Methylene chloride	6	4	ND-3,100	170	1,000

(continued)

TABLE 6-52. (continued)

Toxic Pollutant	Number of		Concentration (μg/l)[a,b]		
	Analyses	Times Detected	Range	Median	Mean
Halogenated aliphatics (continued)					
1,1,2,2-Tetrachloroethane	4	2	ND-32		8.0
Tetrachloroethylene	6	6	ND-109	36	43
1,1,1-Trichloroethane	6	3	ND-22		7.3
Trichloroethylene	6	6	ND-900	230	360
Pesticides and metabolites					
Aldrin	3	1	ND-1.1		0.4
α-BHC	3	0			
β-BHC	3	1	ND-0.02		
δ-BHC	3	1	ND-1.1		0.4
γ-BHC	3	0			
Chlordane	3	1	ND-0.1		
4,4'-DDE	3	1	ND-0.01		
4.4'-DDD	3	1	ND-0.1		
4,4'-DDT	3	1	ND-0.01		
Dieldrin	3	1	ND-0.01		
Endrin	3	1	ND-2.0		0.7
Endrin aldehyde	3	0			
Heptachlor	3	1	ND-0.02		

Source: U.S. Environmental Protection Agency, 1979a.

Notes:

[a]Except asbestos, which is given in fibers/l.

[b]Concentrations were calculated by multiplying the concentrations of the various wastestreams by the normalized percentage of the total flow and then subtracting the concentration in the intake.

ND = Not Detected in sample.

Blanks indicate insufficient data.

TABLE 6-53. CONCENTRATIONS OF TOXIC POLLUTANTS FOUND IN RAW WASTEWATER OF THE PRIMARY TUNGSTEN CATEGORY[a]

	Number of		Concentration (μg/l)[b,c]		
Toxic Pollutant	Analyses	Times Detected	Range	Median	Mean
Metals and inorganics					
Antimony	3	1	ND-700	ND	230
Arsenic	3	3	10-7,200	210	2,500
Asbestos	1	1	6.0×10^9		
Beryllium	3	3	<2.0-29	<10	14
Cadmium	3	3	19-190	<20	76
Chromium	3	3	44-2,000	48	700
Copper	3	3	95-5,000	120	1,700
Cyanide	3	3	0.002-0.14	0.013	0.052
Lead	3	3	<200-20,000	240	6,800
Mercury	3	3	0.20-3.00	1.0	1.4
Nickel	3	3	<50-1,000	92	380
Selenium	3	2	ND-1,000	20	340
Silver	3	3	76-270	86	140
Thallium	3	2	ND-600	200	270
Zinc	3	3	250-1,900	520	890
Phthalates					
Bis(2-ethylhexyl)phthalate	5	5	ND-880	10.0	180
Di-n-butyl phthalate	5	3	ND-23		
Diethyl phthalate	5	0			
Dimethyl phthalate	5	0			
Di-n-octyl phthalate	5	2	ND-1.0		0.2
Aromatics					
Benzene	9	1	ND-3.0		0.7
Chlorobenzene	9	0			
Ethylbenzene	9	1	ND-11		2.2
Nitrobenzene	5	0			
Toluene	9	3	ND-45	3.0	11
1,2,4-Trichlorobenzene	5	0			

(continued)

TABLE 6-53. (continued)

Toxic Pollutant	Number of		Concentration (μg/l)[b,c]		
	Analyses	Times Detected	Range	Median	Mean
Polycyclic aromatic hydrocarbons					
Acenaphthene	5	2	ND-100		21
Acenaphthylene	5	2	ND-110		23
Anthracene	5	2	ND-150		30
Benzo(a)pyrene	5	1	ND-1.0		0.2
Chrysene	5	2	ND-240		48
Fluoranthene	5	1	ND-1.0		0.2
Fluorene	5	2	ND-55		11
Naphthalene	5	2	ND-1,100		220
Phenanthrene	5	0			
Pyrene	5	0			
Polychlorinated biphenyls and related compounds					
Aroclor 1248	5	1	ND-1.0	0.2	0.3
Aroclor 1254	5	1	ND-5.4	0.4	1.4
Halogenated aliphatics					
Bromoform	9	3	ND-48		9.3
Chlorodibromomethane	9	2	ND-38		4.2
Chloroform	9	3	ND-1,800		210
Dichlorobromomethane	9	0			
1,2-Dichloroethane	9	1	ND-8.0		2.1
1,1-Dichloroethylene	9	3	ND-19		4.3
1,2-Trans-dichloroethylene	9	1	ND-2.0		0.2
1,1,2,2-Tetrachloroethane	9	2	ND-35		5.2
Tetrachloroethylene	9	6	ND-69		20
1,1,1-Trichloroethane	9	2	ND-10		1.1
Trichloroethylene	9	3	ND-19		2.9

(continued)

TABLE 6-53. (continued)

Toxic Pollutant	Number of		Concentration (µg/l)[b,c]		
	Analyses	Times Detected	Range	Median	Mean
Pesticides and metabolites					
Aldrin	5	1	ND-7.0		1.4
α-BHC	5	1	ND-0.6		0.1
β-BHC	5	1	ND-0.1		
γ-BHC	5	1	ND-0.2		0.1
Chlordane	5	1	ND-1.2		0.2
4,4'-DDD	5	1	0		
4,4'-DDT	5	1	ND-0.1		
Dieldrin	5	1	ND-0.1		0.1
α-Endosulfan	5	2	ND-15	0.1	3.2
β-Endosulfan	5	2	ND-15		3.1
Endrin	5	1	ND-0.8		0.2
Endrin aldehyde	5	1	ND-0.9	0.2	0.3
Heptachlor	5	1	ND-0.2		0.1
Heptachlor epoxide	5	1	ND-0.2		0.1
Isophorone	9	0			

Source: U.S. Environmental Protection Agency, 1979a.

Notes:

[a]Some numbers in this table do not represent the concentration of the combined total wastewater from the plant but instead represent only one or several wastestreams. This is due to one or more of the streams not having concentration values reported.

[b]Except asbestos, which is given in fibers/l.

[c]All concentrations except those for cyanide and asbestos were calculated by multiplying the concentrations of the various wastestreams by the normalized percentage of the total flow and then subtracting the concentration in the intake.

ND = Not Detected in sample.

Blanks indicate insufficient data.

TABLE 6-54. CONCENTRATIONS OF TOXIC POLLUTANTS FOUND IN RAW WASTEWATER OF THE PRIMARY ZINC CATEGORY

Toxic Pollutant	Number of Analyses	Number of Times Detected	Concentration ($\mu g/l$)[a,b] Range	Median	Mean
Metals and inorganics					
Antimony	4	4	<2.0-2,100	58	550
Arsenic	4	4	3.0-3,000	150	820
Asbestos	2	2	3.2×10^7-4.3×10^7		3.8×10^7
Beryllium	4	4	<2.0-<20	7.5	9.3
Cadmium	4	4	350-44,000	3,400	13,000
Chromium	4	4	<24-610	64	190
Copper	4	4	37-26,000	1,200	7,100
Cyanide	4	4	0.002-0.38	0.007	0.099
Lead	4	4	280-18,000	4,400	6,700
Mercury	4	4	2.9-52	5.4	16
Nickel	4	4	<50-4,300	590	1,400
Selenium	4	4	24-1,200	360	490
Silver	4	4	<25-740	58	220
Thallium	2	2	<20-360		190
Zinc	4	4	8,700-1.7×10^5	160,000	630,000
Phthalates					
Bis(2-ethylhexyl)phthalate	9	7	ND-98	15	28
Butyl benzyl phthalate	9	2	ND-30		3.3
Di-n-butyl phthalate	9	2	ND-26	5.0	3.6
Diethyl phthalate	9	2	ND-18		2.7
Dimethyl phthalate	9	2	ND-22		2.4
Di-n-octyl phthalate	9	0			
Nitrogen compounds					
3,3'-Dichlorobenzidine	9	0			

(continued)

TABLE 6-54. (continued)

Toxic Pollutant	Number of Analyses	Number of Times Detected	Concentration (μg/l)[a,b] Range	Median	Mean
Phenols					
Pentachlorophenol	9	1	ND-8.0		0.9
Aromatics					
Benzene	9	2	ND-24		2.7
Ethylbenzene	9	1	ND-2.0		0.2
Hexachlorobenzene	9	2	ND-100		11
Toluene	9	2	ND-54	7.0	7.5
1,2,4-Trichlorobenzene	9	0			
Polycyclic aromatic hydrocarbons					
Acenaphthylene	9	2	ND-18		2.0
Anthracene	9	1	ND-0.4		
Chrysene	9	2	ND-11		2.2
Fluoranthene	9	2	ND-15		1.7
Fluorene	9	2	ND-14		1.6
Naphthalene	9	0			
Phenanthrene	9	0			
Pyrene	9		ND-15		3.2
Polychlorinated biphenyls and related compounds					
Aroclor 1248	9	0			
Aroclor 1254	9	0			
Halogenated aliphatics					
Bromoform	9	0			
Chloroform	9	3	ND-71	53	16

(continued)

TABLE 6-54. (continued)

Toxic Pollutant	Number of		Concentration (μg/l)[a,b]		
	Analyses	Times Detected	Range	Median	Mean
Halogenated aliphatics (continued)					
1,1-Dichloroethane	9	2	ND-180		20
1,2-Dichloroethane	9	2	ND-22		2.9
1,1-Dichloroethylene	9	2	ND-23		2.6
Methylene chloride	9	5	ND-2,600		350
Tetrachloroethylene	9	1	ND-8.0		0.9
Trichloroethylene	9	2	ND-160	7.2	19
Trichlorofluoromethane	9	2	ND-100		12
Pesticides and metabolites					
Aldrin					
α-BHC	9	0			
β-BHC	9	0			
Chlordane	9	0			
4,4'-DDE	9	0			
4,4'-DDT	9	0			
Dieldrin	9	0			
Heptachlor	9	0			
Heptachlor epoxide	9	0			
Isophorone	9	2	ND-18		2.0

Source: U.S. Environmental Protection Agency, 1979a.

Notes:

[a]Except asbestos, which is given in fibers/l.

[b]All concentrations except those for cyanide and asbestos were calculated by multiplying the concentrations of the various wastestreams by the normalized percentages of the total flow and then subtracting the concentration in the intake.

ND = Not Detected in sample.

Blanks indicate insufficient data.

7. Geohydrologic Settings for Waste Disposal

REGIONAL DELINEATION

Several earth scientists have attempted to divide the United States into groundwater regions. Some of the earliest efforts were completed by Fuller (1905), Ries and Watson (1914), Thomas (1952), and McGuinness (1963). Recent studies by Heath (1982) have resulted in an excellent categorization of groundwater regions within the country. Heath (1982) divided the United States into 14 groundwater regions as shown in Figure 7-1. The 14 regions are:

Southeast Coastal Plain

Atlantic and Gulf Coastal Plain

Piedmont and Blue Ridge

Nonglaciated Central Region

Glaciated Central Region

Northeast and Superior Uplands

High Plains

Western Mountain Ranges

Alluvial Basins

Columbia Lava Plateau

Colorado Plateau and Wyoming Basin

Alaska

Hawaii

Alluvial Valleys.

Table 7-1 shows the principal physical and hydrologic conditions in each of the groundwater regions. This categorization is only partly successful, since no single dominant aquifer exists in some regions. An aquifer widely used in one part of a region may be little used in other parts of the same region. Heath (1982) describes the physical and hydrologic conditions in each region only briefly. Table 7-2 describes the geologic situation and the estimated ranges in values of the most important hydraulic characteristics.

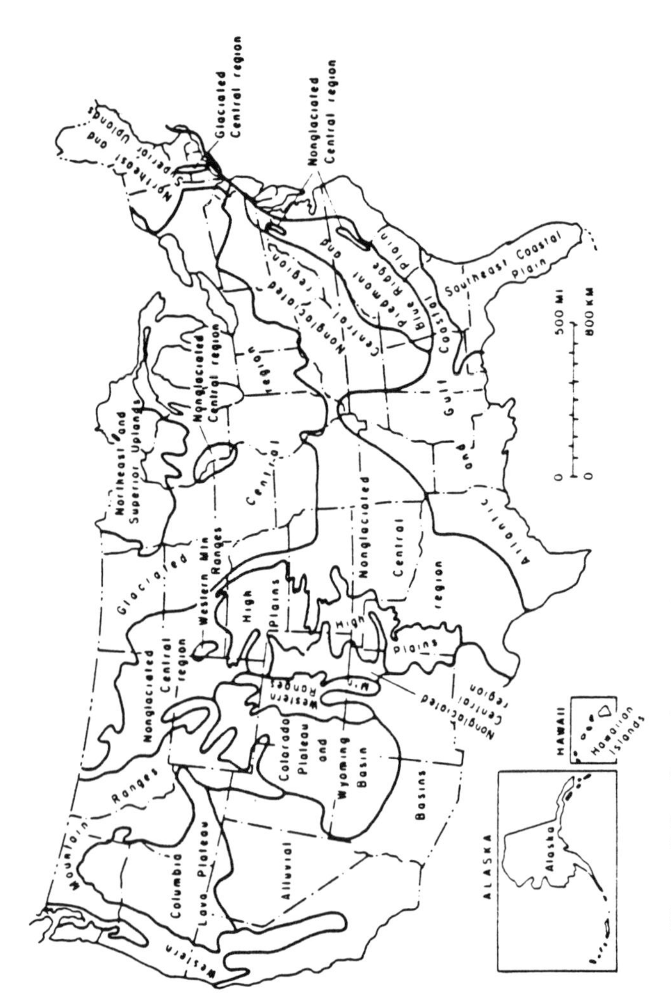

Figure 7-1. Groundwater regions of the United States excluding alluvial valley regions (after Heath, 1982).

TABLE 7-1. THE PRINCIPAL PHYSICAL AND HYDROLOGIC CHARACTERISTICS OF THE GROUNDWATER REGIONS OF THE UNITED STATES (after Heath, 1982)

Region No.	Name	Components of the system															Characteristics of the dominant aquifers																				
																	Water-bearing openings					Composition			Storage and transmission properties							Recharge and discharge conditions					
		Unconfined aquifer			Confining beds				Confined aquifers				Presence and arrangement				Primary			Secondary		Degree of solubility			Porosity			Transmissivity				Recharge			Discharge		
		Hydrologically insignificant	Minor aquifer	Dominant aquifer	Hydrologically insignificant	Thin, discontinuous, or v. leaky	Thick, impermeable	Interlayered with aquifers	Hydrologically insignificant	Not highly productive	Multiple aquifers	A single, dominant aquifer	Single dominant unconfined aquifer	Two interconnected aquifers	Unconfined aquifer, confining bed, confined aquifer	Complex interbedded sequence	Pores in unconsolidated dep.	Pores in semiconsolidated rocks	Tubes and cooling cracks in lava	Fractures and faults	Solution enlarged openings	Insoluble	Mixed soluble and insoluble	Soluble	Large (>0.2)	Moderate (0.01-0.2)	Small (<0.01)	Large (>2500 m^2 day $^{-1}$)	Moderate (250-2500 m^2 day $^{-1}$)	Small (25-250 m^2 day $^{-1}$)	Very small (<25 m^2 day $^{-1}$)	Uplands between streams	Losing streams	Leakage thru confining beds	Springs and surface seepage	Evaporation and basin sinks	Into other aquifers
1	Western Mountain Ranges		X		X					X				X			X			X		X				X				X		X	X		X		
2	Alluvial Basins		X					X			X					X	X					X			X			X					X			X	X
3	Columbia Lava Plateau		X					X			X					X	X		X			X					X	X					X		X		
4	Colorado Plateau and Wyoming Basin	X						X			X					X		X		X		X					X			X		X		X	X		
5	High Plains			X	X				X				X				X					X			X			X					X		X		
6	Nonglaciated Central Region		X					X			X					X		X		X	X		X				X		X			X		X	X		X
7	Glaciated Central Region		X			X					X					X	X	X		X	X		X			X			X					X			X
8	Piedmont and Blue Ridge		X		X					X				X			X			X		X					X				X	X			X		
9	Northeast and Superior Uplands		X			X				X				X			X			X		X					X			X		X			X		
10	Atlantic and Gulf Coastal Plain		X					X			X					X	X				X		X			X			X			X		X	X		X
11	Southeast Coastal Plain		X				X					X			X		X				X			X	X			X				X		X	X		
12	Alluvial Valleys			X		X			X				X				X					X			X			X				X			X		
13	Hawaii			X		X			X				X						X			X			X			X				X	X		X		
14	Alaska			X		X				X						X	X			X			X			X			X				X		X		

TABLE 7-2. COMMON RANGES IN THE HYDRAULIC CHARACTERISTICS OF GROUNDWATER REGIONS OF THE UNITED STATES (all values rounded to one significant figure) (after Heath, 1982)

Region No.	Region	Geologic situation	Common ranges in hydraulic characteristics of the dominant aquifers							
			Transmissivity		Hydraulic conductivity		Recharge rate		Well yield	
			$m^2 day^{-1}$	$ft^2 day^{-1}$	$m\ day^{-1}$	$ft\ day^{-1}$	$mm\ yr^{-1}$	$in\ yr^{-1}$	$m^3 min^{-1}$	$gal\ min^{-1}$
1	Western Mountain Ranges	Mountains with thin soils over fractured rocks, alternating with narrow alluvial and, in part, glaciated valleys	0.5-100	5-1,000	0.0003-15	0.001-50	3-50	0.1-2	0.04-0.4	10-100
2	Alluvial Basins	Thick[1/] alluvial (locally glacial) deposits in basins and valleys bordered by mountains	20-20,000	2,000-200,000	30-600	100-2,000	0.03-30	0.001-1	0.4-20	100-5,000
3	Columbia Lava Plateau	Thick lava sequence interbedded with unconsolidated deposits and overlain by thin soils	2,000-500,000	20,000-5,000,000	200-3,000	500-10,000	5-300	0.2-10	0.4-80	100-20,000
4	Colorado Plateau and Wyoming Basin	Thin[1/] soils over fractured sedimentary rocks	0.5-100	5-1,000	0.003-2	0.01-5	0.3-50	0.01-2	0.04-2	10-1,000
5	High Plains	Thick alluvial deposits over fractured sedimentary rocks	1,000-10,000	10,000-100,000	30-300	100-1,000	5-80	0.2-3	0.4-10	100-3,000
6	Nonglaciated Central region	Thin regolith over fractured sedimentary rocks	300-10,000	3,000-100,000	3-300	10-1,000	5-500	0.2-20	0.4-20	100-5,000
7	Glaciated Central region	Thick glacial deposits over fractured sedimentary rocks	100-2,000	1,000-20,000	2-300	5-1,000	5-300	0.2-10	0.2-2	50-500
8	Piedmont and Blue Ridge	Thick regolith over fractured crystalline and metamorphosed sedimentary rocks	9-200	100-2,000	0.001-1	0.003-3	30-300	1-10	0.2-2	50-500
9	Northeast and Superior Uplands	Thick glacial deposits over fractured crystalline rocks	50-500	500-5,000	2-30	5-100	30-300	1-10	0.1-1	20-200
10	Atlantic and Gulf Coastal Plain	Complexly interbedded sands, silts, and clays	500-10,000	5,000-100,000	3-100	10-400	50-500	2-20	0.4-20	100-5,000
11	Southeast Coastal Plain	Thick layers of sand and clay over semi-consolidated carbonate rocks	1,000-100,000	10,000-1,000,000	30-3,000	100-10,000	30-500	1-20	4-80	1,000-20,000
12	Alluvial Valleys	Thick sand and gravel deposits beneath flood-plains and terraces of streams	200-50,000	2,000-500,000	30-2,000	100-5,000	50-500	2-20	0.4-20	100-5,000
13	Hawaiian Islands	Lava flows segmented by dikes, interbedded with ash deposits, and partly overlain by alluvium	10,000-100,000	100,000-1,000,000	200-3,000	500-10,000	30-1,000	1-40	0.4-20	100-5,000
14	Alaska	Glacial and alluvial deposits in part perennially frozen and overlying crystalline, metamorphic, and sedimentary rocks	100-10,000	1,000-100,000	30-600	100-2,000	3-300	0.1-10	0.04-4	10-1,000

[1/] An average thickness of about five meters was used as the break point between thick and thin.

The classification of groundwater systems by Heath is intended to "enhance the public's understanding of groundwater." His work is also useful for transferring hydrologic knowledge from one area to another. Heath's classification was consulted as a regional starting point for further development of generic geohydrologic settings could be developed. The generic settings identified were strongly influenced by the distribution of locations of actual hazardous waste disposal sites.

GENERIC GEOHYDROLOGIC DISPOSAL AREAS

Several geohydrologic settings have been selected to convey the unsaturated and saturated hydrologic complexity of potential waste disposal sites. Initially, the geohydrologic components for each setting are briefly described and illustrated to introduce the waste disposal site being considered.

The effect of topography, primarily as it may influence infiltration and discharge, is also discussed for each site. Under appropriate conditions of topography, soils, vegetation, and climate, deep percolation can be a significant factor in the water budget of the waste disposal zone. Groundwater flow, in many circumstances, is known to roughly conform to the shape of the surface topography. Therefore, topography can influence the rate (gradient) and direction of groundwater flow. Discharge zones are also related to topography. Topographically low areas, e.g., valley bottoms, are frequently down cut through several water bearing strata and provide an opportunity for discharge to occur.

The geohydrologic setting is also discussed for each waste disposal example. The characteristics of the geologic framework acquired during deposition, metamorphism, postformational movement, or solution weathering are presented and discussed relative to the hydraulics of the geologic units. Whenever possible, the hydraulic characteristics of the aquifer are expressed as hydraulic conductivity and storage. Hydraulic conductivity has been selected because it is a function of the geology rather than the saturated thickness, which varies from site to site. Storage coefficients or specific yields (unconfined conditions) are presented to give a relative idea of the volumes of water that a formation could yield. The relationships of sequences of geologic strata are assessed to highlight the various aquifer conditions (perched, confined, water table) that may be encountered.

With knowledge of the geohydrologic setting and the best waste disposal zone delineated, appropriate monitoring locations are designated. The intent of the monitoring is to detect water moving into and through the waste burial zone and any transportation of waste materials that may occur. All opportunities to monitor waste movement in the vicinity of the burial site are presented.

The generic geohydrologic settings selected for waste disposal discussions in this section are:

Coastal Plains

Glacial Deposits

River Flood Plains

Interior Drainage Basins

Desert and Semiarid Deposits

Fractured Rock Systems.

These settings have been identified as generically representing most waste disposal sites throughout the United States. When the generic setting covers different geologic cross sections, examples are provided.

COASTAL PLAIN

Case A

The waste disposal site evaluated in Case A is similar to Case B (sands over crystalline bedrock with an unconfined aquifer), except that the Case A underlying bedrock is fractured dolomite with solution cavities. In addition, the water table is present in the dolomite beneath the unconsolidated sand deposit. Figure 7-2 depicts the geohydrologic setting for Case A.

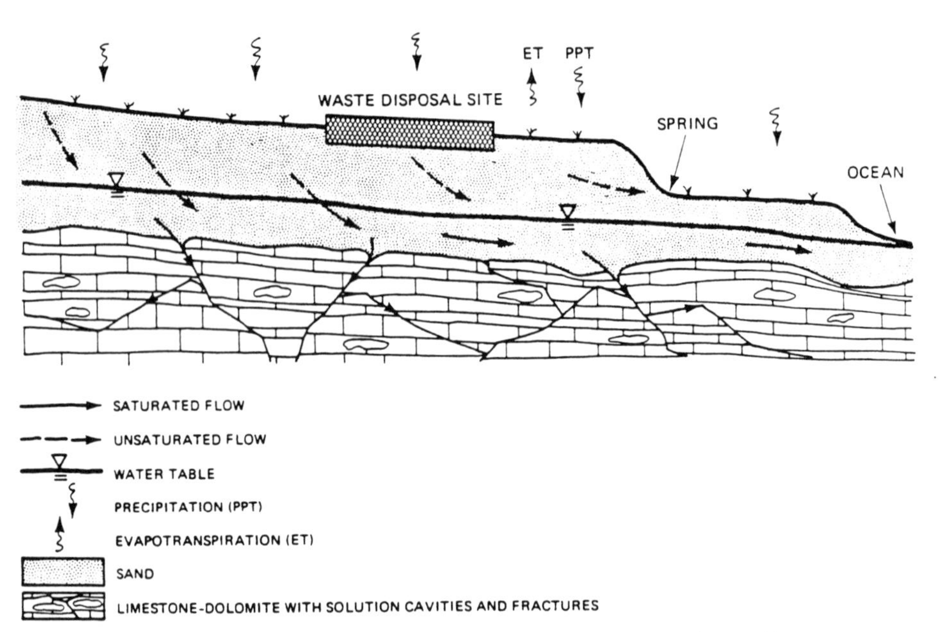

Figure 7-2. Coastal plain geohydrologic setting (Case A).

Topographical Influence on Site Hydrology--

The relatively flat-lying topography and the sandy soils associated with the coastal plain landform allow infiltration to occur at a moderate to high rate. Plans to bury waste in the unconsolidated sand should allow for the effects of water percolating through the disposal zone.

Geohydrologic Setting--

The basement rock specified for this setting is fractured dolomite with some solution cavities. The water table is present in the unconsolidated sand deposit and the dolomite bedrock. Dolomite is not normally a good aquifer because of its low permeability. In this situation, however, the secondary permeability associated with the fractures and solution cavities enhances the hydraulic characteristics of the formation. High values of hydraulic conductivity reported by Chow (1964), assumed to have secondary permeability associated with them, were approximately 250 gal/day/ft^2. For an unconfined water table aquifer, the storage coefficient or specific yield would be about 0.2 (Lohman, 1972).

Hazardous waste disposal facilities have been developed in coastal plain geohydrologic settings in California, Alabama, Mississippi, New Jersey, South Carolina, and New York (U.S. EPA, 1977a). Although this setting is not conducive to the control of leachate, geohydrologic settings do not appear to have been of primary concern during the development of sites before the late 1970's. Leachate, in this generic case, has the potential of surfacing along downgradient springs, thus contaminating surface waters or moving through the fractured bed rock to lower aquifers.

Monitoring for the impacts of waste burial in this geohydrologic setting should include observation of deep percolation into and through the waste zone. An effort should be made to determine if an impermeable safety liner has been installed at the site and its condition. In addition, intensive vadose zone monitoring should focus on the sandy soils above the water table and downgradient and as close to the waste cells as possible. Springs and wells developed in the saturated sand and fractured dolomite should be tested. A generic monitoring program is described in Section 8. Equipment that is effective in sandy materials with relatively high water content (less than 0.8 atmosphere of negative pressure) would probably be appropriate for this setting.

Case B

The Case B setting is a coastal plain with unconsolidated sands overlying shallow crystalline bedrock. Groundwater is present under water table conditions in the sand deposit. Figure 7-3 illustrates the geohydrologic setting.

Topographical Influence on Site Hydrology--

The coastal plain setting can be very complex because of the presence of several depositional environments and the potential for modification by post-depositional erosion and uplift. In the case considered here, the setting has been simplified and, as illustrated in Figure 7-3, the topography is relatively flat with a shallow sequence (30 to 100 feet) of medium- to fine-grained unconsolidated sands.

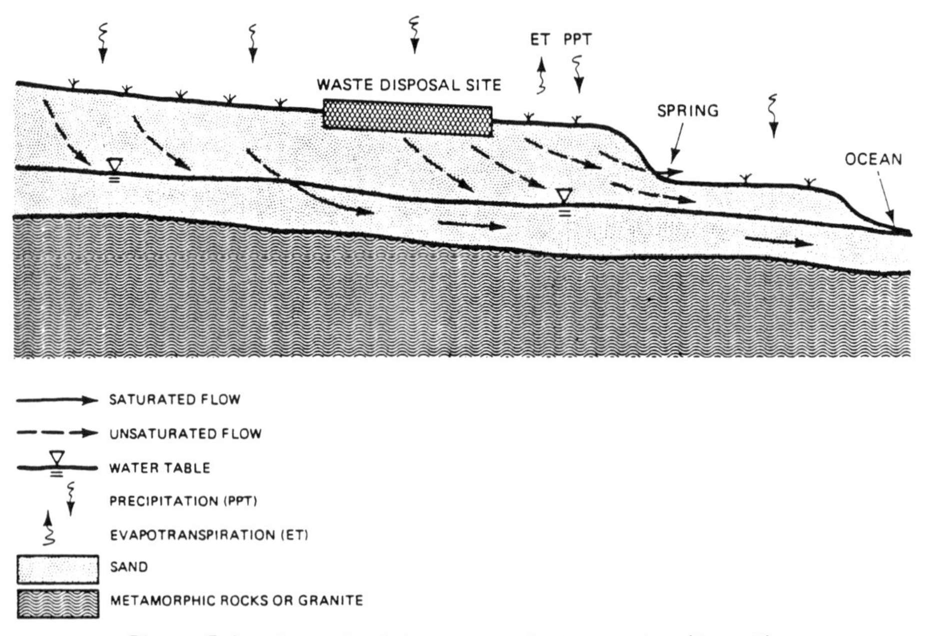

Figure 7-3. Coastal plain geohydrologic setting (Case B).

The association of this setting with the coastal province suggests that precipitation would exceed evapotranspiration demands leaving moisture available for deep percolation, thus providing recharge to groundwater. The gentle topography in a lowland (coastal) setting also enhances the potential for precipitation infiltration. The texture of the sediments would likely allow infiltration to occur at a moderate to high rate. Burial of hazardous waste in this setting must consider the percolation of water through the unconsolidated deposit.

The low topographic position of this potential waste disposal site as part of a coastline also suggests that groundwaters moving through the potential disposal site could be discharged to receiving rivers and/or oceans where the unconsolidated deposits terminate.

Geohydrologic Setting--

Crystalline bedrock depicted beneath the unconsolidated sands generally has an extremely low hydraulic conductivity and, therefore, is considered a perching bed, preventing the downward movement of groundwater from above.

The unconsolidated sands in this example contain a water table aquifer, perched on the relatively impermeable bedrock. The bottom of the burial zone for waste disposal could be reasonably expected to be above the saturated zone to minimize contamination of the water table aquifer. This may not be true, however, for waste disposal sites developed several years ago.

The storage coefficient, or specific yield when considering an unconfined aquifer, for the unconsolidated sands would likely range from approximately 0.20 to 0.32 (Chow, 1964). Here again, the potential for moderately high water content should be anticipated and appropriate vadose zone monitoring equipment selected.

A monitoring program designed to observe hazardous waste buried in the coastal plain setting described here would have to monitor deep percolation through the vadose zone and below where it might influence the water table aquifer. For old sites, multilevel samplers in the water table downgradient and near the waste cells can be used to detect pollutant plume stratification in the saturated zone. At new sites, careful attention should be given to the placement of waste cells or, possibly, recommendations made to change the location of hazardous waste facilities to a more conducive geohydrologic setting. If this setting uses a protective liner, however, leachate detection systems and backup detection units described in Section 8 should be installed to ensure the protection of groundwater resources. Monitoring is not required in the nonpermeable crystalline rock material.

GLACIAL DEPOSITS

Case A

In Case A, the potential waste disposal site is a medium to coarse glacial outwash deposit overlying crystalline metamorphic bedrock. Figure 7-4 illustrates the site-specific geohydrologic setting.

Topographical Influence on Site Hydrology--

Glacial outwash deposits are commonly associated with two landforms dependent upon the nature of the adjacent topography at the time of deposition. When the glacier is confined within narrow and steep valley walls, the corresponding outwash deposit is termed a valley train. An outwash plain is the second landform; it is associated with ice sheets terminating over broad lowlands.

In both forms, the meltwater streams are choked with sediment resulting in a valley or broad lowland deposit primarily laid down by braided streams. The resultant landform is generally flat-lying and is composed of coarser sediments and gravels.

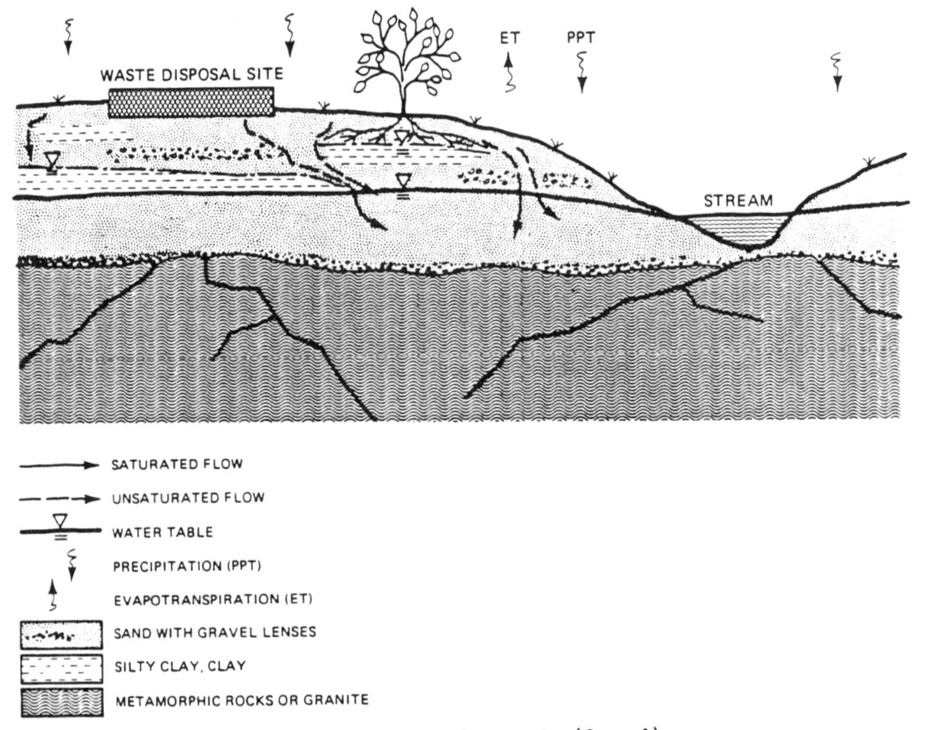

Figure 7-4. Glacial deposit (Case A).

The coarse flat-lying glacial outwash lends itself to recharge by precipitation, although the amount of deep percolation depends upon the local climate, evapotranspiration demand, etc. In addition, streamflow and sheetflow from higher areas are likely sources of recharge to this potential waste disposal site, particularly during periods of flooding. The outwash deposit is characteristically found in a lowland, suggesting that these deposits could be a discharge area for adjacent subcropping bedrock aquifers. In either circumstance, the burial of hazardous waste could have serious consequences for the recharge or discharge of groundwater to the glacial outwash deposit.

Geohydrologic Setting--

The glacial outwash deposit, laid down by braided streams, is characterized by a generally coarse matrix of moderate to well-sorted sands with a lesser amount of interbedded silts and clays. Well-sorted gravel lenses also frequently occur in this type of depositional sequence (Figure 7-4). A coarse basal layer of well-sorted gravels is also commonly associated with outwash deposits.

The lowland deposit being considered may contain groundwater under confined, unconfined, and perched conditions. The generally coarse-textured (sandy) unconsolidated outwash deposit suggests that groundwater could easily occur under water table conditions. The isolated highly permeable gravel zones, particularly the basal gravels, could be under confined groundwater conditions resulting from the adjacent confining beds (less permeable outwash deposit and crystalline bedrock). Where clays and silts have been deposited over a considerable lateral extent, the potential for perched water table conditions exists. The complex nature of the glacial outwash deposit dictates the occurrence of isolated perched groundwater and highly permeable zones throughout the glacial outwash deposit.

The hydraulic characteristics of groundwater flow in glacial outwash deposits have been reported by Chow (1964):

Location	Hydraulic Conductivity (feet/day)	Coefficient
Bristol Co., RI	5,800	0.007
Providence, RI	2,700	0.20

The crystalline metamorphic rocks underlying this type of glacial outwash deposit are usually a relatively impervious boundary. Crystalline rocks do not generally serve as good aquifers because the only permeability is associated with fracturing that may have developed after their initial formation. In this case, the secondary permeability associated with postformational fracturing is assumed minimal.

With respect to burial of waste, the deep percolation of recharge moving through the vadose zone would be the initial transport mechanism (assuming the waste is not liquid) available to mobilize the buried waste. Mobile waste could move laterally along perched water zones, where vegetation could conceivably consume water from the contaminated zone. The more permeable zones, particularly the basal gravels within the saturated zone, would provide the most extensive mechanism for transporting the waste plume. In Case A, the water table gradient is toward a nearby receiving stream. This complex hydrologic setting presents several opportunities for transporting waste. A knowledge of the hydrologic environment, its intricacies, and the chemical mobility of the waste material is essential before plans to bury waste proceed. With the potential for waste movement described here, a rigorous monitoring program should be undertaken before waste burial and monitoring should continue throughout the life of the facility and during postclosure activities.

Groundwater quality monitoring in the complex vadose zone developed in the glacial outwash material will require equipment suitable for saturated and unsaturated sample collection or fluid detection. Water quality samples can be provided by piezometers, multilevel samplers, multiple completion wells, tile drains, and collection pans and manifolds from saturated regions. Unsaturated zones will require suction-type samplers, e.g., suction cup lysimeters, filter candles, cellulose-acetate hollow-fiber filters, and membrane filter samplers. Nonsampling techniques should be employed as primary or backup detection systems as described in Section 8.

Case B

The waste disposal site considered for Case B is a silty clay glacial till with sand and gravel lenses over fractured dolomite bedrock. Figure 7-5 illustrates the site-specific geohydrologic setting to be evaluated for waste disposal.

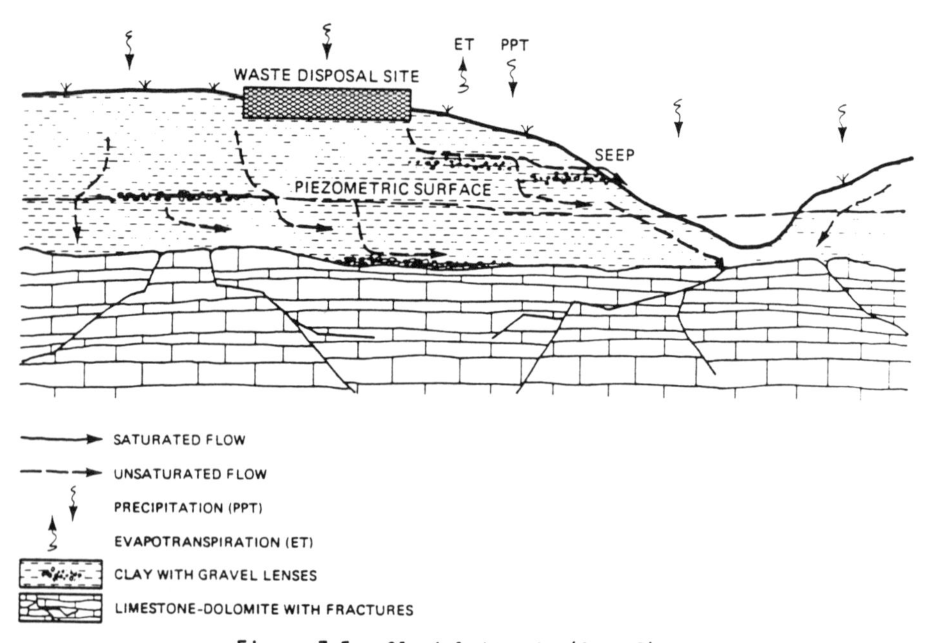

Figure 7-5. Glacial deposit (Case B).

Topographical Influence on Site Hydrology--

Topography is not considered an important hydrologic factor relative to burial of waste in this setting for two reasons: (1) glacial till can be associated with several topographic settings so no generalized landform is involved and (2) the till specified for this example is fine textured, suggesting a lower potential for recharge from precipitation or surface waters.

Geohydrologic Setting--

Glacial till, when composed of silt and clay, generally does not serve as an aquifer but rather acts as a confining bed for adjacent, more permeable zones. The sand and gravel lenses within the silt and clay matrix are capable of transmitting water. In this case, however, the coarse lenses are considered to be of limited extent. Therefore, groundwater quantities from these sand and gravel zones are limited. The stream shown in Figure 7-5 and associated alluvial aquifer are perched on the glacial till deposit. The most permeable zones that transmit groundwater in significant quantities under confined conditions are the basal well-sorted gravel layer and the underlying fractured dolomite bedrock. The gravels and fractured bedrock in this example are in direct hydraulic contact with each other and essentially function as one system.

The hydraulic conductivity of well sorted gravels is reported to range from about $10^{3.5}$ (=3,162) to $10^{6.5}$ (=3,162,277) gal/day/ft^2 (Freeze and Cherry, 1979). The permeability for dolomite was reported by Chow (1964) to have an average value of 36 and 240 gal/day/ft^2 for the Tymochtee and Silurian dolomites, respectively. The respective storage coefficients were 0.00029 and 0.002. Assuming a basal gravel hydraulic conductivity of 100,000 gal/day/ft^2, a gravel depth of 10 feet, and a Silurian dolomite bedrock depth of 100 feet, weighted average transmissivity for the two units would be:

$$\frac{\overbrace{(100{,}000 \text{ gal/day/ft} \times 10 \text{ ft})}^{\text{Gravels}} + \overbrace{(240 \text{ gal/day/ft} \times 100 \text{ ft})}^{\text{Dolomite}}}{110 \text{ ft}} = 9{,}309 \text{ gal/day/ft} .$$

This average transmissivity (9,309 gal/day/ft) indicates that significant amounts of groundwater could be moving through the combined gravel and fractured dolomite confined aquifer system.

Vadose zone monitoring for this case has a different focus from that of Case A. Unsaturated flow conditions would predominate throughout the vadose zone. Deep percolation would be a slow process with lateral migration of pollutants a distinct possibility. Care would be required in developing monitor wells within the confined dolomite and gravel hydraulic unit to ensure intercommunication was not established via the well annular space. Monitoring equipment should be placed as close to the waste cells as possible for early detection of pollutant leakage.

RIVER FLOOD PLAINS

Case A

The geohydrologic setting for Case A is a deposit of loess over fractured limestone-dolomite bedrock, with an adjacent river system that has eroded through both geologic units. The saturated zone occurs within the fractures of the dolomite bedrock and in the silts and gravels above, with the hydraulic gradient being toward the river. Figure 7-6 depicts the geohydrologic components of this potential waste disposal site.

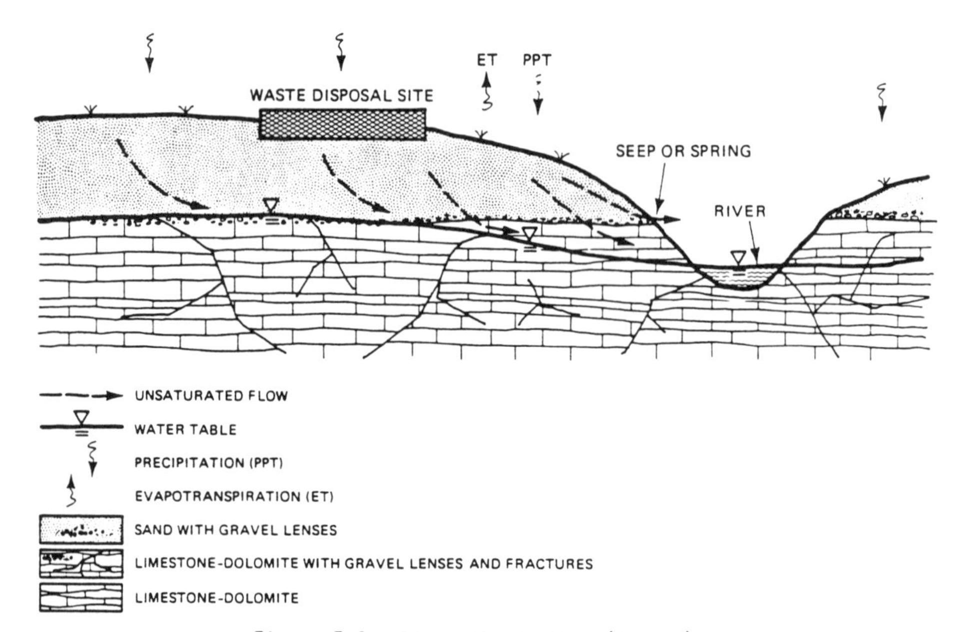

Figure 7-6. River flood plain (Case A).

Topographical Influence on Site Hydrology--

The potential waste disposal site illustrated in Figure 7-6 presents the incised river channel as a discharge zone. The presence of the river as a discharge area, coupled with the steep topography adjacent to the river, is reflected in the water table by the steep hydraulic gradient toward the river. The hydraulic gradient at this site is proportional to the amount of groundwater moving through a cross section of aquifer and, for this reason, larger quantities of water are available for transporting waste materials.

The relatively steep topography adjacent to the river and the fine texture of the loess deposit suggest that runoff would be high and water losses to deep percolation would be minimal over this part of the valley. The flatter portion of the valley would be subjected to greater infiltration rates that could mobilize waste buried in this area. Therefore, the deep percolation aspect of the water balance relative to waste disposal is of greatest concern where the gentle topography encourages infiltration.

Geohydrologic Setting--

The loess deposit and the dolomite are reported by Freeze and Cherry (1979) to have nearly the same range of hydraulic conductivity (0.01 to 10 gal/day/ft^2) while Chow (1964) reports a higher value of 250 gal/day/ft^2 for dolomite at a specific site. This example assumes that the higher reported hydraulic conductivity value is a better estimate for the hydraulic characteristics of the fractured dolomite. The gravels between the silt and dolomite would have the greatest hydraulic conductivity of any of the materials in the example, which could range from 10 thousand to 1 million gal/day/ft^2 (Freeze and Cherry, 1979). Storage coefficients for the gravels and fractured dolomite, when confined by the loess layer, could be on the order of 0.001 (Chow, 1964). Under water table conditions (close to the river), the specific yield would be about 0.2 (an average value for unconfined aquifers, Lohman, 1972).

The potential burial zone should be located above the potentiometric level of the more permeable gravel and fractured dolomite aquifer in order to minimize the mobilization of waste. Placing waste in that part of the valley where the topography is steepest would decrease the amount of deep percolation due to increased runoff at the waste zone, but this location would also be closer to the river. An indepth understanding of the water balance of the immediate site conditions would be essential before considering burying waste close to a river system.

The mass stability of the potential burial site and migration tendency for the river should also be considered before disposing of waste on the valley bottom. Monitoring in this setting should be conducted in the loess layer above and through the waste zone to observe the influence of deep percolation. The combined gravel and fractured dolomite aquifer and the nearby discharge area (river) should be monitored leakage from the disposal zone.

The vadose zone for this site should be monitored directly under new facilities or via slanted access holes under existing or abandoned waste disposal sites. Estimates of flux and pollutant attenuation should be made using depthwise batteries of suction samplers. Access wells or piezometers should be developed in the saturated zone to determine aquifer characteristics and hydraulic gradient and to estimate the amount of water movement toward the river. Pollution dilution potential in the saturated zone would be evaluated based on gradient and water movement data. Surface seeps or springs would be sampled at existing or abandoned sites.

Case B

A valley fill composed of fluvial sand and gravel with underlying metamorphic basement rock is the geologic framework for the potential waste disposal site in Case B. A water table is present within the alluvial fill and a river also flows over the valley fill. Figure 7-7 illustrates the geohydrologic setting.

Topographical Influence on Site Hydrology--

The potential waste disposal site is located on a topographically low valley fill deposit. This position in the topography allows discharge to the valley fill from adjacent subcropping beds that could contribute water to the waste burial zone. The relatively flat valley fill deposit also allows a

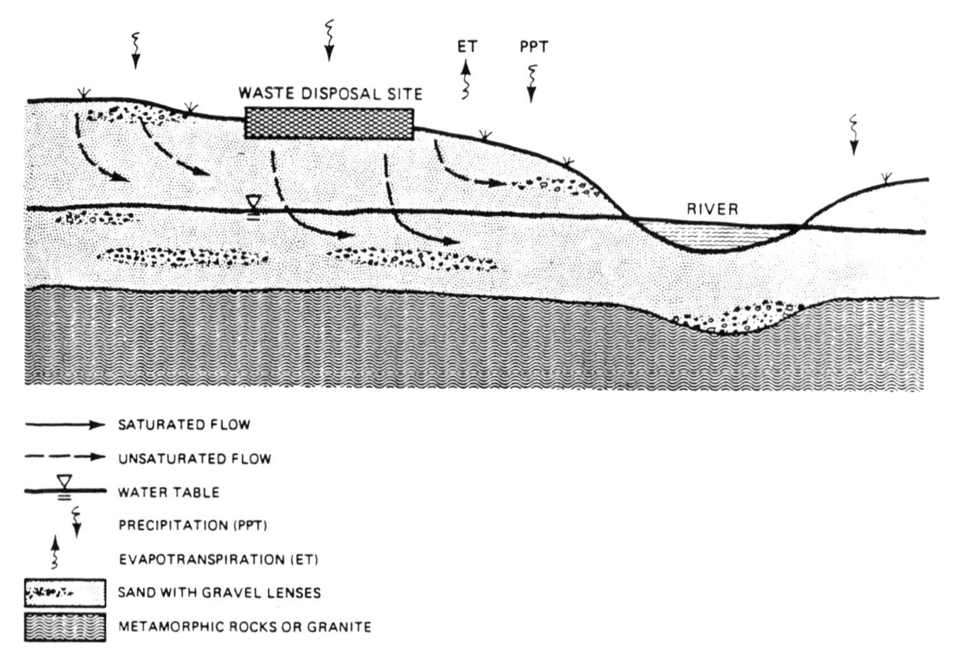

Figure 7-7. River flood plain (Case B).

greater opportunity for infiltration and deep percolation than if the site were located on a hillside where greater runoff would occur. In addition, sheet flow from adjacent uplands and bank storage from the perennial stream can contribute to the water budget associated with the potential disposal site within the valley fill.

Geohydrologic Setting--

The metamorphic bedrock beneath the alluvium is considered to be a perching bed with low hydraulic conductivity. Freeze and Cherry (1979) report that hydraulic conductivities for unfractured metamorphic rock can be as low as 0.0000001 gal/day/ft^2. With such low hydraulic conductivities, the basement rock is not expected to play a significant role in the water budget of the waste disposal site, nor would it provide a zone of transport out of the alluvial zone. Streams associated with alluvial valley fills provide a variety of depositional environments that result in a complex valley fill deposit. Stream-laid deposits are characteristically a well-sorted mixture of interlayered fine-grained (silt and clay) and coarse-grained (sands and gravels) sediments. In many circumstances, coarse gravel deposits, commonly associated with extremely high flows that occurred during periods of glacial retreat, are found at the base of the valley fill. The interbedded fine- and coarse-grained deposits, with hydraulic conductivities as high as 1 million gal/day/ft^2, can transmit significant quantities of groundwater (Freeze and Cherry, 1979). The complex nature of the alluvial valley fill provides circumstances where groundwater can occur under confined (e.g., basal gravels), unconfined, and perched (e.g., over extensive fine-grained deposits) conditions. In this example, groundwater within the alluvial fill occurs under unconfined (water table) conditions. Before burying any waste within the valley fill, the level of the water table and its range of seasonal fluctuation should be established. The burial of waste that is potentially mobilized in water should only be considered above the zone of fluctuation for the water table. Shallow groundwater can also be wicked upward by capillary rise where vegetation can utilize the moisture. Some species of plants are very deep rooted (e.g., sage brush, alfalfa, etc.) and can utilize contaminated waters from depths as great as 60 feet. The potential therefore exists for a vegetation contamination problem as well as a water pollution problem.

The initial monitoring effort should establish storage changes in the vadose zone associated with seasonal influx of water. Multilevel samplers established through the water table would provide seasonal water level fluctuation data and information on pollutant stratification. The shallow water table and permeable sediments dictate near-surface sampling and nonsampling vadose zone monitoring.

INTERIOR DRAINAGE BASINS

The potential waste disposal site evaluated in this example involves a peat deposit overlying silty lake deposits resting on a solution limestone aquifer. Internal drainage occurs in this area of karstic topography. Groundwater in the peat is perched over the silty lacustrine deposits and a water table exists in the limestone aquifer. Figure 7-8 depicts the components of this hydrogeologic setting.

Topographical Influence on Site Hydrology

The basement rock in this geohydrologic setting is karstic limestone. Solution weathering of the limestone has produced the potholes and cavities that influence the surface topography as well as the hydrology of the site. Both the overlying peat and silt beds reflect the sink hole in the limestone, leaving a closed depression. The gentle topography and extremely permeable peat material allow infiltration to occur at an extremely high rate. The little runoff that occurs is internally drained and runs into the pothole, inducing recharge into the peat beds. The pond above the pothole also allows water evaporation, thereby concentrating dissolved constituents that remain after water is lost from the pond.

Geohydrologic Setting

The karstic limestone, which serves as the basement rock for this example, is reported to be extremely permeable. Freeze and Cherry (1979) show hydraulic conductivities for karstic limestone in the range of 1 to 1 million gal/day/ft^2. The associated specific yield for the unconfined aquifer would

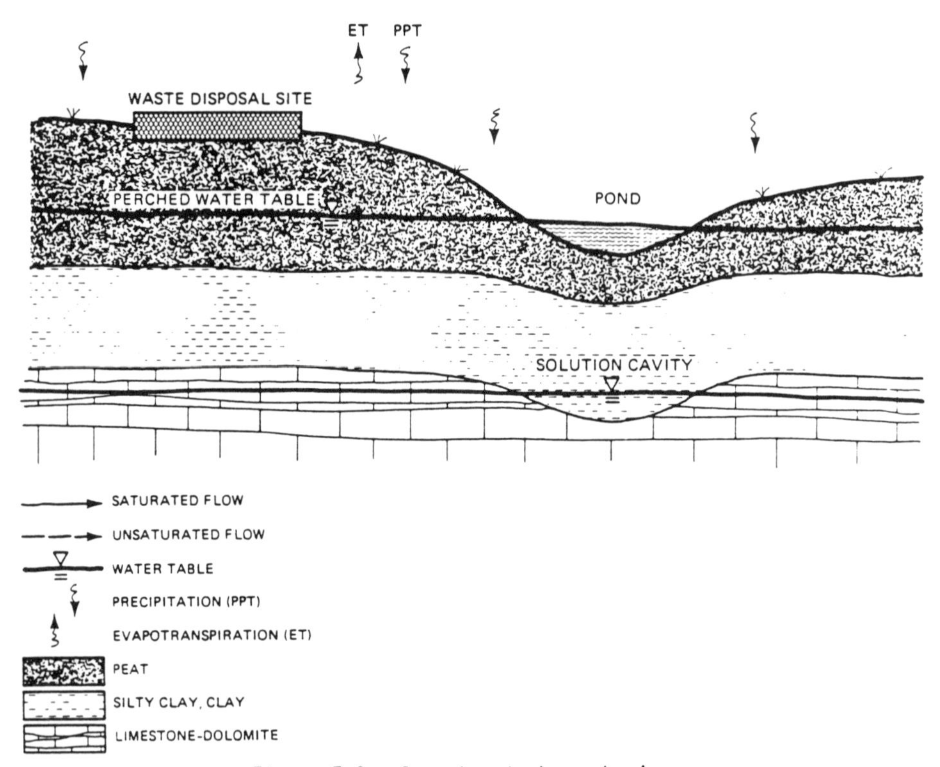

Figure 7-8. Interior drainage basin.

likely be on the order of 0.15 or so because only the solution cavities and fractured portion of the aquifer would be releasing water. A direct relationship exists between the water table configuration in the limestone aquifer and the collapsed solution cavity (pothole). The slightly depressed water table level at this location indicates that groundwater is discharging to some unknown location via the solution cavity.

The lacustrine silts and clays have low permeabilities (less than 0.01 gal/day/ft^2; Cherry and Freeze, 1979) and are considered a perching zone for the water table in the peat. The peat materials are highly permeable and readily accept additions of precipitation.

The presence of two unconfined aquifers overlying each other indicates that groundwater is flowing from the upper zone to the lower zone at a very slow rate and/or that groundwater is discharging rapidly from the karstic limestone aquifer. The silts and clays probably would provide an effective barrier between the two water table aquifers.

The peat materials have been reported by Lapakko and Eger (1981) to have an assimilative quality (cation exchange capacity) for removal of metal ions from solution. The assimilative capacity of peat is low compared with commercial products used to remove metals from mine waters, but the large quantity of peat in the setting described here suggests the peat could effectively adsorb some metals contained in leachate from a waste burial zone.

Burial of waste in the setting described would be difficult without encountering groundwater in the peat. Recharge through the peat is also a definite consideration in the water budget of the waste burial zone.

Vadose zone monitoring would require depthwise batteries of suction samplers above the water table in the peat and throughout the lacustrine silts and clays. Saturated zone monitoring equipment (piezometers, multilevel samplers, wells, etc.) could be installed at shallow depths. Surface water samples should be collected in nearby ponds. Existing waste disposal and abandoned sites should be covered with impermeable material to reduce infiltration and deep percolation.

DESERT AND SEMIARID DEPOSITS

This potential disposal site is alluvium overlying an upper aquifer of fractured oil shale underlain by a clay zone confining layer that separates the upper and lower aquifer. The lower aquifer is marly oil shale with solution cavities and fracturing. An ephemeral stream has down cut into the alluvial deposit. Figure 7-9 illustrates this multiple aquifer system described above.

Topographical Influence on Site Hydrology

The topography of the potential waste disposal site is generally flat with little relief. Little moisture is lost to deep percolation because of the semiarid climate and full utilization of precipitation by vegetation. What little recharge occurs on the site results from ephemeral stream leakage during periods of flooding. In addition, snow drifts build up in some of the protected bottom areas and slow melting in the spring may allow some recharge.

Geohydrologic Setting

The lower aquifer shown in Figure 7-9 is oil shale with enhanced secondary permeability resulting from fracturing and solution cavities. The removal of the evaporite mineral, nahcolite, by groundwater has created a highly permeable zone laced with solution cavities. The transmissivities for this zone are reported to range from 60 to 14,511 gal/day/ft and storage coefficients average about 0.0001 (Slawson, 1980).

The lower aquifer is semiconfined by a lower shale confining bed and the upper semiconfining zone. Figure 7-9 shows the potentiometric surface for the lower aquifer zone as extending through the upper aquifer into the alluvial material.

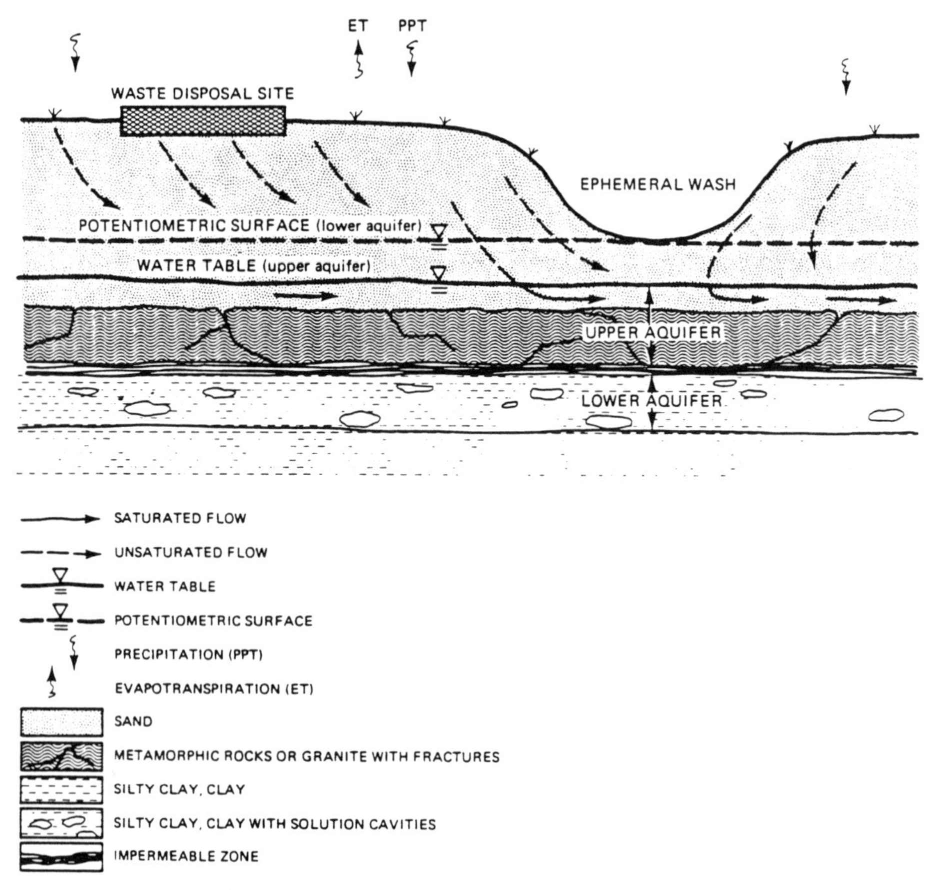

Figure 7-9. Desert and semiarid deposits.

The upper aquifer is a water table aquifer contained in the fractures of the oil shale. Transmissivities for this zone are reported by Slawson (1980) to range from 60 to 7,480 gal/day/ft. The specific yield averages about 0.1. The upper and lower aquifers are slightly connected through the confining layer.

Vadose zone monitoring requires equipment that can operate in very dry soil conditions, e.g., large negative soil-water pressure. Unsaturated sampling methods would generally be ineffective in this geohydrologic setting. Nonsampling methods implanted close to the ground surface should provide adequate leachate detection. Runoff from occasional heavy rains should be kept from entering the waste cells and thereby enhancing leachate production.

FRACTURED ROCK SYSTEMS

The geohydrologic setting for waste disposal presented here is a fractured granite with little other debris or soils above. A water table aquifer is present within the fractured granite. Figure 7-10 illustrates the setting.

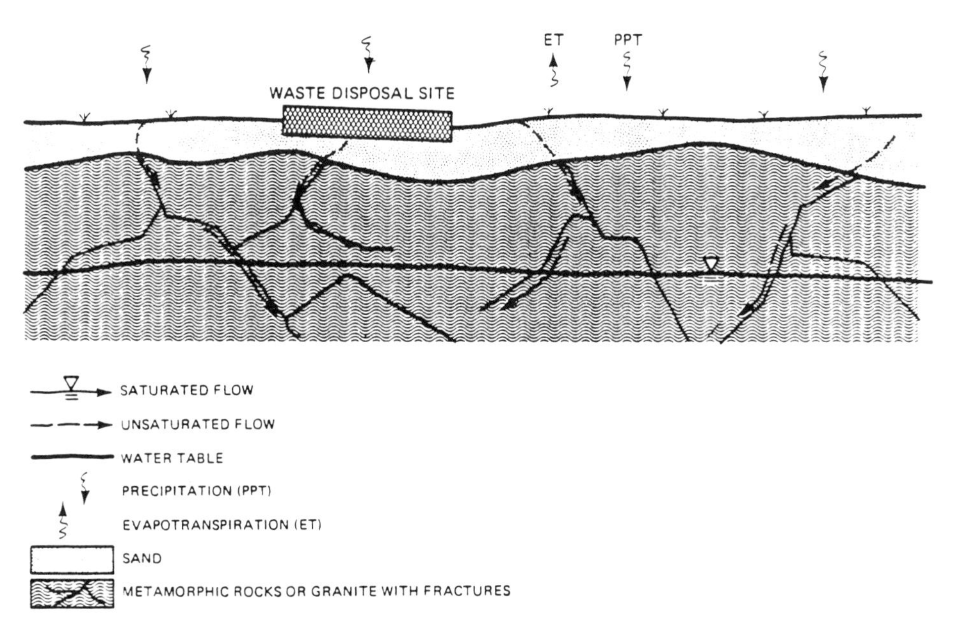

Figure 7-10. Fractured rock system.

Topographical Influence on Site Hydrology

The gentle topography and stoney soils associated with the fractured granite bedrock allow a very high rate of infiltration. Commonly, vegetative communities associated with this soil type reflect xeric moisture conditions, even though relatively high annual precipitation may occur. The porous soils encourage deep percolation at the expense of moisture that would otherwise be available to plants.

Geohydrologic Setting

Figure 7-10 illustrates a granite mass that originally intruded at depth as a batholith. Weathering of the granite bedrock includes exfoliation of granite sheets that allow further expansion and contraction fracturing and solution weathering of the granite to leave the gravelly soils. Soils are immature, shallow, and rocky with partly weathered granite fragments.

The expansion of the granite mass by erosion resulting from unloading cover from the rock has established a strong jointing pattern that provides the secondary permeability for the water table aquifer.

The hydraulic conductivity of the fractured granite bedrock would be expected to fall in the range of 0.1 to 1,000 gal/day/ft^2.

The potential waste burial zone should be located out of the water table to minimize mobilization and transport of waste material via groundwater flow. Deep percolation is the major component of the moisture budget above the water table that must be considered before and after waste burial.

Vadose zone monitoring in fractured rock systems requires a higher density of sampling units to compensate for extreme spatial variability of fluid transport. Monitoring in both saturated and unsaturated regions requires extreme care in the placement of sampling equipment. Suction-type samplers that require good contact with the host rock will be difficult to install. Drilling for deep equipment installation will be hampered by hard drilling media that will increase costs for the monitoring program and thereby limit the number of sampling units utilized.

8. Conceptual Vadose Zone Monitoring Descriptions

Development of a groundwater monitoring program for a hazardous waste disposal facility is an integrative process. Some of the primary elements to be considered, e.g., categorization of waste, waste disposal methods, geohydrologic setting, and monitoring equipments, are discussed elsewhere in this book but are not brought together into a single plan. In general, site and waste characteristics dictate the disposal method and suggest the most effective monitoring program for a given location.

Implementation of the monitoring system requires deployment of the selected monitoring equipment to detect changes in the physical and chemical nature of the groundwater and indicate pollutant migration at a specific point. It does not indicate the system's mass balance characteristics due primarily to the spatial variation of natural soils or geologic formations, e.g., information on the quantity of pollutants migrating from the source. These data can only be estimated by installation of a large enough number of detection units to provide a statistically significant measure of the "average" flux through the test area. Generation of this quantity of data is usually cost prohibitive unless pollution is suspected to be severe enough to justify quantification.

Because of the large number of variables requiring consideration in developing a monitoring system, site-specific examples are not generated for the varied geohydrologic settings, waste disposal methods, and waste categories reviewed in this book. However, generic models of monitoring systems for landfills (including waste piles), surface impoundments, and land treatment facilities are useful in the design of actual monitoring programs. These are developed below for specific geohydrologic settings.

A vadose zone monitoring program for a waste disposal site includes premonitoring activities followed by an active monitoring program. Basically, premonitoring activities consist of: (1) characterization of the of the quantity and quality of potential pollutants, (2) selection or assessment of the disposal method to be used at the site, and (3) evaluation of the geohydrologic setting by assessing hydraulic storage and transmissive properties of the vadose zone, and (4) estimating the effect of geochemical properties on pollutant mobility. The premonitoring program results will provide clues on the potential mobility rates of liquid-borne pollutants through the vadose zone, the storage potential of the region for liquid wastes, and the likelihood of attenuation of specific pollutants. Fuller (1979) discusses premonitoring activities in detail. Geophysical methods that are effective for premonitoring efforts are discussed by Sendlein and Yazicigil (1981).

The active monitoring program will comprise a package of sampling and nonsampling methods selected from an array of possible techniques discussed in Section 4. Additional information on vadose zone monitoring techniques can be found in Wilson (1980). Sampling methods obtain liquids and solids from the vadose zone, whereas nonsampling methods yield inferential evidence for the movement of liquid-borne pollutants. In practice, either the selection or rejection of a particular method of premonitoring and monitoring will depend on a number of practical considerations. Everett et al. (1982) discuss 15 criteria for selecting alternative vadose zone monitoring techniques.

LANDFILLS

A 1977 reconnaissance-level review of case histories for waste disposal facilities throughout the United States indicates that nearly 80 percent were operating as landfills. Approximately 90 percent of the solid waste produced by industrial and municipal activities is deposited in more than 20,000 landfills across the continent. This disproportionately large number of landfills is due primarily to earlier regulations that allowed the disposal of liquid wastes at landfills. Current regulations restrict disposal of liquid wastes in landfills, increasing the numbers of surface impoundments and land treatment programs being developed.

Everett and Hulbert (1980) evaluated landfill facilities for coal strip mining "boom-towns" and developed a solid waste disposal monitoring design. Leachate monitoring in the vadose zone at three sanitary landfills with different geologic environments and disposal practices is reported by Johnson, Cartwright, and Schuller (1981). Additional studies conducted by the U.S. EPA (1977a) defined procedures for groundwater monitoring at solid waste disposal facilities. In the EPA study, cross sections of landfills are given in various geohydrologic settings showing the expected migration of pollutant plumes. Monitoring network procedures focus on saturated conditions below the vadose zone although some vadose zone monitoring equipment is briefly discussed.

Landfill Premonitoring

For this example, the solid waste material being disposed of at the landfill is assumed to be from municipal sources (see Section 6). It is a heterogeneous mixture of organic, inorganic, and living organisms. Although several major environmental factors (e.g., gas production, odor, noise, air pollution, dust, fires, vectors, as well as leachate generation (ASCE, 1976)) must be considered in monitoring the proper operation of a landfill, this subsection focuses only on the potential pollution of groundwater resources from leachate. Earlier studies by Garland and Mosher (1975) cite several examples where groundwater pollution has been caused by landfills. The problems created by migration of landfill leachate into the aquifer used as the water supply of a city are reported in a case history by Apgar and Satherthwaite (1975). These works provide a perspective of potential pollution problems associated with using landfills for disposal.

A wide variety of pollutants can be found in solid wastes deposited in landfills. Methods of measuring these constituents from landfill leachate have been studied by Chain and DeWalle (1975). The type of material accepted at the disposal facility and the industrial base in the surrounding area are clues to the general waste characteristics found at the site. Table 6-1 is a comprehensive listing of parameters associated with leachate from solid wastes. Table 6-2 lists concentrations of these constituents for solid waste leachate studies that have been developed by Steiner et al. (1971) and U.S. EPA (1973a, 1975). Not all of the parameters would be expected at a single hazardous waste disposal site. These classes and concentrations are useful, however, in suggesting indicator parameters that can be used to detect a pollutant plume.

A typical landfill is constructed by using either the area or trench method. With the area method, waste is deposited directly on the ground surface. Runon must be diverted away from the active portion of the landfill. Runoff from the facility must be collected and, if it contains a hazardous waste, treated as specified under the RCRA Part 261. If the landfill material is subject to dispersal by wind, it must be covered or otherwise managed to control hazardous waste. With the trench method, waste is deposited at the end of a trench and covered at the end of each day as required. For this example, hazardous wastes using the trench method is assumed.

Preconstruction data on the vadose zone at the site may have been compiled and should be reviewed as part of the premonitoring effort. This information, along with Soil Conservation Service (SCS) soil maps and well cuttings from geologic formations penetrated by monitor wells, should be examined to determine the thickness, structure, and chemical characteristics of the vadose zone. Water table levels should be plotted to determine existing hydraulic gradients and the thickness of the vadose zone. If vadose zone samples have been saved from site development, several laboratory techniques are available for estimating hydraulic conductivity (K). Saturated K values can be determined using permeameters and from the resultant K values, assuming the hydraulic gradients are unity, the flux at the site can be estimated from Darcy's equation. Saturated K values for different layers can also be determined using the relationship between grain size and K developed from grain-size distribution curves. Additional laboratory methods for measuring unsaturated K values include the long soil columns, pressure plate methods, and other column techniques.

Storage potential for water-borne pollutants can be inferred from storage coefficients developed from pump tests in perched water systems within the vadose zone. These data can be generated at minimal cost if wells and observation piezometers are available from preliminary work at the site. Aerial photographs should be reviewed for evidence of springs and seeps caused by modifications of the water table under landfills and potential threats to surface water quality at the site. Rainfall data and ambient surface water quality for the site should be known. Surface structures designed to control runon and runoff from the waste material should be inspected to ensure compliance with the RCRA.

Data on the chemical characteristics of the vadose zone are vital in developing and understanding pollutant attenuation. In particular, the percentage of colloidal-sized particles, e.g., clay minerals, and pH of a representative soil-moisture extract should be determined. In porous geologic materials, these particles have the capability to exchange ionic constituents adsorbed on the particle surfaces. The nature of the surface charge of these particles is a function of pH. At high pH, a negatively charged surface occurs, while at low pH, a positively charged surface is developed. The tendency for adsorption of anions or cations therefore depends on the pH of the vadose zone soil-water solution. This information will be useful in estimating pollutant attenuation based on expected leachate constituents given in Table 6-2. In addition, the redox (reduction-oxidation) potential of perched groundwater, soil-moisture extract, and/or leachate should be measured. Commercially available "Eh probes" can be used to obtain these data. With the pH and Eh of the vadose zone known, Eh-pH diagrams can be constructed showing stability fields for major dissolved species and solid phases. These diagrams are useful in understanding the occurrence and mobility of minor and trace elements. Cloke (1966) describes in detail the construction of the Eh-pH diagram with addition theory presented by Guenther (1975).

Ambient groundwater quality is an important data base to establish during the premonitoring effort. At existing sites, care must be taken to ensure that water unaffected by the landfill has been sampled to determine background quality data. Of particular interest as indicator parameters are TDS, COD, conductivity, and BOD_5 that have been found in high concentrations in solid waste leachate studies (see Table 6-2). In addition, temperature, color, chlorine, and iron are listed as indicator parameters by U.S. EPA (1977a).

The geohydrologic setting for the landfill is assumed to be a glaciofluvial deposit composed of stratified sand and gravel with some clay lenses. Underlying the glacial deposits is a fractured crystalline metamorphic bedrock. The deposits are located within a river flood plain with the predominant hydraulic gradient toward the river. Section 7 gives further details of the geohydrologic setting.

Active and New Landfill Monitoring

The continued operation of a landfill by both the area and trench methods requires additional land for disposal of new waste material. Therefore, each site is composed of a combination of existing and projected waste cells. Monitoring activities for these areas will differ because cost-effective placement of a specific type of monitoring equipment for new cells may not be possible for existing cells. For example, a resistivity network installed under the protective liner of the landfill to determine leachate migration through the liner could be easily incorporated in the earthwork required for development of a new waste cell. This type of installation would not be possible under an existing cell. Segregation of wastes by toxicity may dictate alternate monitoring networks around selected waste cells. Depending on the toxicity level of the waste material, its vertical proximity with groundwater aquifers and the use of aquifer nondestructive monitoring methods (e.g., geophysical techniques) may be preferable to sampling methods that require boreholes through the potentially polluted strata. These boreholes have been known to short circuit liquid-borne wastes to the water table through the annular well space of an improperly completed monitor well.

A generic monitoring design for an active and new hazardous waste landfill is shown in Figure 8-1. Design elements include both nonsampling and sampling methods. Nonsampling methods include (1) neutron moderation probes, (2) tensiometers, and (3) a resistivity network underlying new waste cells. Surface geophysical methods can be used as required for detection or definition of pollutant plumes. Sampling methods include (1) multiple completion wells, (2) multilevel samplers, (3) suction samplers, (4) piezometers, and (5) gas samplers. Technical descriptions, field implementation, and range of applications and limitations of these techniques are discussed in Sections 4 and 5. The following paragraphs describe use of the equipment for monitoring a landfill.

Nonsampling Methods--

If no hydrogeologic information is available for the landfill before implementing the active monitoring program, data collection could begin with slant-drilling neutron probe access tubes under the waste cells. Although the neutron probe is a nonsampling method, soil samples from the drilling operation should be collected and analyzed for particle size distribution, cation exchange capacity (CEC), and pollutant concentrations. Detailed lithologic logs could be constructed by a well-site geologist based on borehole cuttings. Impermeable lenses with potential for ephemeral perched groundwater should be carefully noted on cross sectional diagrams constructed for the site. If stable formations are penetrated under the landfill, dual-packer pressure permeability tests can be made to determine vadose zone hydraulic properties. Alternate methods of estimating saturated hydraulic conductivity (K) in the absence of water table conditions are by air-entry permeameter, infiltration gradient, or the double tube method. Water samples can be collected for chemical analyses if perched groundwater zones are penetrated. These preliminary holes should be drilled with caution to avoid establishing a hydraulic connection between potentially contaminated areas and the water table.

Following the appropriate data collection effort, a seamless stainless steel or aluminum neutron probe access tube is inserted in the hole. Unsaturated water content measurements at specific depths under the landfill should be made. Subsequent readings will indicate if changes in storage capacity are occurring over time. While these data tell nothing of the energetics of the water movement, they can be used to infer pollutant migration.

Tensiometers can be installed near or below waste cells to measure soil suction values that are not in excess of -0.8 bar. If soil-water pressure is found to exceed this operational limit, alternate monitoring methods can be employed (e.g., psychrometers). Several tensiometers should be installed at increasing depths throughout the region of interest to characterize the direction of unsaturated water movement in the vadose zone. Soil suction measurements should also be used to adjust suction samplers for optimal efficiency.

For new waste cells or at new landfills, several types of monitoring equipment can be cost-effectively placed below protective liners for leachate

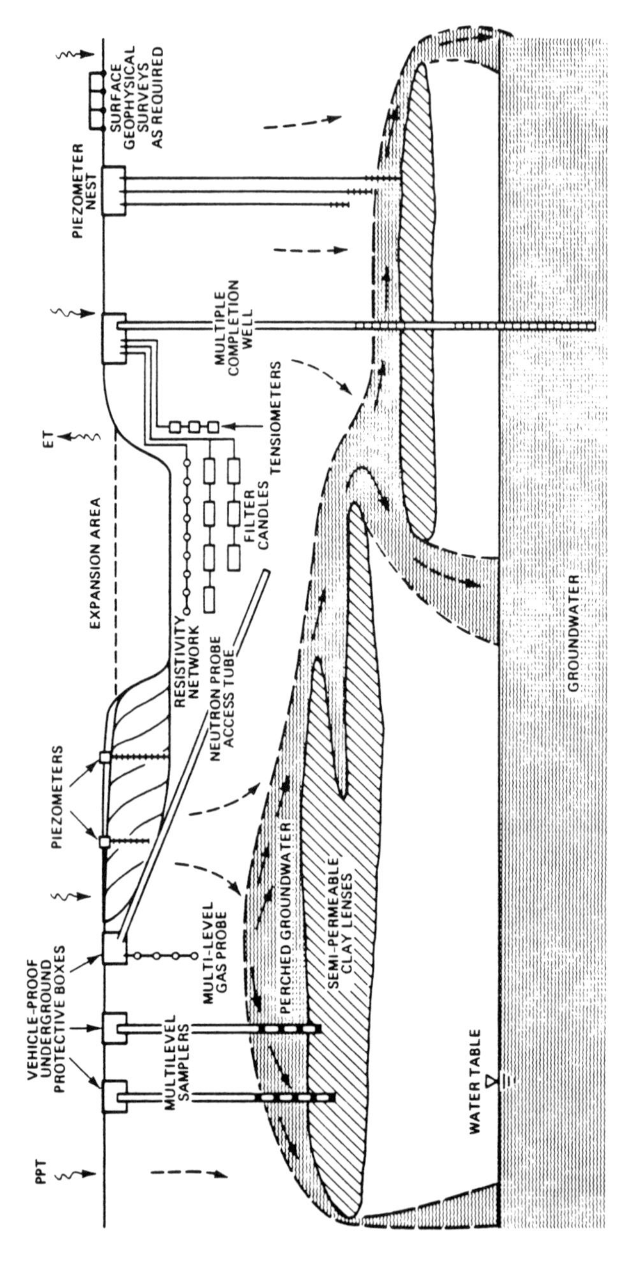

Figure 8-1. Generic monitoring design for existing hazardous waste landfill.

detection. The generic landfill monitoring design presented here has a resistivity network installed below the liner for leachate detection. Alternate types of nonsampling methods could include heat dissipation sensors, resistivity blocks, or thermocouple psychrometers. Each of these monitoring methods should have multiple-depth installations to monitor movement of fluids through the vadose zone. In addition, gamma ray attenuation methods are a viable nonsampling technique.

Kaufmann et al. (1981) describes a resistivity network developed from a modified use of the Wenner array for detection of leachate migration through a protective liner. The system is designed and constructed to measure the apparent resistivities of soils beneath the waste disposal facility. Interpretation of the data is similar to surface applications; a decrease in apparent resistivity indicates the leakage of conductive contaminants through the base of the waste cell. Monitoring with this type of system can be non-labor-intensive if done under continuous computer surveillance.

Surface geophysical techniques have been employed at existing landfills with varying degrees of success. If shallow water table or perched water table conditions exist at the site and detectable variations in conductivity between the leachate and ambient groundwater quality are present, the four-probe (Wenner array) electrode configuration can successfully determine the lateral migration of pollutant plumes. Current studies are underway to determine if this technique can be used to detect leaks in landfill liners. Magnetometers have been useful in locating abandoned waste cells and landfills that contain metallic objects. Acoustic emission (AE) monitoring is conceptually applicable to the detection of leaks in landfills. Ongoing studies by the EPA should determine if this technique, which measures soil mechanical changes rather than changes in soil dielectric properties, has potential for future application to hazardous waste landfills.

Sampling Methods--

Direct sampling in the soil and soil-water region underlying the landfill should accompany installation of the nonsampling equipment.

Single or multiple completion wells are an effective method of obtaining regional samples from saturated zones under the waste cells. Data on aquifer characteristics can be determined from pumping tests. Water quality samples drawn from the cone of depression surrounding a pumping well will have a greater probability of capturing leachate emanating from the waste cells because of the larger area from which the sample is collected. These samples will be diluted compared to point source samples, e.g., suction cup lysimeters, etc., and this should be kept in mind when interpreting the water chemistry data.

Monitoring wells installed should be beneath and downgradient of the landfill. Experience has shown that surface completion of wells and nonsampling equipment installations should be protected by vehicle-proof underground boxes. This completion technique will not inhibit vehicle movement in the landfill facility. An artist's conception of a landfill with vadose zone monitoring equipment is shown in Figure 8-2.

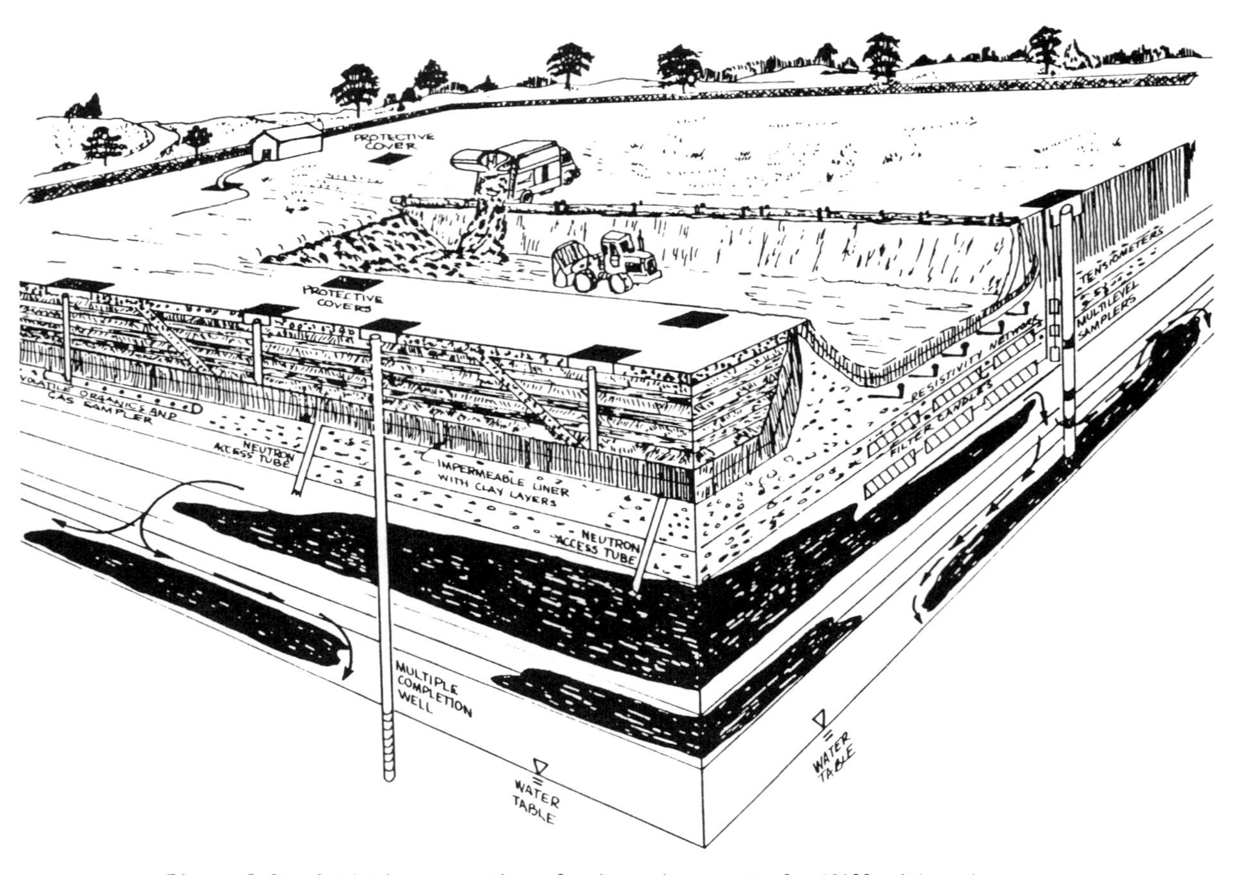

Figure 8-2. Artist's conception of a hazardous waste landfill with vadose zone monitoring equipment.

Clusters of multilevel samplers installed by slant-drilling under the facility can be used to obtain liquid samples from saturated zones as well as information on the hydraulic gradient underlying the hazardous waste landfill. Multilevel samplers determine stratification of liquid-borne pollutants and can detect a pollutant plume that might go unnoticed due to dilution in a standard monitor well. In addition, multilevel samplers can be used to establish the water quality of infiltration surges occurring during wet months or following the spring thaw.

In situ samples of pore fluid are obtained from filter candles installed horizontally under the resistivity network. Installation of these units represents "planned redundance" in the monitoring design. The vacuum applied to suction samplers is determined via tensiometers to minimize sample filtering or segregation during collection. The horizontal position of the samplers increases the area sampled. However, these units are still providing pore fluid from a very limited area. Multidepth batteries of these units provide data on pollutant attenuation with depth. Alternate sampling techniques could include cellulose-acetate hollow-fiber filters, membrane filter samplers, or suction lysimeters installed under the resistivity network.

Piezometers that do not penetrate the safety liner can be placed in the landfill to collect leachate samples and estimate liquid volume within the waste cells. With appropriate assumptions about storage coefficients within the waste piles, changes in water level can be used to infer water movement through the liner by the water budget method. These data are useful in estimating the size of a pollutant plume or magnitude of a given groundwater pollution problem.

Single piezometers or nests can be installed near the waste cells for sampling saturated layers in the vadose zone. With the recent development of small-diameter (1.50- and 1.78-inch) air-actuated sampling pumps, piezometers can provide representative water quality samples. Some sampling methods in small-diameter casings (e.g., bailing or swabbing) have been shown to produce nonrepresentative samples unless extreme care is taken to standardize sample collection techniques (Kelly, 1982).

Gas samplers can be installed under new cells or near existing cells. This approach is important where nonconductive fluid may leak through the liner and a resistivity network would fail to detect the leak. A number of sampling equipment configurations are possible. Figure 8-3 shows a multilevel gas probe. Gas vapor sampling pipes installed under new waste cells are shown in Figure 8-4. Gas, temperature, and moisture probes can also be laid down within waste cells as they are developed.

SURFACE IMPOUNDMENT

The following generic model is useful for comparative evaluation. When developing a site-specific vadose zone monitoring package for a hazardous waste surface impoundment, two separate cases for generic surface impoundments will be examined: Case 1, a new unused facility, and Case 2, an active site. This format is necessary because access for instrument implacement differs between the two sites. The procedure for developing both sites is identical.

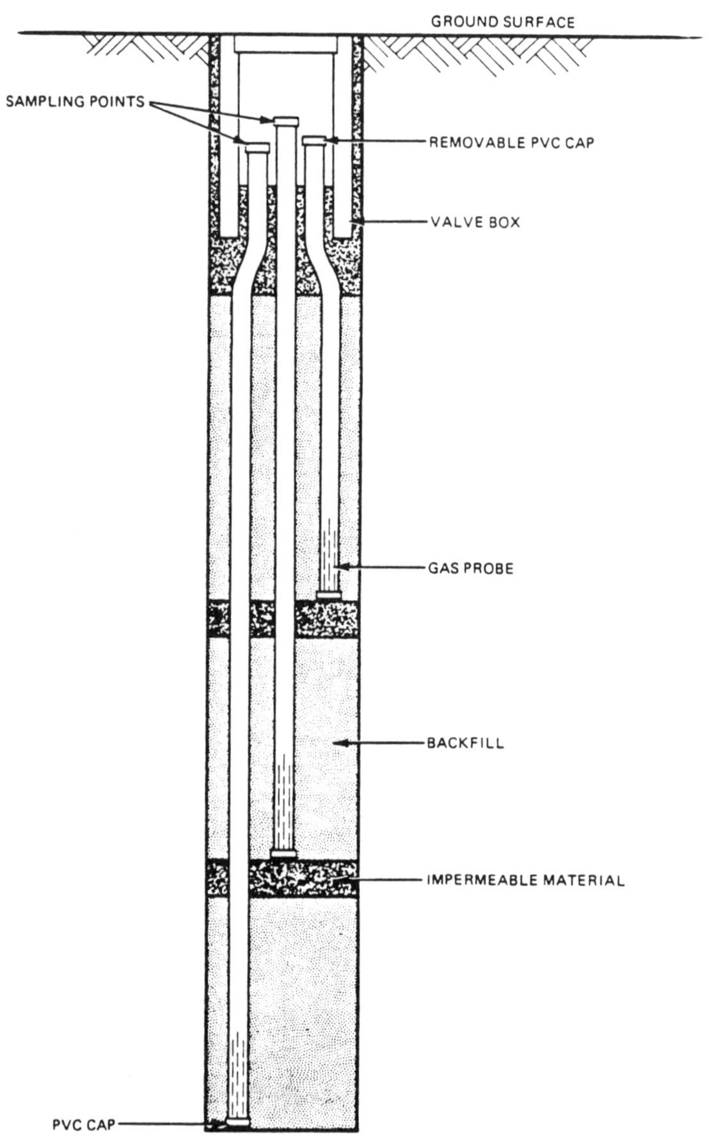

Figure 8-3. Multilevel gas probe.

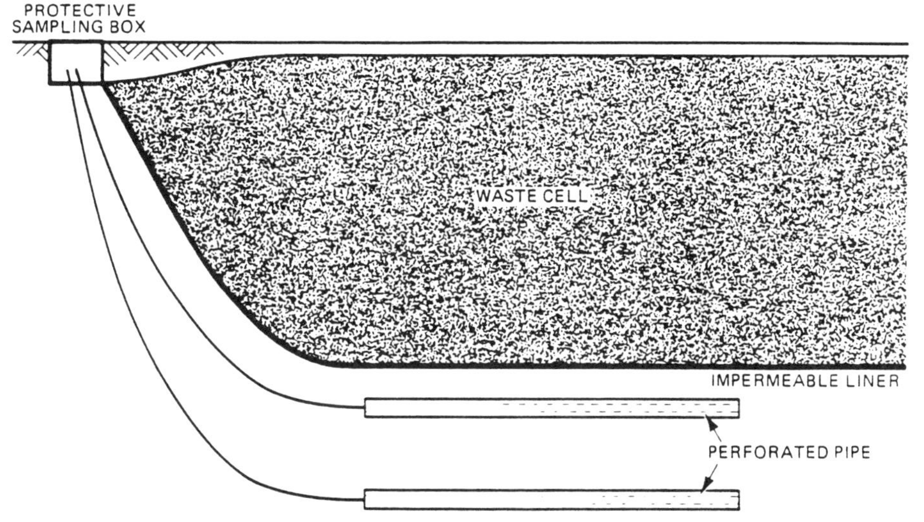

Figure 8-4. Gas vapor sampling pipes beneath a landfill.

A premonitoring effort was undertaken to characterize the wastewater and vadose zone properties. Subsequently, this generic model for a specific site was prepared.

It was assumed at the outset that vadose zone monitoring would be a component of a comprehensive monitoring program, including source and groundwater monitoring. A comprehensive program is best developed by following a sequential process such as that presented by Todd et al. (1976) and Everett (1980).

Tinlin (1976) developed monitoring plans for percolation and lined ponds and reports on a number of case histories on leachate formation and migration from hazardous waste surface impoundments. Included in this work are illustrative examples of pollution occurring from brine disposal in Arkansas; plating waste contamination in Long Island, New York; pollution potential of an oxidation pond near Tucson, Arizona; and multiple-source nitrate pollution in the Fresno-Clovis, California metropolitan area. Cole (1972) presents case histories on pollution from saline waters, nitrates, phenolic substances, and petrochemicals throughout Europe.

Surface Impoundment Premonitoring

The source of wastewater being disposed of in this generic surface impoundment is assumed to be from a primary aluminum metal manufacturing facility. Table 6-45 summarizes potential pollutants commonly found in wastewaters from such a plant. The table shows that a wide range of potentially dangerous constituents emanates from the raw primary aluminum wastewater, including organics, inorganics, and heavy metals.

The well construction program for the groundwater monitoring package is presumed to yield valuable information for the vadose zone monitoring effort. In particular, drill cuttings obtained (for example, during cable tool drilling) in 10-foot increments can be used to determine the predominant grain size and thus the storage and fluid transmission properties of the vadose zone. Clay-sized fractions can be separated from the drill cuttings for determination of cation exchange capacity (CEC). During construction of the wells, U.S. Bureau of Reclamation (USBR) borehole tests can be conducted to determine the depthwise distribution of hydraulic conductivity values. Finally, the depth to groundwater in the principal water-bearing formation can be determined, thus defining the overall storage capacity of the vadose zone.

Table 8-1 summarizes the results of the hypothetical premonitoring effort. The overburden geology consists of a depth of layered alluvium, with relatively fine material in the upper 40 feet. Perched groundwater detected at 40 feet extends to about 60 feet, in fairly coarse-size deposits. Finer material is noted at depths of 60 to 80 feet, underlain by coarser material to 100 feet. The water table is noted at 92 feet.

TABLE 8-1. SUMMARY OF RESULTS OF A PREMONITORING PROGRAM AT THE SITE OF A GENERIC HAZARDOUS WASTEWATER IMPOUNDMENT FOR PRIMARY ALUMINUM INDUSTRY

Lithology

Depth (ft)	Predominant Grain Size	CEC (meq/l)
0-10	Silty clay	29
10-20	Silty sand	15
20-30	Clayey-fine sand	23
30-40	Clayey sand	25
40-50	Gravelly sand (perched groundwater)	10
50-60	Gravelly sand (saturated)	11
60-70	Silty gravel (dry)	15
70-80	Clayey gravel	18
80-90	Gravel sand	7
90-100	Poorly-graded gravel (saturated)	9

Depth to groundwater depth = 92 feet

Saturated hydraulic conductivity from borehole test at 20- to 30-foot depth = 0.0004 in./hr

Estimated flux of wastewater (assuming unit hydraulic gradient) = 0.0004 in./hr

Estimated velocity of wastewater (assuming average field capacity at 10 percent) 0.004 in./hr or about 35 in./yr

The highest CEC is found in the material within the upper 10 feet of the profile. However, the CEC of material in the region from 20 to 40 feet is also high. The base of the impoundment is 20 feet below land surface. Consequently, if leakage should occur below the impoundment liner, considerable attenuation of cationic pollutants will occur in the region from 20 to 40 feet.

Saturated hydraulic conductivity, K_{sat}, values by the USBR borehole method average 0.0004 in./hr for the material underlying the impoundment. Thus, assuming vertical flow under a unit hydraulic gradient, the calculated flux value corresponding to this K_{sat} value is 0.0004 in./hr. Furthermore, if an average, drained water content of 10 percent is assumed, the vertical velocity of percolating water under a unit hydraulic gradient is 0.004 in./hr or about 35 in./yr.

New Surface Impoundment Monitoring

The monitoring package depicted in Figure 8-5 can be developed based on the results of the premonitoring program. To accommodate measurement of intake rates in the impoundment, a stilling well with a water stage recorder and an access platform is installed. During operation of the impoundment, intake rates can be determined by instantaneous rate methods in which wastewater level declines, measured in the lagoon during a brief shutdown period, are related to The lagoon's volume.

Figure 8-5 shows horizontal filter candle units installed beneath the impoundment to sample percolating water in case of failure at the liner. Filter candles contact a larger area than point samplers such as lysimeters. The individual units are laid in sheet metal troughs. Because of the importance of detecting wastewater movement below a failed liner, back-up units are installed beneath the shallower units. In addition, other units are installed at the same depths at other locations under the pond. As with the generic landfill model, such duplication can be regarded as "planned redundancy" to avoid the problems envisioned in the Hazardous Waste Regulations (Section 1). Inlet and outlet lines terminate in aboveground shelters containing vacuum pumps, sample bottles, and other appurtenances for obtaining either discrete or continuous samples.

A second array for direct sampling of groundwater consists of monitor wells installed in perched groundwater lenses and the main aquifer. The wells are particularly valuable in case water movement beneath the pond occurs at matric pressures below the limit of the filter candles (i.e., about -0.8 atmosphere). Ideally, the monitor wells should be of large enough diameter to permit the installation of permanent submersible pumps (Schmidt, personal communication, 1981). Sampling should be initiated as soon as the wells are completed to obtain baseline water quality values.

Indirect methods selected for detecting percolation of wastewater from the lagoon consist of tensiometers, heat-dissipation sensors, and access wells for neutron moisture logging. Tensiometers and heat dissipation units are useful in estimating both storage changes and hydraulic head gradients. In addition, the tensiometers will indicate the appropriate negative pressure to apply to the filter candles during sampling to avoid affecting unsaturated

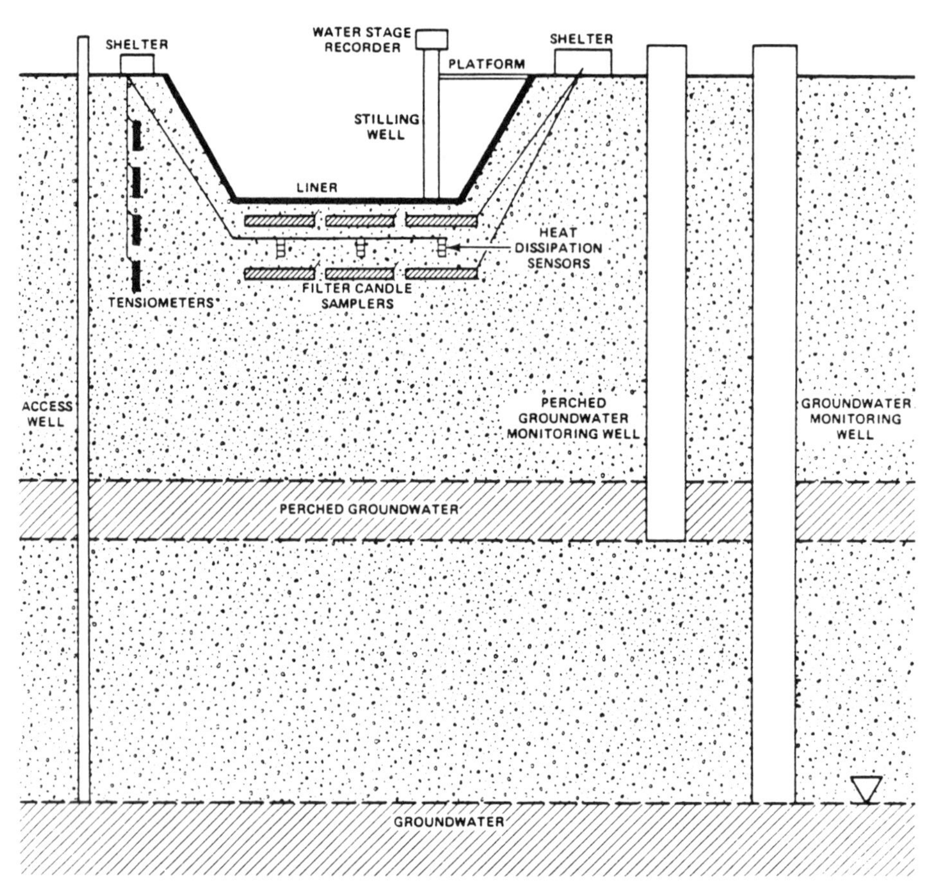

Figure 8-5. Water quality monitoring design for a new surface impoundment.

flow paths. The heat-dissipation sensors provide information on hydraulic gradients at negative pressures below the operating range of tensiometers. The selection of heat dissipation sensors in this example was arbitrary. To conform with the principle of planned redundancy, batteries of psychrometers, hygrometers, and/or electric resistance blocks can also be installed. The tensiometer readings would be recorded manually by a field technician. However, signals from the other matric-potential sensors could be recorded at the plant.

Neutron moisture logging in the access wells will show the lateral spread of wastewater in the vadose zone in case of liner failure if storage changes occur in the geologic profile. The wells can also be used to monitor changes in groundwater levels and obtain water samples from the vicinity of the water table.

Active Surface Impoundment Monitoring

It will not be possible to install monitoring units beneath the base of active and abandoned impoundments. Consequently, such units must be installed at the periphery of the facility. Figure 8-6 shows a conceptualized scheme. Again, a stilling well and water stage recorder are installed to aid in determining intake rates by the instantaneous rate method. A cheaper technique would be to position stilling wells on the sides of the lagoon. Flux and velocity in the vadose zone can be estimated for active ponds. Dividing this value by representative water content values gives an estimate of velocity.

Sampling techniques consist of clustered suction cup lysimeters in a common borehole and perched groundwater and main aquifer monitoring wells. Nonsampling techniques consist of tensiometers and access wells for neutron logging. As with a new lagoon, installation of other nonsampling units such as psychrometers and/or hygrometers is advisable for monitoring in the dry range where tensiometers become inoperative.

The previous subsection described a conceptualized vadose zone monitoring package for one disposal method, namely, a wastewater impoundment. Obviously, a wide variety of monitoring packages is possible (Everett, 1980), depending upon variables such as waste type, disposal method, site-specific conditions such as topography, climate, and geohydrology, and the selection criteria listed in Everett, Wilson, and McMillion (1982). In view of the criticality of the waste disposal problem and the importance of vadose zone monitoring as an early warning of pollutant movement, the authors feel that a concerted effort should be undertaken to develop a systematic approach for matching the selection variables. The possibility of new and innovative methods for vadose zone monitoring should be undertaken. For example, techniques for counting radiation decay may be more appropriate for detecting leakage from nuclear waste disposal areas than methods presented here.

LAND TREATMENT FACILITIES

As defined in Hazardous Waste and Consolidated Permit Regulations (U.S. EPA, 1980b), a land treatment facility is "... that part of a facility at which hazardous waste is applied onto or incorporated into the soil surface." The objective of land treatment is to enhance the microbial decomposition of waste pollutants, or to otherwise retard their mobility by soil physical/chemical reactions. According to Ross and Phung (1980), land treatment of wastes involves three steps following application and incorporation: (1) mixing the waste with surface soil to aerate the mass and expose waste to soil microorganisms, (2) adding nutrients or amendments (optional), and (3) remixing the soil and waste periodically to maintain aerobic conditions. Crops may be grown concurrently in waste disposal land treatment areas if that care is taken to prevent introducing pollutants such as cadmium in the food chain.

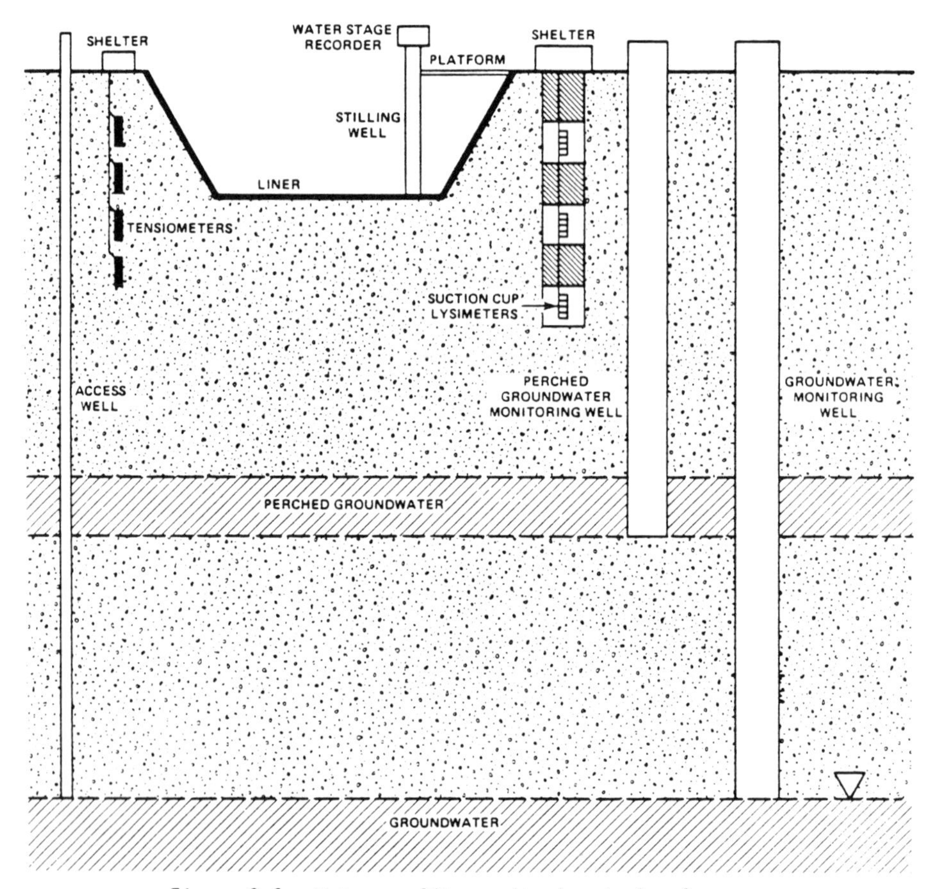

Figure 8-6. Water quality monitoring design for an active surface impoundment.

Techniques for selecting, managing, and operating land treatment facilities were reviewed by Phung et al. (1978), Ross and Phung (1980), Huddleston (1980), and Miller (1980). EPA requirements for surface water control, record keeping, waste analyses, monitoring, use for food chain crops, and closure are included in "Hazardous Waste Management System: Standards Applicable to Generators of Hazardous Waste" (U.S. EPA, 1980c). In contrast to the monitoring requirements for hazardous waste impoundments and landfills, vadose zone monitoring is required by the EPA at land treatment areas. The designated techniques are pore-water and solids sampling.

Figure 8-7 illustrates a conceptual representation of a land treatment facility. As shown, the facility contains a lagoon for storing incoming wastes. Conceivably, such a lagoon would be lined to minimize seepage and would include monitoring facilities, such as described previously for impoundments. The field is diked to prevent uncontrollable runoff. Whatever runoff occurs is collected in a sump. The sketch also illustrates concurrent waste application and disking.

According to Phung et al. (1978), the quantity of industrial waste disposed of in land treatment facilities will increase in time because of more rigorous air and water pollution regulations. For example, certain liquid wastes once deposited in landfills will be applied to land treatment areas. Estimates of the amount of industrial wastewater and sludge suitable for land application in 1975, 1980, and 1985 are:

	1975	1980	1985
Wastewater (10^5 acre-feet/yr)	6.0	6.8-7.5	7.6-9.4
Sludge (10^6 t/yr, dry wt)	7.2-7.5	8.8-9.1	10.8-11.1

As stated by Ross and Phung (1980), the "suitability" of a specific industrial waste for land treatment depends on characteristics such as pH, BOD, concentration of chemical elements (heavy metals), soluble salts and hazardous chemicals, bulk density of waste solids, and volatility. Because of such limitations, the values of wastewater and sludge for 1975, 1980, and 1985 represent only 1 percent and 3 percent, respectively, of waste generated or projected for generation in the United States. Industries producing waste suitable for land treatment are (Ross and Phung, 1980):

Food and kindred products

Textile finishing

Wood preserving

Paper and allied products

Organic fibers (noncellulosic)

Drugs and pharmaceuticals

Soap and detergents

Organic chemicals

Petroleum refining

Leather tanning and finishing.

Land Treatment Premonitoring

Premonitoring activities for land treatment of hazardous wastes are identical to those for solid waste and impoundment site selection, e.g., characterization of wastes and vadose zone properties. Ideally, these activities should be staged as follows: (1) identify waste properties that will affect the mobility of pollutants in the vadose zone, (2) identify properties of

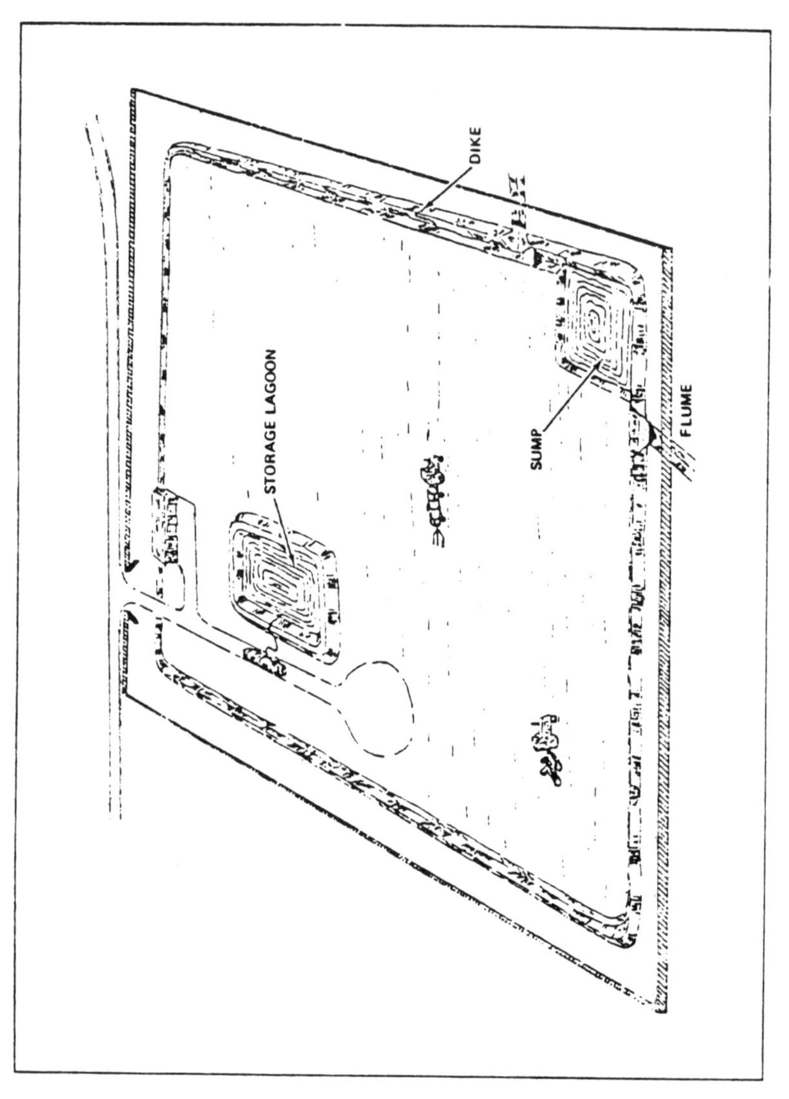

Figure 8-7. Artist's conception of land cultivation site.

vadose zone solids that will affect the mobility of waste pollutants, and (3) conduct tests to evaluate waste and soil interactions promoting pollutant attenuation in the vadose zone.

During a survey of industrial wastewater, sludges, and residues, Geraghty and Miller, Inc. (Miller, 1980) found that information on the composition of such waste is limited, in part because of the proprietary nature of industrial processes. Examples of pollutants that may be found in industrial wastewater, sludge, and residues are listed in Section 6.

A first step in site selection and premonitoring is thus to inventory the potential sources and completely analyze a representative sample of the waste (the requirements for sample collection and analyses are critical but cannot be elaborated upon here). The fate of major and trace inorganics found in the waste analyses can be qualitatively predicted from relationships among constituents and parameters. For example, the mobility of certain cationic trace inorganics is inhibited at pH values above neutrality (Fuller, 1977). For organics, additional laboratory tests may be required to identify properties governing their fate in the environment. These properties include water solubility, chemical structure, octanol-water partition coefficient, polarity, and vapor pressure (Howard, Saxena, and Sikka, 1978).

As an example of clues available for one of these properties, namely chemical structure, Hutzinger (1981) indicated that biodegradability of an aromatic compound decreases with increasing numbers of chlorine substitutions on the ring.

A general assessment of the physical and chemical properties of the soils at a potential or active land treatment site can be derived by reviewing published soil survey information. Soil surveys include maps depicting the areal distribution of mapping types and detailed descriptions of soil properties such as texture, structure, and infiltration rates. In addition to published information, the local Soil Conservation Service (SCS) office may contain unpublished soils information, possibly dealing with the actual disposal field. Onsite tests may be required to determine local variations in mapping units that affect hydraulic and attenuation properties. A soil survey specialist should assist in this task.

After a surface survey has been completed with hand-operated equipment, deeper cores should be obtained in each mapping unit using power equipment. Statistical approaches for determining the number of cores to account for variations in soil physical properties are outlined by Warrick and Nielsen (1980). In the selection of landforms, taking cores to a depth of 8 to 12 feet is generally recommended (Huddleston, 1980). Within the context of vadose zone monitoring, however, a few holes should be extended to the water table. The drilling and sampling program can be coupled with monitor well installation. During drilling, careful logs should be maintained showing changes in strata, presence of perched groundwater, and depth to water table. Core samples should be placed in appropriate containers and labelled for subsequent laboratory testing.

Core samples are used to determine baseline physical and chemical conditions at a new site and the extent of pollutant movement in active sites. The parameters and concentrations that should be characterized are listed in Tables 6-1 and 6-2. In addition to the factors listed, other properties that should be determined for specifically estimating pollutant mobility include clay content, specific surface, hydrous oxides, iron and aluminum oxides, and organic matter content (Fuller, 1979). These properties are related to the cation exchange capacity and specific adsorption potential. The mobility of excessive concentrations of cationic heavy metals in a waste is known to be reduced as the opportunity for adsorption increases.

The interactions between solids and wastes are generally more complex than can be estimated from information on solid and waste properties separately. Laboratory approaches for determining physical and chemical interactions include soil thin layer chromatography (TLC), batch tests, and column tests using samples of vadose zone solids and waste from the proposed disposal site. Techniques are available for estimating the chemical, photochemical, and biological degradation of waste pollutants in soils (Howard, Saxena, and Sikka, 1978).

Conducting permeameter tests on intact soil cores is advisable to determine saturated and unsaturated hydraulic conductivities. Ideally, waste fluid should be used for these tests. A state-of-the-art review of hydraulic testing on cores is included in a recent ASTM publication (Zimmie and Riggs, 1981).

For new sites, field plots can be constructed for more definitive information on interactions between soils and chemical wastes. A sufficient number of plots should be selected to ensure that each mapping unit is adequately represented. Draining-profile tests explained in detail by Wilson (1980) can be used to characterize hydraulic conductivity and flux characteristics of each mapping unit. Suction lysimeters should be installed for evaluating the in situ mobility of waste pollutants. Access tubes should also be installed for neutron moisture logging to determine storage properties of the upper vadose zone. Incidentally, the extent that neutron logging is influenced by neutron moderation in chemical wastes (e.g., by hydrogen chlorine in aromatic compounds) can also be determined in these test plots.

Active and New Land Treatment Monitoring

Generic designs for vadose zone monitoring at active and new land treatment facilities are presented in this subsection. The premise is that the active site is reaching capacity and that an adjacent area is being developed to meet future demands. The hydrologic setting is a glaciofluvial deposit similar to that for the solid waste landfill. Section 7 presents details of this setting.

Federal requirements for monitoring of land treatment areas are described in detail by U.S. EPA (1980b) in "Standards Applicable to Owners and Operators of Hazardous Waste Treatment, Storage, and Disposal Facilities." As stated in these "standards," the objective of monitoring at land treatment areas is to "... provide waste-specific, constituent specific, and site-specific evidence that the treatment objective is being met." A comprehensive monitoring program to demonstrate treatment includes waste sampling, soil core sampling, soil pore liquid sampling, and groundwater sampling. This subsection focuses on soil pore liquid monitoring.

The recently issued final permitting standards (EPA, 1982) for owners and operators of hazardous waste treatment, storage, and disposal facilities specify conduct of "unsaturated" zone monitoring at all land treatment areas "... to determine whether hazardous constituents migrate out of the treatment zone" (40 CFR Part 278). The techniques designated for such monitoring include soil coring and soil pore liquid samplers. Important requirements of unsaturated (vadose) zone monitoring include: (1) determining background levels for each hazardous constituent selected for monitoring, (2) subsurface soil monitoring and soil-pore sampling immediately below the treatment zone, (3) determining whether a statistically significant change over background levels occurs for specified hazardous constituents, and (4) notifying the Regional Director when statistically significant changes in the specified constituents are observed. In the context of the regulations (S.264.271), the maximum depth of the treatment zone must be (1) no more than 1.5 meters (5 feet) below the initial soil surface, and (2) at least 1 meter (3 feet) above the seasonal high water table.

Although a number of techniques are available for indirectly monitoring the movement of pollutants beneath waste disposal facilities, soil core sampling and suction-cup lysimeters, or samplers, remain the principal methods for directly sampling pore fluids in unsaturated media. Lysimeters have been used for many years by agriculturists for monitoring the flux of solutes beneath irrigated fields (Biggar and Nielsen, 1976). Similarly, they have been used to detect the deep movement of pollutants beneath land treatment areas (Parizek and Lane, 1970) and beneath sanitary landfills (Johnson, Cartwright, and Schuller, 1981).

Guidance documents are currently in preparation by EPA on the subject of land treatment disposal. The following procedure presents an approach which allows a user to randomly select a vadose zone monitoring location in both the background and the active site (U.S. EPA, 1983). The location, within a given uniform area of a land treatment unit (i.e., active portion monitoring), at which a soil-pore liquid monitoring device should be installed can be determined using the following procedure:

1. Divide the land treatment unit into uniform soil areas. A uniform soil area at an active land treatment unit is defined as a soil series upon which similar hazardous wastes have been applied at similar application rates. A certified professional soil scientist should be consulted in completing this step.

2. Map each uniform area by establishing two baselines at right angles to each other which intersect at an arbitrarily selected origin, for example, the southwest corner. Each baseline should extend to the boundary of the uniform area.

3. Establish a scale interval along each baseline. The units of this scale may be feet, yards, miles, or other units depending on the size of the uniform area. Both baselines must have the same scale.

4. Draw two random numbers from a random numbers table (usually available in any basic statistics book). Use these numbers to locate one point along each of the baselines.

5. Locate the intersection of two lines drawn perpendicular to these two baseline points. This intersection represents one randomly selected location for collection of one soil-pore liquid device. If this location at the intersection is outside the uniform area, disregard and repeat the above procedure.

6. For soil-pore liquid monitoring, repeat the above procedure as many times as necessary to obtain three locations for installation of soil-pore liquid monitoring devices within each uniform area. In addition, there should be no less than one soil-pore liquid monitoring device (location) per 4 acres of uniform area. Monitoring at these same randomly selected locations will continue throughout land treatment unit life (i.e., devices do not have to be relocated at every sampling event).

Locations for monitoring on background areas should be randomly determined using the following procedure:

1. Consult a certified professional soil scientist in determining an acceptable background area. The background area must have characteristics (i.e., at least soil series classification) similar to those present in the uniform area of the land treatment unit it is representing.

2. Map an arbitrarily selected portion of the background area by establishing two baselines at right angles to each other which intersect at an arbitrarily selected origin.

3. Complete steps 3, 4, and 5 as defined above.

4. For soil-pore liquid monitoring, repeat the above procedure as necessary to obtain two locations for soil-pore liquid monitoring devices within each background area.

While sampling depths should be within 30 cm (12 in.) of the bottom of the treatment zone (Everett and Wilson, 1983), seasonality is the rule with soil-pore liquid sample timing (i.e., scheduled sampling cannot be on a preset date but must be geared to precipitation events). Assuming that sampling is done soon after leachate-generating precipitation or snowmelt, the frequency also varies depending on site conditions. As a starting point, sampling should be done quarterly. The timing of sampling should be geared to the waste application schedule as much as possible.

For land treatment units at which wastes are applied infrequently (i.e., only once or twice a year) or where leachate-generating precipitation is highly seasonal, quarterly sampling and analysis of soil-pore liquid may be unnecessary. Because soil-pore liquid is instituted primarily to detect fast-moving hazardous constituents, monitoring for these constituents many months after waste application may be useless. If fast-moving hazardous constituents are to migrate out of the treatment zone, they will usually migrate at least within 90 days following waste application, unless little precipitation or snowmelt has occurred. Therefore, where wastes are applied infrequently or leachate generation is seasonal, soil-pore liquid may be monitored less frequently (semiannually or annually).

The background concentrations of hazardous constituents in the soil-pore liquid should be established by installing two monitoring devices at random locations for each soil series present in the treatment zone. Samples should be taken on at least a quarterly basis for at least 1 year and can be composited to give one sample per quarter. Analysis of these samples should be used to calculate an arithmetic mean and variance for each hazardous constituent. After background values are established, additional soil-pore liquid samples should occasionally be taken to determine if the background values are changing over time.

The generic monitoring program presented here is based on the premise that the owner or operator has chosen a mixture of nonsampling methods in addition to the required sampling program. These nonsampling methods do not get at specific constituents and are not required in the regulations (EPA, 1983). In addition, the owner or operator is assumed sufficiently environmentally conscious to elect to use alternative sampling methods to back up the suction cup samplers that are inoperative at soil-water pressures below -0.8 bar.

Nonsampling Methods--

Nonsampling approaches to monitoring at an active land treatment area include conducting a water budget analysis and installing nonsampling units. The water budget approach using soil moisture accounting is a simple, rapid method for estimating the volume of deep percolation below a selected soil depth. The method developed by Thornthwaite and Mather (1957) is used most often and a computer program designated WATBUG is available to simplify calculations (Wilmott, 1977). Inflow components that must be measured include precipitation and wastewater application. Outflow components requiring measurement include runoff and crop evapotranspiration (ET). The change in storage equals water content change in the depth of interest. All terms are equated to deep percolation.

Figure 8-8 shows a generic assortment of monitoring units at a land treatment station. Elements of the design include (1) an access well for neutron moisture logging, (2) implantable EC probes for detecting changes in salinity, (3) tensiometers for measuring matric potential down to -0.8 bar, and (4) thermal dissipation sensors for measuring matric potential below -0.8 bar. Technical descriptions, limitations, and field applications of these techniques are described in other sections of this book.

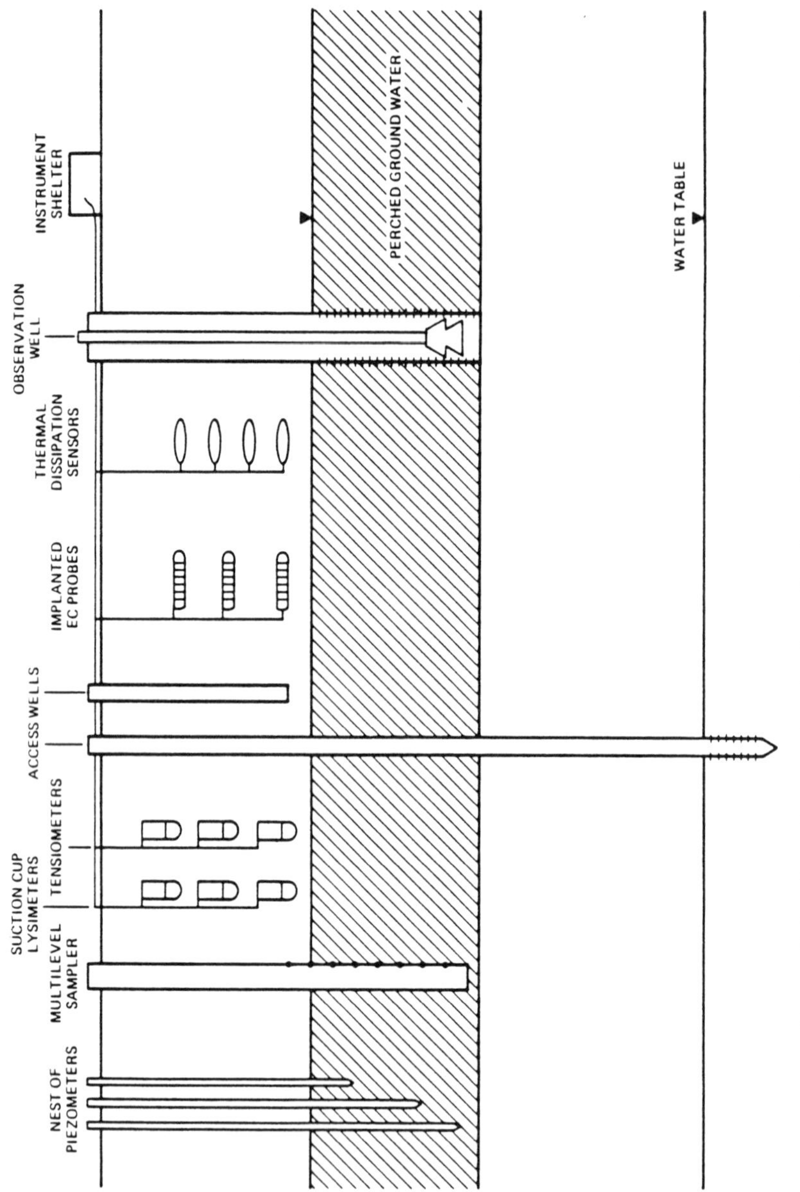

Figure 8-8. Generic assortment of monitoring units at land treatment area.

In order to prevent operational inconvenience and sampling bias, the monitoring system is designed and installed so that the aboveground portion of the device is located at least 10 meters from the sampling location. If the aboveground portion of the device is located immediately above the sampling device, the sampling location will often be avoided because of operational difficulties. Thus, samples collected at this location will be biased and not representative of the treated area (Figure 8-9).

Figure 8-8 shows that two access wells have been installed at the station, one shallow and one deep. In practice, a large number of shallow ($\leq$10 feet) access wells could be installed throughout the site to determine changes in water content during application and drying cycles. If all of the water content changes appear to be occurring in the shallow depth, deeper units may not be required. If extensive deep percolation is occurring, however, deeper access wells will be needed, perhaps extending to the water table. This method is useful in determining site-specific water balances and soil-water flux. This moisture logging accounts for changes in water content, a capacity factor. In fact, water (and pollutants) may flow through specific zones without a change in water content being registered on moisture logs. Thus, backup facilities are required.

A depthwise battery of implantable EC probes is included. These units are based on the four-probe Wenner array for detecting electrical resistivity in the field of measurement. Changes in resistivity indicate that conductive fluids are moving past the units. Alternative techniques that could be used for detecting changes in salinity include salinity sensors, portable EC probes, and a portable four-electrode array.

Tensiometer units are included for sensing the soil-water matric potential. Tensiometer readings are useful in determining the correct vacuum to be applied to suction cup samples (to avoid influencing the flow field). Soil-water pressure gradients detectable by a battery of units are indicative of the flux of pollutants. Consequently, these units serve to support the results obtained from moisture logging. For shallow units where the gravitational component does not limit the matric potential readings, inexpensive hydraulic switches can be used to attach many tensiometers to a single pressure transducer (Rawlins, 1976). Deeper units will require the use of integral pressure transducers. The latter units also have the advantage of rapid response time and no aboveground components except lead wires running to a shelter.

Inasmuch as tensiometer units fail at negative pressures below -0.8 bar, the generic system also includes a battery of thermal dissipation sensors, operative in the dry range. Other units that could be used to extend the range of tensiometers for detecting changes in matric potential include electrical resistance and capacitance blocks, thermocouple psychrometers and hygrometers, and osmotic tensiometers (Rawlins, 1976).

Sampling Methods--

Sampling methods at an active site include soil sampling, pursuant to the requirements of EPA (1980b), and the generic methods depicted in Figure 8-8.

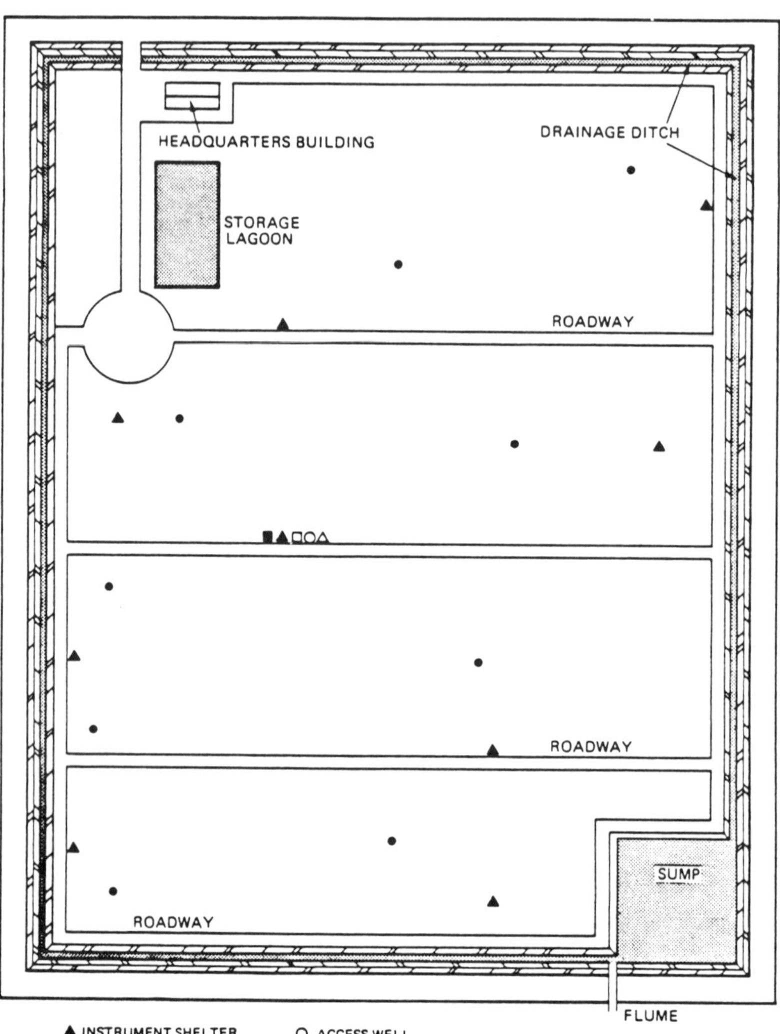

Figure 8-9. Plan view of monitoring units for land treatment area.

Soil sampling includes using hand augers and samplers such as the Veihmeyer tube. Deeper methods require the use of power equipment including flight augers and hollow stem augers with wire-line samplers (Kaufmann, 1981).

Generic methods depicted in Figure 8-8 include: (1) suction cup samplers, (2) multilevel samplers in perched groundwater, (3) a depthwise array of piezometers, (4) an observation well, and (5) a screened well point at the end of the access well.

Suction cup lysimeters are installed to ensure detecting the vertical movement of pollutants. For in situ shallow sampling, the vacuum-pressure unit will be adequate. Filter-candle-type samplers could be used instead of suction cup units.

Multilevel samplers are installed in a shallow body of perched groundwater are useful for determining the vertical extent of a plume. A horizontal transect of these units is useful in detecting the lateral dimensions of a plume. They are also used to measure hydraulic gradients. As shown in the figure, some of the sampling points extend above the water table to facilitate sampling during a water table rise (useful for collecting pollutants adsorbed in the vadose zone). A battery of piezometer units could be used in place of or to supplement the multilevel samplers. Piezometers are useful for determining saturated hydraulic conductivity values.

Figure 8-8 shows a standard observation well installed in a body of perched groundwater. Such wells permit extracting large volumes of perched groundwater for analysis. Results may indicate the integrated water quality flowing through the vadose zone. Pumps installed in the observation wells facilitate sampling. Pumps are also useful for conducting pumping tests when determining the hydraulic properties of the vadose zone.

References

Alemi, M.H., D.R. Nielsen, and J.W. Biggar, 1976. Determining the Hydraulic Conductivity of Soil Cores by Centrifugation, J. Soil Sci. Soc. Amer., 40, 42-218.

Alesii, B.A., 1976. Cyanide Mobility in Soils, unpublished M.S. thesis, U. of Arizona, Tucson.

Alexander, M., 1961. Introduction to Soil Microbiology, John Wiley and Sons, New York.

Allison, L.E., 1971. A Simple Device for Sampling Ground Waters in Auger Holes, Soil Sci. Soc. Amer. Proc., 35, 844-845.

American Society of Civil Engineers (ASCE), 1976. Sanitary Landfill, ASCE Solid Waste Management Committee of the Environmental Engineering Division, Headquarters of the Society of New York.

Anderson, K.E., (ed), 1977. Water Well Handbook, Missouri Water Well Drillers Assocation, Rolla, Missouri.

Apgar, M.A., and D. Langmuir, 1971. Ground-Water Pollution Potential of a Landfill Above the Water Table, Ground Water, 9(6), 76-93.

Apgar, M.A., and W.B. Satherthwaite, Jr., 1975. Ground Water Contamination Associated with The Llangollen Landfill, New Castle County, Delaware, Proc. Res. Symp., Gas and Leachate from Landfills, New Brunswick, New Jersey, U.S. Environmental Protection Agency, National Environmental Research Center, Cincinnati, Ohio.

Ayers, R.S., and R.L. Branson, (ed), 1973. Nitrates in the Upper Santa Ana River Basin in Relation to Groundwater Pollution, Bull. 861, California Agric. Exp. Station

Ayers, R.S., and D.W. Westcot, 1976. Water Quality for Agriculture, Irrigation and Drainage Paper No. 29, UN Food and Agriculture Organization.

Bailey, G.W., and J.L. White, 1970. Factors Influencing the Adsorption, Desorption, and Movement of Pesticides in Soil, in Residue Reviews, 32, Springer-Verlag, New York.

Bear, J., 1972. Dynamics of Fluids in Porous Media, American Elsevier, New York, 1972.

Bear, J., D. Zaslavsky, and S. Irmay, 1968. Physical Principles of Water Percolation and Seepage, UN Educational, Scientific and Cultural Organization, 465 pp.

Bertrand, A.R., 1965. Rate of Water Intake in the Field, in Methods of Soil Analyses, C.A. Black, (ed), Agronomy No. 9, 197-208, Amer. Soc. Ag., Madison, Wisconsin.

Beven, K., and P. Germann, 1982. Macropores and Water Flow in Soils, Water Resources Res., 18(5), 1311-1325.

Bianchi, W., 1967. Measuring Soil Moisture Tension Changes, Agric. Engineering, 43(7), 398, 399, 404.

Bianchi, W.C., C.E. Johnson, and E.E. Haskell, 1962. A Positive Action Pump for Sampling Small Bore Wells, Soil Sci. Soc. Amer. Proc., 26(1), 86-87.

Biggar, J.W., and D.R. Nielsen, 1976. Spatial Variability of the Leaching Characteristics of a Field Soil, Water Resources Res., 12(1), 78-84.

Black, C.A., (ed), 1965. Methods of Soil Analyses, (in two parts), Agronomy No. 9, Amer. Soc. Agron., Madison, Wisconsin.

Black and Veatch, 1981a. Interim Report on the Impact of Pollutants on the Blue Plains Treatment Plant, Dept. Environmental Services, Government of the District of Columbia, Washington, D.C.

Black and Veatch, 1981b. Second Interim Report, Industrial Pretreatment Program, Washington Suburban Sanitary Commission, Laurel, Maryland.

Blake, G.R., 1965. Bulk Density, in Methods of Soil Analyses, C.A. Black, (ed), Agronomy No. 9, 374-390, Amer. Soc. Agron., Madison, Wisconsin.

Bodman, G.B., and E.A. Coleman, 1944. Moisture and Energy Conditions During Downward Entry of Water into Soils, Soil Sci. Soc. Amer. Proc., 8, 116-122.

Bohn, H.L., B.L. McNeal, and G.A. O'Connor, 1979. Soil Chemistry, Wiley Interscience, New York.

Bordner, R., J. Winter, and P. Scarpino, 1978. Microbiological Methods for Monitoring the Environment, EPA-600/8-78-017, U.S. Environmental Laboratory, Environmental Monitoring and Support Laboratory.

Borg, I.Y., R. Stone, H.B. Levy, and L.D. Ramspott, 1976. Information Pertinent to the Migration of Radionuclides in Ground Water at the Nevada Test Site, Part I: Review and Analysis of Existing Information, Lawrence Livermore Laboratory, Livermore, California, NTIS.

Bouma, J., F.G. Baker, and P.L.M. Veneman, 1974. Measurement of Water Movement in Soil Pedons Above the Water Table, Info. Circ. No. 27, U. of Wisconsin-Extension, Geological and Natural History Survey.

Bouwer, H., 1978. Groundwater Hydrology, McGraw-Hill, New York, 480 pp.

Bouwer, H., 1980. Deep Percolation and Ground-Water Management, in Proc. of the Deep Percolation Symp., Arizona Dept. of Water Resources Report No. 1, 13-19.

Bouwer, H., and R.D. Jackson, 1974. Determining Soil Properties, in Drainage for Agriculture, J. van Schilfgaarde, (ed), Agronomy No. 17, 611-675, Amer. Soc. Agron., Madison, Wisconsin.

Bouwer, H., and R.C. Rice, 1963. Seepage Meters in Recharge and Seepage Studies, Journal Irrigation and Drainage Division, ASCE, 89, (IRI), 17-42.

Bower, C.A., and L.V. Wilcox, 1965. Soluble Salts, in Methods of Soil Analyses, C.A. Black, (ed), Agronomy No. 9, 933-951, Amer. Soc. Agron., Madison, Wisconsin.

Bouyoucas, G.J., 1960. Measuring Soil Moisture Tension, Agricultural Engineering, Michigan, 41, 40-41.

Brakensiek, D.L., 1979. Empirical and Simplified Models of the Infiltration Process, in Infiltration Research Planning Workshop, Part 1, State of the Art Reports ARM-NC-4, Sciences and Education Administration, U.S. Dept. of Agric.

Brakensiek, D.L., H.B. Osborn, and W.J. Rawls, 1979. Field Manual for Research in Agricultural Hydrology, Agricultural Handbook No. 224, Sciences and Education Adm., U.S. Dept. of Agric.

Broadbent, F.E., 1973. Organics, in Proc. Joint Conference on Recycling Municipal Sludges and Effluents on Land, sponsored by Environmental Protection Agency, U.S. Dept. of Agric., and Nat. Assoc. of State Univ. and Land-Grant Colleges, Champaign, Illinois, July 9-13.

Brown, R.W., 1970. Measurement of Water Potential With Thermocouple Psychrometers: Construction and Applications, USDA Forest Service Res. Paper INT-80, USDA Forest Service Intermountain Forest and Range Experiment Station, Ogden, Utah.

Cartwright, K., and M.R. McComas, 1968. Geophysical Surveys in the Vicinity of Sanitary Landfills in Northeastern Illinois, Ground Water, 6(5), 23-30.

Cartwright, K., and F.B. Sherman, 1972. Electrical Earth Resistivity Surveying in Landfill Investigations, in Proc. 10th Ann. Engineering and Soils Engineering Symp., 77-92, Moscow, Idaho.

Cary, J.W., 1973. Soil Water Flowmeters with Thermocouple Outputs, Soil Sci. Soc. Amer. Proc., 37, 176-181.

Chain and DeWalle, 1975. Compilation of Methodology for Measuring Pollution Parameters of Landfill Leachate, U. of Illinois, U.S. Environmental Protection Agency, Cincinnati, Ohio.

Chang, A.C., and A.L. Page, 1979. Hydraulic and Water Quality Considerations of Artificial Groundwater Recharge, in Proc. Twelfth Biennial Conference on Ground Water, Sacramento, September 20-21, 1979, California Water Resources Center, U. of California, Davis, Report No. 45, 88-99.

Chapman, H.D., 1965. Cation Exchange Capacity, in Methods of Soil Analyses, C.A. Black, (ed), Agronomy No. 9, 891-900, Amer. Soc. Agron., Madison, Wisconsin.

Childs, E.C., 1969. An Introduction to the Physical Basis of Soil Water Phenomena, Wiley Interscience, New York.

Chow, T.L., 1977. A Porous Cup Soil-Water Sampler with Volume Control, Soil Sci., 124, 173-176.

Chow, V.T., 1964. Handbook of Applied Hydrology, McGraw-Hill, New York.

Cloke, P.L., 1966. The Geochemical Application of Eh-pH Diagrams, J. Geol. Educ., 4, 140-148.

Coelho, M.A., 1974. Spatial Variability of Water Related Soil Physical Properties, unpublished M.S. thesis, The U. of Arizona, Tucson.

Cole, J.A., (ed), 1972. Groundwater Pollution in Europe, Proc. conference organized by the Water Research Association in Reading, England, Water Information Center, Inc., Port Washington, New York.

Cooley, R.L., J.F. Harsh, and D.C. Lewis, 1972. Principles of Ground-Water Hydrology, Hydrologic Engineering Methods for Water Resources Development, Vol. 10, Hydrologic Engineering Center, U.S. Army Corps of Engineers, Davis, California.

Corey, J.C., S.F. Peterson, and M.A. Wakat, 1971. Measurement of Attenuation of ^{137}Cs and ^{241}Am Gamma Rays for Soil Density and Water Content Determinations, Soil Sci. Soc. Amer. Proc., 35, 215-219.

Davis, J.B., 1956. Microbial Decomposition of Hydrocarbons, Industrial and Engineering Chemistry, 48(9), 1444-1448.

Davis, S.N., and R.J.M. De Wiest, 1966. Hydrogeology, John Wiley and Sons, New York.

Day, P.R., G.H. Bolt, and D.M. Anderson, 1967. Nature of Soil Water, in Irrigation of Agricultural Lands, R.M. Hagan, H.R. Haise, and T.W. Edminster, (ed), Agronomy No. 9, 193-208, Amer. Soc. Agron., Madison, Wisconsin.

Dazzo, F.B., and D.F. Rothwell, 1974. Evaluation of Porcelain Cup Water Samplers for Bacteriological Sampling, Applied Microbiology, 27(6), 1172-1174.

Dirksen, C., 1974a. Water Flux Measurements in Unsaturated Soils, in Flow--Its Measurement and Control in Science and Industry, 1, 479-486, Inst. Soc. Amer., Pittsburgh, Pennsylvania.

Dirksen, C., 1974b. Field Test of Soil Water Flux Meters, in Trans. Amer. Soc. Agric. Eng., 17(6), 1038-1042.

Donnan, W.W., 1957. Field Methods, in Drainage of Agricultural Lands, J.N. Luthin, (ed), Amer. Soc. Agron., Madison, Wisconsin.

Duke, H.R., and H.R. Haise, 1973. Vacuum Extractors to Assess Deep Percolation Losses and Chemical Constituents of Soil Water, in Soil Sci. Soc. Amer. Proc., 37, 963-964.

Dunlap, W.J., and J.F. McNabb, 1973. Subsurface Biological Activity in Relation to Ground Water Pollution, EPA-660/2-73-014, U.S. Environmental Protection Agency, Corvallis, Oregon.

Dunlap, W.J., J.F. McNabb, M.R. Scalf, and R.L. Cosby, 1977. Sampling for Organic Chemicals and Microorganisms in the Subsurface, EPA-600/2-77-176, U.S. Environmental Protection Agency.

Ellis, B.G., 1973. The Soil as a Chemical Filter, in Recycling Treated Municipal Waste Water and Sludge Through Forest and Cropland, W.E. Sopper and L.T. Kardos, (ed), Pennsylvania State Press.

Enfield, C.G., J.J.C. Hsieh, and A.W. Warrick, 1973. Evaluation of Water Flux Above a Deep Water Table Using Thermocouple Psychrometers, Soil Sci. Soc. Amer. Proc., 37(6), 968-970.

England, C.B., 1974. Comments on "A Technique Using Porous Cups for Water Sampling at Any Depth in the Unsaturated Zone" by Warren W. Wood, Water Resources Res., 10(6), 1049.

Estes, J.E., D.S. Simonett, L.R. Tinney, C.E. Ezra, B. Bowman, and M. Roberts, 1978. Remote Sensing Detection of Perched Water Tables, Technical Completion Report, Contribution No. 175, California Water Resources Center, U. of California, Davis, California, 1978.

Evans, D.D., T.W. Sammis, A.W. Warrick, 1976. Transient Movement of Water and Solutes in Unsaturated Soil Systems, Phase II, Project Completion Report, OWRT Project No. B040-ARIZ, Dept. of Hydrology and Water Resources, U. of Arizona, Tucson.

Evans, D.D., and A.W. Warrick, 1970. Time of Transit of Water Moving Vertically for Ground Water Recharge, Symposium on Saline Water, Contribution No. 13, Committee on Desert and Arid Zone Research, Southwest and Rocky Mountain Division Amer. Acad. Adv. Sci., 87-97.

Everett, L.G., (ed), 1979. Groundwater Quality Monitoring of Western Coal Strip Mining: Identification and Priority Ranking of Potential Pollution Sources, EPA-600/7-79-024, U.S. Environmental Protection Agency, Las Vegas, Nevada.

Everett, L.G., 1980. Groundwater Monitoring, General Electric Co. Technology Marketing Operations, Schenectady, New York.

Everett, L.G., 1981. Monitoring in the Vadose Zone, Ground Water Monitoring Rev., 1(2), 44-51.

Everett, L.G., and M. Hulbert, 1980. Groundwater Quality Monitoring Designs for Municipal Pollution Sources: Preliminary Designs for Coal Strip Mining Communities, U.S. Environmental Protection Agency, EMSL, Las Vegas, Nevada.

Everett, L.G., and L.G. Wilson, 1983. Unsaturated Zone Monitoring at Hazardous Waste Land Treatment Units, Guidance Manual to EPA, Office of Solid Waste, Washington, D.C.

Everett, L.G., K.D. Schmidt, R.M. Tinlin, and D.K. Todd, 1976. Monitoring Groundwater Quality: Methods and Costs, EPA-600/4-76-023, U.S. Environmental Protection Agency, Environmental Monitoring and Support Laboratory, Las Vegas, Nevada.

Everett, L.G., L.G. Wilson, and L.G. McMillion, 1982. Vadose Zone Monitoring Concepts for Hazardous Waste Sites, Ground Water, 20(3), May-June, 1982.

Fenn, D., E. Cocozza, J. Isbister, O. Briads, B. Yare, and P. Roux, 1977. Procedures Manual for Ground Water Monitoring at Solid Waste Disposal Facilities, EPA/530/SW-611, U.S. Environmental Protection Agency.

Fenn, D.G., K.J. Hanley, and T.V. DeGeare, 1975. Use of the Water Balance Method for Predicting Leachate Generation from Solid Waste Disposal Sites, EPA/530/SW-618, U.S. Environmental Protection Agency, Office of Solid Waste.

Freeze, R.A., 1969. The Mechanism of Natural Ground-Water Recharge; 1 - One-Dimensional Vertical, Unsteady, Unsaturated Flow Above a Discharging GroundWater Flow System, Water Res. Research, 5(1), 153-171.

Freeze, R.A., and J.A. Cherry, 1979. Groundwater, Prentice-Hall, Englewood Cliffs, New Jersey.

Frissel, M.J., P. Poelstra, K. Harmsen, and G.H. Bolt, 1974. Tracing Soil-Moisture Migration with ^{36}Cl, ^{60}Co and Tritium, in Isotope and Radiation Techniques in Soil Physics and Irrigation Studies 1973, International Atomic Energy Agency, Vienna,.

Fuller, M.L., 1905 Underground Waters of Eastern United States, Water-Supply Paper 114, U.S. Geological Survey.

Fuller, W.H., 1977. Movement of N-Selected Metals, Asbestos, and Cyanide in Soil: Application to Waste Disposal Problems, EPA-600/2-77-020, U.S. Environmental Protection Agency.

Fuller, W.H., 1978. Investigation of Landfill Leachate Pollutant Attenuation by Soils, EPA-600/2-78-158, U.S. Environmental Protection Agency, Cincinnati, Ohio.

Fuller, W.H., 1979. Premonitoring Waste Disposal Sites, in Establishment of Water Quality Monitoring Programs, L.G. Everett and K.D. Schmidt, (ed), 85-95.

Gairon, S., and A. Hadas, 1973. Measurement of the Water Status in Soils, in Arid Zone Irrigation, B. Yaron, E. Danfois, and Y. Vaadia, (ed), 215-226, Springer-Verlag, New York.

Gardner, W.H., Water Content, in 1965. Methods of Soil Analyses, C.A. Black, (ed), Agronomy No. 9, 82-125, Amer. Soc. Agron., Madison, Wisconsin.

Garland, G.A., and D.C. Mosher, 1975. Leachate Effect from Improper Land Disposal, J. Wastewater, 6, 42-48.

Geraghty & Miller, Inc., 1977. The Prevalence of Subsurface Migration of Hazardous Chemical Substances at Selected Industrial Waste Land Disposal Sites, EPA/530/SW-634, U.S. Environmental Protection Agency.

Gibb, J.P., R.M. Schuller, and R.A. Griffin, 1981. Procedures for the Collection of Representative Water Quality Data from Monitoring Wells, Cooperative Groundwater Report 7, Illinois State Water Survey.

Gilbert, R.G., C.P. Gerba, R.C. Rice, H. Bouwer, C. Wallis, and J.L. Melnick, 1976. Virus and Bacteria Removal from Wastewater by Land Treatment, Applied and Environmental Microbiology, 32(3), 333-338.

Green, W.H., and G.A. Ampt, 1911. Studies on Soil Physics: I. Flow of Air and Water Through Soils, J. of Agric. Research., 4, 1-24.

Grier, H.E., W. Burton, and C. Tiwari, 1977. Overland Cycling of Animal Waste, in Land as a Waste Management Alternative, Ann Arbor Science, pp 693-702.

Gronowski, R., 1979. Innovations Simplify Shallow Well Monitoring, Water Well J., 33(7), 57.

Guenther, W.B., 1975. Chemical Equilibrium: A Practical Introduction for the Physical and Life Sciences, Plenum Press, New York.

Guma'a, G.S., 1978. Spatial Variability of In-Situ Available Water, unpublished Ph.D. dissertation, U. of Arizona, Tucson.

Hall, W.A., 1955. Theoretical Aspects of Water Spreading, Amer. Soc. Agric. Eng., 36(6), 394-397, 1955.

Hamon, W.R., 1979. Infiltrometer Using Simulated Rainfall for Infiltration Research, in Infiltration Research Planning Workshop, Part 1, State of the Art Reports, ARM-NC-4, Sciences and Education Administration, The U.S. Dept. of Agric.

Hansen, E.A., and A.R. Harris, 1974. A Groundwater Profile Sampler, Water Resources Res., 10(2), 375.

Hansen, E.A., and A.R. Harris, 1975. Validity of Soil-Water Samples Collected With Porous Ceramic Cups, Soil Sci. Soc. Amer. Proc., 39, 528-536.

Hansen, E.A., and A.R. Harris, 1980. An Improved Technique for Spatial Sampling of Solutes in Shallow Groundwater Systems, Water Resources Res., 16(4), 827-829.

Hantush, M.A., 1967. Growth and Decay of Ground-Water Mounds in Response to Uniform Percolation, Water Resources Res., 3(1), 277-234.

Hart, Fred C., Associates, 1979. Draft Environmental Impact Statement on the Proposed Guidelines for the Landfill Disposal of Solid Waste, U.S. Environmental Protection Agency, Office of Solid Waste.

Heath, R.C., 1982. Classification of Ground-Water Systems of the United States, Ground Water.

Hem, J.D., 1970. Study and Interpretation of the Chemical Characteristics of Natural Water, Water Supply Paper 1473, U.S. Geological Survey.

Hillel, D., 1971 Soil and Water Physical Principles and Processes, Academic Press, New York.

Hoffman, G.J., C. Dirksen, R.D. Ingvalson, E.V. Maas, J.D. Oster, S.L. Rawlins, J.D. Rhoades, and J. van Schilfgaarde, 1978. Minimizing Salt in Drain Water by Irrigation Management, in Agricultural Water Management, 1, 233-252, Elsevier Scientific Publishing, Amsterdam.

Holmes, J.W., S.A. Taylor, and S.J. Richards, 1967. Measurement of Soil Water, in Irrigation of Agricultural Lands, R.M. Hagan, H.R. Haise, and T.W. Edminster, (ed), Agronomy No. 9, 275-298, Amer. Soc. Agron., Madison, Wisconsin.

Horton, R.E., 1935. Surface Runoff Phenomena: Part 1, Analysis of the Hydrograph, Pub 101, Horton Hydrology Laboratory, Pub 101, Ann Arbor, Michigan, Edwards Bros., Inc.

Hounslow, A., J. Fitzpatrick, L. Cerrillo, and M. Freeland, 1978. Overburden Mineralogy as Related to Ground-Water Chemical Changes in Coal Strip Mining, EPA-600/7-78-156, U.S. Environmental Protection Agency, Ada, Oklahoma.

Howard, P.H., J. Saxena, and H. Sikka, 1978. Determining the Fate of Chemicals, Envir. Sci. and Technology, 12(4), 398-407.

Howe, Charles R., 1982. Earth-Resistivity Method is Used to Detect Contaminants, Johnson Driller's Journal, First Quarter.

Hsieh, J.J.C., C.G. Enfield, 1974. Communications in Soil Survey and Plant Analysis, Steady State Methods of Measuring Unsaturated Hydraulic Conductivity, 5(2), 123-129.

Huddleson, R.L., 1980. Solid Waste Disposal: Land Farming, in Indus. Wastewater and Solid Waste Engineering, V. Cavaseno, (ed), 275-280, McGraw-Hill Co.

Huibregtse, K.R., and J.H. Moser, 1976. Handbook for Sampling and Sample Preservation of Water and Wastewater, EPA-600/4-76-049, U.S. Environmental Protection Agency, Quality Assurance Branch, Cincinnati, Ohio.

Hutzinger, O., 1981. Environmental and Toxicological Chemistry at the University of Amsterdam: Five Years of Philosophy and Practice of Environmental Health Chemistry, in Environmental Health Chemistry, J.D. McKinney, (ed), Ann Arbor Science, 15-58.

Jackson, D.R., F.S. Brinkley, and E.A. Bondietti, 1976. Extraction of Soil Water Using Cellulose-Acetate Hollow Fibers, J. Soil Sci. Soc. Amer., 40, 327-329.

Jackson, R.D., R.J. Reginato, and C.H.M. van Bavel, 1965. Comparison of Measured and Calculated Hydraulic Conductivities of Unsaturated Soils, Water Resources Res., 1, 375-379.

Jakubick, A.T., 1976. Migration of Plutonium in Natural Soils, in Transuranium Nuclides in the Environment, International Atomic Energy Agency, Vienna.

Jensen, M.E., 1973. Consumptive Use of Water and Irrigation Water Requirements, ASCE.

Johnson, E.E., Inc. 1966. Ground Water and Wells, E.E. Johnson, Inc., St. Paul, Minnesota.

Johnson, T.M., K. Cartwright, and R.M. Schuller, 1981. Monitoring of Leachate Migration in the Unsaturated Zone in the Vicinity of Sanitary Landfills, Ground Water Monitoring Rev.

Jurinak, J.J., and J. Santillan-Medrano, 1974. The Chemistry and Transport of Lead and Cadmium in Soils, Res. Report No. 18, Agric. Exp. Station, Utah State U.

Kaufman, R.F., T.A. Gleason, R.B. Ellwood, and G.P. Lindsey, 1981. Ground-Water Monitoring Techniques for Arid Zone Hazardous Waste Disposal Sites, Ground Water Monitoring Rev.

Keeney, D.R., and R.E. Wildung, 1977. Chemical Properties of Soils, in Soils for Management of Organic Wastes and Waste Water, p 9, Soil Sci. Soc. Amer., American Soc. of Agron., Crop Science Soc. Amer., Madison, Wisconsin.

Kelly, K.D., 1982. Bailing and Construction Considerations for Deep Aquifer Monitoring Wells on Western Oil Shale Leases, Groundwater, 20(2), 179-185.

Keys, W.S., 1967. The Application of Radiation Logs to Ground-Water Hydrology, in International Atomic Energy Agency Symposium on Use of Isotopes in Hydrology, 477-488.

Keys, W.S., and L.M. MacCary, 1971. Application of Borehole Geophysics to Water Resources Investigations, in Technical Techniques of Water-Resources Investigations of the U.S. Geological Survey, Book 2, 126 pp.

Kirkham, D., 1947. Studies of Hillside Seepage in the Iowa Drift Area, Soil Sci. Soc. Amer. Proc., 12, 73-80.

Kirkham, D., and W.L. Powers, 1972. Advanced Soil Physics, Wiley Interscience, New York.

Klute, A., 1965. Laboratory Measurement of Hydraulic Conductivity of Unsaturated Soil, in Methods of Soil Analyses, C.A. Black, (ed), Agronomy No. 9, 253-261, Amer. Soc. Agron., Madison, Wisconsin.

Klute, A., 1969. The Movement of Water in Unsaturated Soils, in The Progress of Hydrology, Proc. First Int. Seminar for the Hydrology Prof., National Science Foundation Science Seminar, Dept. of Civil Eng., U. of Illinois, July 13-25.

Kohout, F.A., 1960. Cyclic Flow of Salt Water in Biscayne Aquifer of South Eastern Florida, J. of Geophys. Rev., 65, 2133-2141.

Kokotov, Y.A., R.F. Popova, and A.P. Urbanyuk, 1961. Sorption of Long-Lived Fission Products on Soils and Clay Minerals, Radiokhimiya, 3(2), 199-206.

Korte, N.E., J. Skopp, W.H. Fuller, E.E. Niebla, and B.A. Alesii, 1976. Trace Element Movement in Soils: Influence of Soil Physical and Chemical Properties, Soil Sci., 122(5), 350-359.

Kraatz, D.B., 1977. Irrigation Canal Lining, FAO Land and Water Development Series, No. 1, UN Food and Agriculture Organization.

Kraijenhoff van deLeur, D.A., 1962. Some Effects of the Unsaturated Zone on Nonsteady Free-Surface Groundwater Flow as Studied in a Scaled Granular Model, J. Geophys. Res., 67(11), 4347-4362.

Lapakko, K., and P. Eger, 1981. Trace Metal Removal from Mining Stock Pile Runoff Using Peat, Wood Chips, Tailing, Till, and Zeolite, in Symposium on Surface Mining Hydrology, Sedimentology and Reclamation, U. of Kentucky.

LaRue, M.E., D.R. Nielsen, and R.M. Hagan, 1968. Soil Water Flux Below a Ryegrass Root Zone, Agron. J., 60, 625-629.

Law Engineering Testing Company, 1982. Lysimeter Evaluation, American Petroleum Institute.

Leenheer, J.A., and E.W.D. Huffman, Jr., 1976. Classification of Organic Solutes in Water by Using Macroreticular Resins, J. Res. U.S. Geol. Survey, 4(6), 737-751.

Leonard, R.A., G.W. Bailey, and R.R. Swank, 1976. Transport, Detoxification Fate, and Effects of Pesticides in Soil and Water, in Land Application of Soils Material, Soil Cons. Soc. Amer.

Levin, M.J., and D.R. Jackson, 1977. A Comparison of In-Situ Extractors for Sampling Soil Water, J. Soil Sci. Soc. Amer., 41, 535-536.

Lissey, A., 1967. The Use of Reducers to Increase the Sensitivity of Piezometers, J. Hydrology, 5(2), 197-205.

Lohman, S.W., 1972. Ground-Water Hydraulics, Professional Paper No. 708 U.S. Geological Survey.

Lund, L.J., and P.F. Pratt, 1977. Variability of Nitrate Leaching Within Defined Management Units, in Proc. National Conference on Irrigation Return Flow Quality Management, Colorado State U.

Lusczynski, N.J., 1961. Filter-Press Method of Extracting Water Samples for Chloride Analysis, Water Supply Paper 1544-A, U.S. Geological Survey.

Luthin, J.M., (ed), 1957. Drainage of Agricultural Lands, Amer. Soc. Agron., Madison, Wisconsin.

Luthin, J.M., and P.R. Day, 1955. Lateral Flow Above a Sloping Water Table, Soil Sci. Soc. Amer. Proc., 19, 406-410.

Mann, J.F., 1976. Wastewaters in the Vadose Zone of Arid Regions: Hydrologic Interactions, Ground Water, 14(6), 367-372.

Mather, J.R., and P.A. Rodriguez, 1978. The Use of the Water Budget in Evaluating Leaching Through Solid Waste Landfills, Final Report on OWRT Project No. A-040-DEL, Water Resources Center, U. of Delaware, Newark Delaware.

Matlock, W.G., and P.R. Davis, 1972. Groundwater in the Santa Cruz Valley Technical Bull. 194, Agric. Exp. Station, The U. of Arizona, Tucson.

Matlock, W.G., G.C.A. Morin, and J.E. Posedly, 1976. Well Cuttings Analysis in Ground-Water Resources Evaluation, Ground Water, 14(5), 272-277.

McGuinness, C.L., 1963. The Role of Ground Water in the National Water Situation, Water-Supply Paper 1800, U.S. Geological Survey.

McMichael, F.C., and J.E. McKee, 1966. Wastewater Reclamation at Whittier Narrows, Water Quality Publication No. 33, State of California.

McMillion, L.G., and J.W. Keeley, 1968. Sampling Equipment for Groundwater Investigations, Ground Water, 6(2), 9-11.

McNeal, B.L., 1974. Soil Salts and Their Effect on Water Movement, in Drainage for Agriculture, J. van Schilfgaarde, (ed), Agronomy No. 17, 409-433, Amer. Soc. Agron., Madison, Wisconsin.

McWhorter, D.B., and J.A. Brookman, 1972. Pit Recharge Influenced by Subsurface Spreading, Ground Water, 10(5), 6-11.

Meinzer, O.E., 1942. Ground Water, in Hydrology, Oscar E. Meinzer, (ed), 385-477, Dover Publications, New York.

Merrill, S.D., and S.C. Rawlins, 1972. Field Measurement of Soil Water Potential with Thermocouple Psychrometers, Soil Sci., 113(2), 102-109.

Meyer, W.R., 1963. Use of a Neutron Moisture Probe to Determine the Storage Coefficient of an Unconfined Aquifer, Prof. Paper 450-E, 174-176, U.S. Geological Survey.

Miller, D.W., 1980. Waste Disposal Effects on Ground Water, Premier Press, Berkeley, California.

Miller, R.H., 1973. The Soil as a Biological Filter, in Recycling Treated Municipal Waste Water and Sludge Through Forest and Cropland, W.E. Sopper and L.T. Kardos, (eds), Pennsylvania State U. Press.

Miller, G.R., and N.B. Barch, 1981. Impacts of Municipal Treatment Plants on Toxic Pollutants in Wastewater, in The Environmental Professional, Vol. 3, 79-84.

Millington, R.J., and J.P. Quirk, 1959. Permeability of Porous Media, Nature, 183, 387-388.

Mooij, H., and F.A. Rovers, 1976. Recommended Groundwater and Soil Sampling Procedures, EPS-4-EC, Environmental Protection Service, Canada.

Morrison, R.D., and T.C. Tsai, 1981. Modified Vacuum-Pressure Lysimeter for Vadose Zone Sampling, Calscience Research, Inc., Huntington Beach, California.

Mortland, M.M., and W.D. Kemper, 1965. Specific Surface, in Methods of Soil Analyses, C.A. Black, (ed), Agronomy No. 9, 532-543, Amer. Soc. Agron., Madison, Wisconsin.

Mualem, Y., 1976. A Catalog of the Hydraulic Properties of Unsaturated Soils, Report on Research Project 442, Technion, Israel Institute of Technology.

Murrmann, R.P., and F.R. Koutz, 1972. Role of Soil Chemical Processes in Reclamation of Wastewater Applied to Land, in Wastewater Management by Disposal on the Land, Cold Regions Research and Eng. Lab. Specialty Report 171, U.S. Army Corps of Engineers.

Myhre, D.L., J.O. Sanford, and W.F. Jones, 1969. Apparatus and Techniques for Installing Access Tubes in Soil Profiles to Measure Soil Water, Soil Science, 108(4), 296-299.

Nelson, J.L., 1959. Soil Column Studies with Radiostrontium: I. Effects of Temperature of Species of Accompanying Ion, H.W. 62035, Hanford Atomic Products Operation, Richland, Washington.

Nielsen, D.R., and J.W. Biggar, 1973. Analyzing Soil Water and Solute Movement Under Field Conditions, in Soil-Moisture and Irrigation Studies II, International Atomic Energy Agency.

Nielsen, D.R., J.W. Biggar, and K.T. Erh, 1973. Spatial Variability of Field Measured Soil-Water Properties, Hilgardia, 43(7), 215-260.

Nommik, H., 1965. Ammonium Fixation and Other Reactions Involving a Nonenzymatic Immobilization of Mineral Nitrogen in Soil, in Soil Nitrogen, W.V. Barthalemew and F.E. Clark, (ed), Agronomy No. 10.

Norris, S.E., 1972. The Use of Gamma Logs in Determining the Character of Unconsolidated Sediments and Well Construction Features, Ground Water, 10(6), 14-21.

Ongerth, H.J., D.P. Spath, J. Crook, and A.F. Greenberg, 1973. Public Health, Aspects of Organics in Water, J. Amer. Water Works Assoc., 65, 495.

Oster, J.D., and L.S. Willardson, 1971. Reliability of Salinity Sensors for the Management of Soil Salinity, Agron. J., 63, 695-698.

Paetzold, R.F., 1979. Measurement of Soil Physical Properties, in Infiltration Research Planning Workshop, Part 1, State of the Art Reports, ARM-NC-4, Sciences and Education Admin., U.S. Dept. of Agric.

Parizek, R.R., L.J. Kardos, W.E. Sopper, E.A. Myers, D.E. Dairs, M.A. Farrell, and J.B. Nesbitt, 1967. Waste Water Renovation and Conservation, Pennsylvania State Studies, No. 23, The Pennsylvania State U. Press.

Parizek, R.R., and B.E. Lane, 1970. Soil-Water Sampling Using Pan and Deep Pressure-Vacuum Lysimeters, J. Hydrology, 11, 1-21.

Phene, C.J., G.J. Hoffman, and S.L. Rawlins, 1971. Measuring Soil Matric Potential In-Situ by Sensing Heat Dissipation Within a Porous Body: I Theory and Sensor Construction, in Soil Sci. Soc. Amer. Proc., 35, 27-33.

Phene, C.J., S.L. Rawlins, and G.J. Hoffman, 1971. Measuring Soil Matric Potential In-Situ by Sensing Heat Dissipation Within a Porous Body: II Experimental Results, in Soil Sci. Soc. Amer. Proc., 35, 225-229.

Philip, J.R., 1969. Theory of Infiltration, in Advances in Hydroscience, 5, 216-269.

Phung, T., L. Barker, D. Ross, and D. Bauer, 1978. Land Cultivation of Industrial Wastes and Municipal Solid Wastes: State of the Art Study, Volume 1: Technical Summary and Literature Review, EPA-600/2-78-140a, U.S. Environmental Protection Agency.

Pickens, J.F., J.A. Cherry, R.M. Coupland, G.E. Grisak, W.F. Merritt, and B.A. Risto, 1981. A Multi-Level Device for Ground-Water Sampling, Ground Water Monitoring Rev., Vol. 1, No. 1.

Pickens, J.F., J.A. Cherry, G.E. Grisak, W.F. Merritt, and G.A. Risto, 1978. A Multilevel Device for Ground-Water Sampling and Piezometric Monitoring, Ground Water, 16(5), 322-327.

Pickens, J.F., and G.E. Grisak, 1979. Reply to the Preceding Discussion of Vanhof et al. at "A Multilevel Device for Groundwater Sampling and Piezometric Monitoring", Ground Water, 17(4), 393-397.

Pratt, P.F., A.C. Chang, J.P. Martin, A.L. Page, and C.F. Kleine, 1978. Removal of Biological and Chemical Contaminants by Soil Systems in Association with Ground Water Recharge by Spreading or Injection of Treated Municipal Waste Water, in Health Aspects of Wastewater Recharge, A State-of-the-Art Review, Water Information Center, Huntington, New York.

Pratt, P.F., W.W. Jones, and V.E. Hunsaker, 1972. Nitrate in Deep Soil Profiles in Relation to Fertilizer Rates and Leaching Volume, J. Envir. Quality, 1(1), 97-102.

Pratt, P.F., L.J. Lund, and J.M Rible, 1977. An Approach to Measuring Leaching of Nitrate from Freely Drained Irrigated Fields, in Nitrogen and the Environment, Academic Press.

Pratt, P.F., J.E. Warneke, and P.A. Nash, 1976. Sampling the Unsaturated Zone in Irrigated Field Plots, J. Soil Sci. Soc. Amer., 40, 277-279.

Quirk, J.P., and R.K. Schofield, 1955. The Effect of Electrolyte Concentration on Soil Permeability, J. Soil Sci., 6, 1963-1978.

Ralston, D.R., 1967. Influences of Water Well Design on Neutron Logging, unpublished M.S. thesis, U. of Arizona, Tucson.

Rawlins, S.L., 1976. Measurement of Water Content and the State of Water in Soils, in Water Deficits and Plant Growth, T.T. Kozlowski, (ed), Vol. 4, Soil Water Measurement, Plant Responses, and Grading for Drought Resistance, 1-55, Academic Press, New York.

Rawlins, S.L., and F.N. Dalton, 1967. Psychrometric Measurement of Soil Water Potential Without Precise Temperature Control, Soil Sci. Soc. Amer. Proc., 31, 297-301.

Reeve, R.C., 1965. Hydraulic Head, in Methods of Soil Analyses, C.A. Black, (ed), Agronomy No. 9, 180-196, Amer. Soc. Agron., Madison, Wisconsin.

Reginato, R.J., 1974. Count Rate Instability in Gamma Ray Transmission Equipment, Soil Sci. Soc. Amer. Proc., 38, 156-157.

Reginato, R.J., and R.D. Jackson, 1971. Field Measurement of Soil-Water Content by Gamma-Ray Transmission Compensated for Temperature Fluctuations, Soil Sci. Soc. Amer. Proc., 35, 529-533.

Reginato, R.J., and C.H.M. van Bavel, 1964. Soil Water Measurement with Gamma Attenuation, Soil Sci. Soc. Amer. Proc., 28(6), 721-724.

Reitemeier, R.F., 1946. Effect of Moisture Content on the Dissolved and Exchangeable Ions of Soils and Arid Regions, Soil Sci., 61(3), 195-214.

Reynolds, S.G., 1970. Gravimetric Method of Soil Moisture Determination, Part 1, A Study of Equipment and Methodological Problems, J. Hydrology, 2(3) ,258-273.

Rhoades, J.D., 1979a. Monitoring Soil Salinity: A Review of Methods, in Establishment of Water Quality Monitoring Programs, Amer. Water Resources Assoc.

Rhoades, J.D., 1979b. Salinity Management and Monitoring, in Proceedings of the Twelfth Biennial Conference on Ground Water, Sacramento, September 20-21, 1979, California Water Resources Center, U. of California, Davis.

Rhoades, J.D., 1979c. Inexpensive Four-Electrode Probe for Monitoring Soil Salinity, J. Soil Sci. Soc. Amer.

Rhoades, J.D., and L. Bernstein, 1971. Chemical, Physical and Biological Characteristics of Irrigation and Soil Water, in Water and Water Pollution Handbook, Vol. 1, L.L. Ciaccio, (ed), 141-222, Marcel Dekker, Inc., New York.

Rhoades, J.D., and A.D. Halvorson, 1977. Electrical Conductivity Methods for Detecting and Delineating Saline Seeps and Measuring Salinity in Northern Great Plains Soils, ARS W-42, U.S. Dept. of Agric.

Rhoades, J.D., and J. van Schilfgaarde, 1976. An Electrical Conductivity Probe for Determining Soil Salinity, Soil Sci. Soc. Amer. Proc., 40, 647-651.

Rhoades, J.D., M.T. Kaddah, A.D. Halvorson, and J.R. Prather, 1977. Establishing Soil Electrical Conductivity-Salinity Calibrations Using Four-Electrode Cells Containing Undisturbed Soil Cores, Soil Sci., 123, 137-141.

Rhodes, D.W., 1957a. Adsorption of Plutonium by Soil, Soil Sci., 84, 465-472.

Rhodes, D.W., 1957b. The Effect of pH on the Uptake of Radioactive Isotopes from Solution by a Soil, Soil Sci. Soc. Amer. Proc., 21, 389-392.

Rible, J.M., P.A. Nash, P.F. Pratt, and L.H. Lund, 1976. Sampling the Unsaturated Zone of Irrigated Lands for Reliable Estimates of Nitrate Concentrations, J. Soil Sci. Soc. Amer., 40, 566-570.

Richards, L.A., 1949. Methods of Measuring Soil Moisture Tension, Soil Sci., Vol. 68, 95-112.

Richards, L.A., (ed), 1954. Diagnosis and Improvement of Saline and Alkali Soils, Agric. Handbook 60, U.S. Dept. of Agric.

Richards, L.A., 1965. Physical Condition of Water in Soils, in Methods of Soil Analyses, C.A. Black, (ed), Agronomy No. 9, 128-152, Amer. Soc. Agron., Madison, Wisconsin.

Richards, L.A., 1966. A Soil Salinity Sensor of Improved Design, Soil Sci. Soc. Amer. Proc., 30, 333-337.

Ries, H., and T.L. Watson, 1914. Engineering Geology, John Wiley and Sons, New York.

Robertson, J.M., C.R. Toussaint, and M.A. Jorque, 1974. Organic Compounds Entering Ground Water from a Landfill, EPA-660/2-74-077, U.S. Environmental Protection Agency.

Rolston, D.E., and F.E. Broadbent, 1977. Field Measurement of Denitrification, EPA-600/2-77-233, U.S. Environmental Protection Agency.

Rose, C.W., 1966. Agricultural Physics, Pergamon Press, 1966.

Ross, D.E., and H.T. Phung, 1980. Soil Incorporation (Land Farming) of Industrial Wastes, in Toxic and Hazardous Waste Disposal, Vol. Four, R.P. Pojasek, (ed), 291-308, Ann Arbor Science.

Runnells, D.D., 1976. Wastewaters in the Vadose Zone of Arid Regions: Geochemical Interactions, Ground Water, 14(6), 374-385.

Saines, M., 1981. Errors in Interpretation of Ground-Water Level Data, Ground Water Monitoring Rev., 56-61, Spring.

Schmidt-Collerus, J.J., 1974. The Disposal and Environmental Effects of Carbonaceous Solid Wastes from Commercial Oil Shale Operations, U. of Denver Res. Inst., Denver, Colorado.

Schmugge, T.J., T.J. Jackson, and H.L. McKim, 1980. Survey of Methods for Soil Moisture Determination, Water Resources Res., 16(6), 961-979.

Schuller, R.M., J.P. Gibb, and R.A. Griffin, 1981. Recommended Sampling Procedures for Monitoring Wells, Ground Water Monitoring Rev., 1(1).

Schwille, F., 1967. Petroleum Contamination of the Subsoil - A Hydrological Problem, in The Joint Problems of Oil and Water Industries, P. Hepple, (ed), Elsevier, Amsterdam.

Sciences and Education Administration, 1979. Infiltration Research Planning Workshop, Part 1, State of the Art Reports, ARM-NC-4, U.S. Dept. of Agric.

Scott, V.H., and J.C. Scalmanini, 1978. Water Wells and Pumps: Their Design, Construction, Operation and Maintenance, Bull. 1889, Div. Agric. Sci., U. of California.

Sendlein, V.A., and H. Yazicigil, 1981. Surface Geophysical Methods for Ground Water Monitoring, Part I, Ground Water Monitoring Rev.

Shaffer, K.A., D.D. Fritton, and D.E. Baker, 1979. Drainage Water Sampling in a Wet, Dual-Pore Soil System, J. Envir. Quality, 8(2), 241-246.

Shuval, H.E., and N. Gruener, 1973. Health Considerations in Renovating Wastewater for Domestic Use, Envir. Sci. and Tech., 7, 600.

Sidle, R.C., 1979. Infiltration Measurements and Soil Hydraulics Characteristics, in Infiltration Research Planning Workshop, Part 1, State of the Art Reports, ARM-NC-4, Sciences and Education Admin., U.S. Dept. of Agric.

Silka, L.R., and T.L. Swearingen, 1978. A Manual for Evaluating Contamination Potential of Surface Impoundments, EPA-570/9-78-003, Ground Water Protection Branch, Office of Drinking Water, U.S. Environmental Protection Agency.

Simonson, R.W., 1957. What Soils Are, in Soil, The Yearbook of Agriculture, U.S. Dept. of Agric.

Simpson, T.W., and R.L. Cunningham, 1982. The Occurrence of Flow Channels in Soils, J. Env. Qual., 11(1), 29-30.

Slawson, G.C., (ed), 1980. Groundwater Quality Monitoring of Western Oil Shale Development: Monitoring Program Development, EPA-600/7-80-089, U.S. Environmental Protection Agency, Las Vegas, Nevada.

Slawson, G.C., K.E. Kelly, E.W. Hoylman, and L.G. Everett, 1982. Groundwater Recommendations for In Situ Oil Shale Development, U.S. Environmental Protection Agency, Las Vegas, Nevada.

Smith, J.L., and D.M. McWhorter, 1977. Continuous Subsurface Injection of Liquid Organic Wastes, in Land as a Waste Management Alternative, Ann Arbor Science, pp 646-656.

Soil Moisture Equipment Corporation, 1978. Operating Instructions for the Model 1900 Soil Water Sampler, Santa Barbara, California.

Soiltest, Inc., 1976. Soil Testing Equipment, General Catalog.

Sopper, W.E., and L.T. Kardos, (eds), 1973. Recycling Treated Municipal Waste Water and Sludges Through Forest and Cropland, Pennsylvania State U. Press.

Sowers, N.M., 1979. Geological Factors Affecting Geohydrology of Tongue River Valley Sheraton Wyoming, M.S., U. of Wyoming, Laramie.

Stallman, R.W., 1967. Flow in the Zone of Aeration, in Advances in Hydroscience, Vol. 4, Ven Te Chow, (ed), Academic Press, New York.

Steele, R.G.D., and J.H. Torrie, 1960. Principles and Procedures of Statistics, McGraw-Hill, New York.

Steiner, R.C., A.A. Fungaroli, R.J. Schoenberger, and P.W. Purdom, 1971. Criteria for Sanitary Landfill Development, Public Works, 102(2), 77-79.

Stephens, D.B., and S.P. Neuman, 1980. Analysis of Borehole Infiltration Tests Above the Water Table, Tech. Report No. 35, Department of Hydrology and Water Resources, U. of Arizona.

Stevenson, C.D., 1978. Simple Apparatus for Monitoring Land Disposal Systems by Sampling Percolating Soil Waters, Envir. Sci. and Engr, 12(3), 329-331.

Stollar, R.L., and P. Roux, 1975. Earth Resistivity Surveys - A Method for Determining Ground-Water Contamination, Ground Water, 13(2), 145-150.

Sverdrup and Parcel, 1981. Priority Pollutant Analysis for Industrial Pretreatment Program, Metropolitan St. Louis Sewer District, St. Louis, Missouri.

Taylor, G.S., and J.N. Luthin, 1969. The Use of Electronic Computers to Solve Subsurface Drainage Problems, Water Resources Research, 5(1), 144-152.

Thomas, G.W., and R.E. Phillips, 1979. Consequences of Water Movement in Macropores, J. Envir. Quality, 8(2), 149-152.

Thomas, H.E., 1952. Ground-Water Regions of the United States--Their Storage Facilities, U.S. 83rd Congress, House Interior and Insular Affairs Committee, The Physical and Economic Foundation of Natural Resources, Vol. 3.

Thomas, R.E., 1973. The Soil as a Physical Filter, in Recycling Treated Municipal Wastewater and Sludge Through Forest and Cropland, W.E. Sopper and L.T. Kardos, (eds), Pennsylvania State U. Press.

Thornthwaite, C.W., and J.R. Mather, 1957. Instructions and Tables for Computing Potential Evapotranspiration and Water Balance, Drexel Institute of Technology Laboratory of Climatology, Publications in Climatology, Volume X, No. 3, Centerton, New Jersey.

Tinlin, R.M., 1976. Monitoring Groundwater Quality: Illustrative Examples, General Electric Company-Tempo, Santa Barbara, California.

Todd, D.K., R.M. Tinlin, K.D. Schmidt, and L.G. Everett, 1976. Monitoring Groundwater Quality: Monitoring Methodology, EPA-600/4-76-026, U.S. Environmental Protection Agency, Environmental Monitoring and Support Laboratory, Las Vegas, Nevada.

Topp, G.C., and W. Zebchuck, 1979. The Determination of Soil-Water Desorption Curves for Soil Cores, Canadian J. of Soil Sci., 59, 19-26.

Trescott, P.C., and G.F. Pinder, 1970. Air Pump for Small-Diameter Piezometers, Ground Water, 8(3), 10-15.

Tsai, T.C., Morrison, R.D., and Stearns, R.J., 1980. Validity of the Porous Cup Vacuum/Suction Lysimeter as a Sampling Tool for Vadose Waters, unpublished report, Calscience Research Inc., Huntington Beach, California, 1-11.

U.S. Bureau of Reclamation, 1977. Ground Water Manual, A Water Resources Technical Publication, U.S. Dept. of the Interior.

U.S. Environmental Protection Agency, 1973a. An Environmental Assessment of Potential Gas and Leachate Problems at Land Disposal Sites, EPA SW-110, Office of Solid Waste.

U.S. Environmental Protection Agency, 1973b. Development Document for Proposed Effluent Limitations Guidelines and New Performance Standards for the Basic Fertilizer Chemicals Segment of the Fertilizer Manufacturing Point Source Category, EPA-440/1-73/011, Office of Air and Water Programs.

U.S. Environmental Protection Agency, 1973c. Development Document for Proposed Effluent Limitations Guidelines and New Performance Standards for the Feedlots Point Source Category, EPA-440/1-73/004, Office of Air and Water Programs.

U.S. Environmental Protection Agency, 1974a. Development Document for Proposed Effluent Guidelines and New Source Performance Standards for the Dairy Product Processing Point Source Category, EPA-440/1-73/021, Office of Air and Water Programs.

U.S. Environmental Protection Agency, 1974b. Development Document for Effluent Limitations Guidelines and New Source Performance Standards for the Petroleum Refining Point Source Category, EPA-440/1-74/014-a, Washington, D.C.

U.S. Environmental Protection Agency, 1975. Gas and Leachate from Land Disposal of Municipal Solid Waste: Summary Report, Municipal Environmental Research Laboratory, Cincinnati, Ohio.

U.S. Environmental Protection Agency, 1977a. Procedures Manual for Ground Water Monitoring at Solid Waste Disposal Facilities, EPA SW-611, Office of Solid Waste.

U.S. Environmental Protection Agency, 1977b. Interim Final Supplement for Pretreatment to the Development Document for the Petroleum Refining Industry Existing Point Source Category, EPA-440/1-76/083A, Washington, D.C.

U.S. Environmental Protection Agency, 1978a. Hazardous Waste Guidelines and Regulations, Federal Register, 43(243), Part IV, 58946-59027.

U.S. Environmental Protection Agency, 1978b. Draft Development Document Including the Data Base for the Review of Effluent Limitations Guidelines (BATEA), New Source Performance Standards, and Pretreatment Standards for the Petroleum Refining Point Source Category, Washington, D.C.

U.S. Environmental Protection Agency, 1978c. Technical Study Report BATEA-NSPS-PSES-PSN-Textile Mills Point Source Category, draft contractor's report, Contracts 68-01-3289 and 68-01-3884, Effluent Guidelines Division, Washington, D.C.

U.S. Environmental Protection Agency, 1979a. Draft Development Document for Effluent Limitations Guidelines and Standards for the Nonferrous Metals Manufacturing Point Source Category, Office of Water and Waste.

U.S. Environmental Protection Agency, 1979b. Draft Development Document for Inorganic Chemicals and Manufacturing Point Source Category - BATEA, NSPS, and Pretreatment Standards, Contract 68-01-4492, Effluent Guidelines Division, Washington, D.C.

U.S. Environmental Protection Agency, 1979c. Development Document for Proposed Effluent Limitations Guidelines, New Source Performance Standards, and Pretreatment Standards for the Leather Tanning and Finishing Point Source Category, Effluent Guidelines Division, Washington, D.C.

U.S. Environmental Protection Agency, 1979d. Draft Application Forms for EPA Programs, Federal Register, 44(116), Part III, 34346-34416.

U.S. Environmental Protection Agency, 1979e. Landfill Disposal of Solid Waste, Proproposed Guidelines, Federal Register, 44(59), Part II, 18138-18148.

U.S. Environmental Protection Agency, 1979f. Development Document for Proposed Effluent Limitations Guidelines, New Source Performance Standards, and Pretreatment Standards for the Petroleum Refining Point Source Category, EPA-440/1-79/014-b, Washington, D.C.

U.S. Environmental Protection Agency, EPA-600/8-80-042b, 1980a. Treatability Manual, Volume II: Industrial Descriptions, Office of Research and Development.

U.S. Environmental Protection Agency, 1980b. Hazardous Waste and Consolidated Permit Regulations, Federal Register, 45(98), 33066-33588.

U.S. Environmental Protection Agency, 1980c. Hazardous Waste Management System: Standards Applicable to Generators of Hazardous Waste, Federal Register, 45(98), 33140-33148.

U.S. Environmental Protection Agency, 1982. Hazardous Waste Management System: Permitting Requirements for Land Disposal Facilities, Federal Register, 47(143), 32274-32382.

U.S. Environmental Protection Agency, 1983. RCRA Guidance Document, Land Treatment, Office of Solid Waste, Washington, D.C.

U.S. Environmental Protection Agency, no date. Manual of Water Well Construction Practices, EPA-570/9-75-001, Office of Water Supply.

U.S. Environmental Protection Agency, U.S. Army Corps of Engineers, and U.S. Department of Agriculture, 1977. Process Design Manual for Land Treatment of Municipal Wastewater, EPA-625/1-77-008 (COE EM/110-1-501).

U.S. Salinity Laboratory Staff, 1977. Minimizing Salt in Return Flow Through Irrigation Management, EPA-600/2-77-134, U.S. Environmental Protection Agency, Ada, Oklahoma.

van Bavel, C.H.M., 1963. Neutron Scattering Measurement of Soil Moisture: Development and Current Status, Proc. Int. Symp. Humidity and Moisture, Washington, D.C., 171-184.

Vanhof, J.A., K.U. Weyer, and S.H. Whitaker, 1979. Discussion of "A Multilevel Device for Ground-Water Sampling and Piezometric Monitoring by J.F. Pickens, J.A. Cherry, G.E. Grisak, W.F. Merritt, and B.A. Rizto," Ground Water, 17(4), 391-393.

van Schilfgaarde, J., 1970. Theory of Flow to Drains, in Advances in Hydroscience, 6, 43-106.

Vomicil, J.A., 1965. Porosity, in Methods of Soil Analyses, C.A. Black, (ed), Agronomy No. 9, 299-314, Amer. Soc. of Agron., Madison, Wisconsin.

Wagner, G.H., 1962. Use of Ceramic Cups to Sample Soil Water Within the Profile, Soil Science, 94, 379-386.

Walker, W.H., 1973. Where Have All the Toxic Chemicals Gone?, Ground Water, 11(2), 11-20.

Warner, D.L., 1969. Preliminary Field Studies Using Earth Resistivity Measurements for Delineating Zones of Contaminated Ground Water, Ground Water, 7(1), 9-16.

Warrick, A.W., and A. Amoozegar-Fard, (in press) 1980. Area Prediction of Water and Solute Flux in the Unsaturated Zone, Final Report on Grant No. R-804-751-010, U.S. Environmental Protection Agency.

Warrick, A.W., and D.R. Nielsen, 1980. Spatial Variability of Soil Physical Properties in the Field, in Applications of Soil Physics, D. Hillel, (ed), 319-344, Academic Press, New York.

Watson, K.K., 1967. A Recording Field Tensiometer with Rapid Response Characteristics, J. Hydrology, 5, 33-39.

Weeks, E.P., 1978. Field Determination of Vertical Permeability to Air in the Unsaturated Zone, USGS Professional Paper No. 1051.

Wheeler, M.L., 1976. Moisture and Solute Transport in Porous Media, in Actinides in the Environment, A.M. Friedman, (ed), ACS/Symposium Series 35, American Chemical Society.

Willardson, L.S., B.D. Meek, and M.J. Huber, 1973. A Flow Path Ground Water Sampler, Soil Sci. Soc. Amer. Proc., 36, 965-966.

Willmott, C.J., 1977. WATBUG: A FORTRAN IV Algorithm for Calculating the Climatic Water Budget, Contribution No. 23, U. of Delaware Water Resources Center.

Wilson, L.G., 1971. Observations on Water Content Changes in Stratified Sediments During Pit Recharge, Ground Water, 9(3), 29-40.

Wilson, L.G., June 1980. Monitoring in the Vadose Zone: A Review of Technical Elements and Methods, EPA-600/7-80-134, U.S. Environmental Protection Agency, Las Vegas, Nevada

Wilson, L.G., and K.J. DeCook, 1968. Field Observations on Changes in the Subsurface Water Regime During Influent Seepage in the Santa Cruz River, Water Resources Research, 4(6), 1219-1234.

Wilson, L.G., and J.N. Luthin, 1963. Effect of Air Flow Ahead of the Wetting Front on Infiltration, Soil Science, 96, 136-143.

Wilson, L.G., P.S. Osborne, and D.J. Percious, 1968. Dilution of an Industrial Waste Effluent with River Water in the Vadose Region During Pit Recharge, Paper No. 68-727, presented at the Winter Meetings, Amer. Soc. Agric. Eng.

Wilson, L.G., and K.D. Schmidt, 1979. Monitoring Perched Ground Water in the Vadose Zone, in Establishment of Water Quality Monitoring Programs, Amer. Water Resources Assoc.

Wilson, L.G., and G.G. Small, 1973. Pollution Potential of a Sanitary Land fill Near Tucson, Hydraulic Engineering and the Environment, Proc. 21st Annual Hydraulics Division Specialty Conference, ASCE.

Winograd, I.J., 1974. Radioactive Waste Storage in the Arid Zone, EOS, Trans. Amer. Geophys. Union, 55(10), 884-894.

Wolff, R.G., 1967. Weathering Woodstock Granite Near Baltimore, Maryland, American J. Sci., 265, 106-117.

Wood, W.W., 1973. A Technique Using Porous Cups for Water Sampling at Any Depth in the Unsaturated Zone, Water Resources Res., 9(2), 486-488.

Wood, W.W., and D.C. Signor, 1975. Geochemical Factors Affecting Artificial Recharge in the Unsaturated Zone, in Trans. Amer. Soc. of Agric. Eng., 18(4), 677-683.

Zimmie, T.F., and C.O. Riggs, (eds), 1981. Permeability and Ground Water Contaminant Transport, ASTM, Philadelphia, Pennsylvania.

Zohdy, A.A.R., G.P. Eaton, and D.R. Mabey, 1974. Application of Surface Geophysics to Ground-Water Investigations, in Techniques of Water Resources Investigations of The United States Geological Survey, Chapter D1, Book 2, Collection of Environmental Data.

Appendix A—Conversion Factors

U.S. Customary to SI (Metric)

U.S. customary unit			SI	
Name	Abbreviation	Multiplier	Symbol	Name
acre	acre	0.405	ha	hectare
acre-foot	acre-ft	1,233	m^3	cubic metre
cubic foot	ft^3	28.32 0.0283	l m^3	litre cubic metre
cubic feet per second	ft^3/s	28.32	l/s	litres per second
degrees Fahrenheit	^oF	$0.555(^oF-32)$	^oC	degrees Celsius
feet per second	ft/s	0.305	m/s	metres per second
foot (feet)	ft	0.305	m	metre(s)
gallon(s)	gal	3.785	l	litre(s)
gallons per acre per day	gal/acre•d	9.353	l/ha•d	litres per hectare per day
gallons per day	gal/d	4.381×10^{-5}	l/s	litres per second
gallons per minute	gal/min	0.0631	l/s	litres per second
horsepower	hp	0.746	kW	kilowatt
inch(es)	in.	2.54	cm	centimetre(s)
inches per hour	in./hr	2.54	cm/h	centimeters per hour
mile	mi	1.609	km	kilometre

(continued)

CONVERSION FACTORS (continued)

U.S. customary unit			SI	
Name	Abbreviation	Multiplier	Symbol	Name
miles per hour	mi/h	0.45	m/s	metres per second
million gallons	Mgal	3.785	Ml	megalitres (litres x 10^6)
		3,785	m^3	cubic metres
million gallons per acre	Mgal/acre	8,353	m^3/ha	cubic metres per hectare
million gallons per day	Mgal/d	43.8	l/s	litres per second
		0.044	m^3/s	cubic metres per second
parts per million	ppm	1	mg/l	milligrams per litre
pound(s)	lb	0.454 453.6	kg g	kilogram(s) gram(s)
pounds per acre per day	lb/acre•d	1.12	kg/ha•d	kilograms per hectare per day
pounds per square inch	lb/in.^{2}	0.069	kg/cm^2	kilograms per square centimetre
		0.69	N/cm^2	Newtons per square centimetre
square foot	ft^2	0.0929	m^2	square metre
square inch	in.^{2}	6.452	cm^2	square centimetre
square mile	mi^2	2.590 259	km^2 ha	square kilometre hectare

Appendix B—Glossary

adjusted sodium adsorption ratio: A modified form of the sodium adsorption ratio that accounts for the interrelationships among sodium, calcium, and magnesium ions in an applied water or matrix solution, and for the tendency for calcium and magnesium ions to precipitate or dissolve in the presence of carbonate and bicarbonate.

adsorption isotherm: A graphical representation of the amount of adsorbate (solute) adsorbed by an adsorbent as a function of the equilibrium concentration of adsorbate (Bohn, McNeal, and O'Connor, 1979).

anisotropy: A porous sytem is anisotropic with respect to the hydraulic conductivity (or other properties) if values of the hydraulic conductivity are not equal in all directions away from a given point.

antecedent water: Water stored in a soil before to infiltration of applied water.

average linear velocity: The flux (specific discharge) divided by the water filled porosity; accounts for variations in the cross sectional areas through which flow occurs in a porous system.

capillary fringe: The basal region of the vadose zone comprising sediments that are saturated, or nearly saturated, near the water table, gradually decreasing in water content with increasing elevation above the water table.

cascading water: Perched groundwater that enters a well casing via cracks or uncovered perforations, trickling or pouring down the inside of the casing.

cation exchange capacity (CEC): The total capacity of a porous system to adsorb cations from a solution.

clay micelle: An individual clay particle.

Debye Huckel theory: A model used to calculate the activty coefficients of single ions.

distribution coefficient: A representation of the partitioning of solids between liquid and solid phases in a porous system (Freeze and Cherry, 1979).

electrical conductivity: The inverse of the electrical resistivity of a solution containing electrolytes; measured via a wheatstone bridge and a standardized cell containing platinum electrodes, exactly 1 cm^2 in surface area, located of 1 cm apart.

field capacity (specific retention): The amount of water retained by a soil (or aquifer) against the force of gravity.

fillable porosity: The volume of water that an unconfined aquifer stores during a unit rise in water table per unit surface area.

flux: The volume of water crossing a unit area of porous material per unit time.

gravitational head: The component of total hydraulic head relating to the position of a given mass of soil water relative to an arbitrary datum.

groundwater recharge: The flux of water across a water table.

groundwater zone: The geohydrological region underlying a water table.

hydraulic conductivity: The proportionality factor in Darcy's equation; generally, a measure of the ease with which water moves through a porous system.

hydraulic gradient: The change in total hydraulic head of water per unit distance of flow.

infiltration capacity: The maximum rate at which water enters a soil.

isotropy: A porous system is isotropic with respect to hydraulic conductivity (or other properties) if values of the hydraulic conductivity are equal in all directions away from a point.

Klinkenberg effect: An effect that accounts for the greater permeability of a porous system to air movement than water movement; in particular, the velocity of air in the immediate vicinity of grains is finite because of slippage; in contrast, the velocity of water near grains is zero.

Langelier saturation index: A representation of the degree of supersaturation or undersaturation of water with respect to calcium carbonate.

matric potential: The energy required to extract water from a soil against the capillary and adsorptive forces of the soil matrix.

matric suction: For an isothermal soil system, matric suction is the pressure difference across a membrane separating soil solution, in-place, from the same in bulk (Hillel, 1971).

matrix: The solid framework of a porous system.

osmotic potential: The component of the total soil-water potential associated with dissolved ions.

perched groundwater: A saturated groundwater body in the vadose zone, commonly developed at the interface between regions of varying texture.

percolation: The movement of water through the vadose zone, in contrast to infiltration at the land surface and recharge across a water table.

porosity: The fraction of a given volume of a porous system not occupied by solid grains.

pressure head: The head of water at a point in a porous system, negative for unsaturated systems, positive for saturated systems; quantitatively, it is the water pressure divided by the specific weight of water.

recharge: The movement of water and pollutants across the water table.

saturation extract: A solution extracted from a solids sample wetted with distilled water until a prescribed end point has been reached.

sodium adsorption ratio: A relationship between the concentration of sodium in an applied water or matrix solution and the concentrations of calcium and magnesium.

soil bulk density: The mass of dry soil in a volume of bulk soil.

soil peds: A soil structural unit, such as a crumb, prism, granule, or block.

soil-water characteristic curve: A graphical representation of the change in water content of a soil or porous media with changing soil-water pressure.

soil-water pressure: The pressure on the water in a soil-water system, as measured by a piezometer for a saturated soil, or by a tensiometer for an unsaturated soil.

specific discharge: See flux.

specific yield: The volume of water that an unconfined aquifer releases from storage per unit surface area of aquifer per unit drop in the water table elevation.

tortuosity: A general term used to describe variations in the flow paths of water in a porous system caused by the presence of solid grains; quantitatively, it is the ratio of the distance travelled by a mass of water flowing through pores of a system to the length of the system (Hillel, 1971).

total hydraulic head: The sum of gravitational head and pressure head in water within a porous system.

total soil-water potential: The sum of the energy-related components of a soil-water system; e.g., the sum of the gravitational, matric, and osmotic components.

transmissivity: The flow of water in gal/day through a cross section of aquifer 1-foot thick and 1-mile wide under a hydraulic gradient of 1 foot mile at field temperature times the aquifer thickness in feet.

vadose zone: The hydrogeological region extending from the soil surface to the top of the principal water table.

water content: The amount of water stored within a porous matrix, expressed on either a volumetric (volume per unit volume) or mass (mass per unit mass) of solid.

water content profile: A log of the change in water content with depth in a profile through the vadose zone.

Appendix C—Monitoring Methods* Index

* Each method is referenced many times in the book.

ABOUT THE AUTHORS

Dr. **Lorne G. Everett** is Manager of the Natural Resources Program for Kaman Tempo, formerly General Electric's Center for Advanced Studies, Santa Barbara, California. His current hydrology interests are related to the design of groundwater quality monitoring programs for coal strip mining, oil shale extraction, uranium mine abandonment, and hazardous waste disposal areas. In addition, he oversees programs relating to minerals, industrial, and agricultural development.

After completing his Ph.D. in Hydrology at the University of Arizona in 1972, Dr. Everett was invited to join the faculty in the Department of Hydrology. Prior to his current position, Dr. Everett was the Manager of Tempo's Water Resources Program and a principal investigator in developing a national groundwater quality monitoring methodology for the U.S. Environmental Protection Agency. Dr. Everett recently completed a major EPA contract to develop groundwater quality monitoring guidelines for all western coal strip mine operations and for surface and in situ extraction of shale oil. He has written fundamental EPA manuals on soil core monitoring and soil pore-liquid monitoring at hazardous waste disposal sites. Dr. Everett was asked to develop and present training programs to all 10 EPA regions on groundwater monitoring permit requirements for RCRA.

Dr. Everett has worked under contract to the U.S. Department of Justice in managing testimony relative to water resource decisions. He has testified before Congress on national legislation relative to water monitoring. Dr. Everett was invited by the American Water Resources Association to be the Technical Chairman of a special symposium on water quality monitoring. He has published over 75 professional papers, book chapters, and reports. He is the principal author of the book *Establishment of Water Quality Monitoring Programs* and his handbook entitled *Groundwater Monitoring* is distributed internationally.

Dr. **L. Graham Wilson** received his B.S. in Agriculture from the University of British Columbia in 1951 and his M.S. in Irrigation Science and Ph.D. in Soil Physics from the University of California at Davis in 1957 and 1962, respectively. From 1956 to 1960, he was an assistant specialist in Irrigation at the University of California at Davis, working on drainage-salinity problems in irrigated regions of Northern California; the San Joaquin Valley, and the Imperial Valley. Since 1962, Dr. Wilson has been a Hydrologist at the Water Resources Research Center, the University of Arizona, Tucson, Arizona.

While at the University of Arizona, he has conducted research on natural and artificial recharge, and on the nature of water flow and pollutant movement in the vadose zone and groundwater underlying waste disposal operations. He has been associated with a number of studies on vadose zone and groundwater monitoring and has designed monitoring networks for recharge pits, oxidation ponds, sanitary landfills, and coal strip mining and oil shale processing operations. Dr. Wilson has written numerous journal articles and technical documents on vadose zone monitoring at hazardous waste disposal sites. Recently, as a consultant to Kaman Tempo, he assisted in developing a guidance document for vadose zone monitoring at land treatment operations.

Mr. **Edward W. Hoylman** is a Senior Hydrogeologist with the Natural Resources Program of Kaman Tempo, Santa Barbara, California. Currently, he is involved in developing groundwater and vadose zone monitoring programs for hazardous waste disposal sites. Specifically, his applications are in the use of geophysical tools, computer software, and the application of vadose zone monitoring equipment.

After completing his M.S. in Geology at the University of California at Los Angeles, Mr. Hoylman held several positions related to Hydrogeology. He has worked for the Aerogeophysics Company, Inc.; Drilling Fluids Specialists, Inc.; Petrolog Geologic Well Logging Service, Inc.; Great Basin Petroleum, Inc.; and as a Faculty Instructor in Hydrology. Mr. Hoylman has written several fundamental papers and coauthored two books on the subject of groundwater monitoring. He was extensively involved in a major EPA investigation to develop vadose zone monitoring guidelines for hazardous waste disposal sites.

Other Noyes Publications

HANDBOOK OF LAND TREATMENT SYSTEMS FOR INDUSTRIAL AND MUNICIPAL WASTES

by

Sherwood C. Reed
U.S. Army Corps of Engineers
Cold Regions Research
and Engineering Laboratory

Ronald W. Crites
George S. Nolte and Associates

The basic technology for land treatment of industrial and municipal wastes is defined in this book. The performance of land treatment systems in removing or retaining waste constituents is documented, and environmental effects are discussed. Detailed procedures are provided, for example, for selecting potential treatment sites, for determining infiltration rates, and for calculating aerosol transport of pathogens and nitrogen loss in storage ponds.

The book covers a range of topics from small onsite soil absorption to large agricultural utilization systems, from municipal wastewater to industrial wastes and sludges. Also included are cost data and energy conservation considerations.

Though a previously known method, land treatment was seldom used to solve waste treatment problems prior to the 1970s. Considerable recent research, however, has shown the treatment capability and reliability of land treatment and also that the capital and operating costs and energy usage can be significantly less than for more familiar mechanical systems. Land treatment can be a viable and cost effective choice for many industries, for small towns, and for moderately large cities or sections of large metropolitan areas.

Data tables, procedures, equations, work examples, and over 65 figures and illustrations enhance the presentation of this excellent and thorough reference work. A condensed table of contents giving **chapter titles and selected subtitles** is listed below.

ISBN 0-8155-0991-X (1984)

427 pages

Other Noyes Publications

GROUNDWATER CONTAMINATION AND EMERGENCY RESPONSE GUIDE

by

J.H. Guswa
W.J. Lyman
Arthur J. Little, Inc.

A.S. Donigian, Jr., T.Y.R. Lo, E.W. Shanahan
Anderson-Nichols & Co., Inc.

Pollution Technology Review No. 111

An overview of groundwater hydrology; a technology review of equipment, methods, and field techniques; and a methodology for estimating groundwater contamination under emergency response conditions are provided in this book. It describes the state of the art of the various techniques used to identify, quantify, and respond to groundwater pollution incidents.

Interest in the causes and effects of groundwater contamination has increased significantly in the past decade as numerous incidents have brought the potential problems to public attention. Protection of our groundwater resources is of critical importance, thus making the book both timely and relevant.

Part I assesses methodology for investigating and evaluating known or suspected instances of contamination. Part II surveys groundwater fundamentals, state-of-the-art equipment, monitoring methods, and treatment and containment technologies. It will serve as a desk reference and guidance manual. Part III details possible emergency response actions at toxic spill and hazardous waste disposal sites.

A condensed table of contents listing **part and selected chapter titles** is given below.

ISBN 0-8155-0999-5 (1984)

490 pages

Other Noyes Publications

TRIHALOMETHANE REDUCTION IN DRINKING WATER
Technologies, Costs, Effectiveness, Monitoring, Compliance

Edited by

Gordon Culp

Pollution Technology Review No. 114

This book evaluates technologies for the effective reduction of trihalomethanes (THMs) in drinking water. It is based on studies by *Culp/Wesner/Culp; Temple, Barker & Sloane, Inc.; Malcolm Pirnie, Inc.*, and the *U.S. Environmental Protection Agency.* In addition to the treatment technologies described, their costs and effectiveness are evaluated. Monitoring methods and compliance with federal drinking water regulations are also covered.

The first part of the book will serve as a guidance manual for those planning changes in water treatment systems for THM control. It discusses THM formation, necessary steps which must be followed for compliance with THM maximum contaminant levels (MCLs), and procedures to ensure preservation of the finished water.

The second part of the book describes best available treatment methods as well as potentially available treatment methods. Cost analyses with ranges of possible costs are included.

Part III covers monitoring and compliance as they pertain to promulgated federal drinking water regulations. Suggestions and recommendations for implementation of the TTHM (total THM) amendment, issued in February 1983, are provided.

A condensed table of contents listing **parts and chapter titles and selected subtitles** is given below.

ISBN 0-8155-1002-0 (1984)

251 pages